Plant Conservation in the Tropics

perspectives and practice

Plant Conservation in the Tropics

perspectives and practice

Edited by

Mike Maunder • Colin Clubbe • Clare Hankamer • Madeleine Groves

Published by The Royal Botanic Gardens, Kew, 2002

© Copyright 2002
The Board of Trustees of
the Royal Botanic Gardens, Kew
and authors of text and illustrations
of individual papers

First published 2002

Production Editor: S. Dickerson

Graphic design and page
make up by Media Resources,
Information Services Department,
Royal Botanic Gardens, Kew

ISBN 1 84246 014 5

Printed in Great Britain
by The Cromwell Press Ltd

Contents

Section II. Species Conservation and Sustainable Use: Perspectives and Practice

Appendices

Global Biodiversity Hotspots

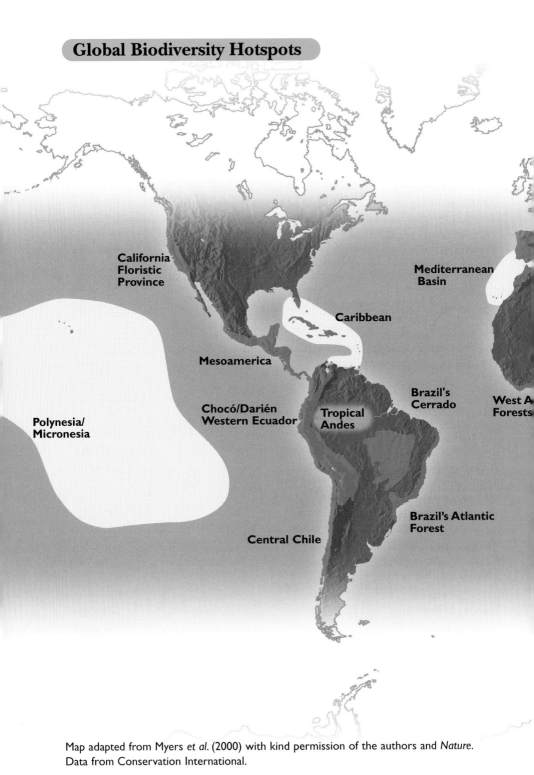

California
Floristic
Province

Mediterranean
Basin

Caribbean

Mesoamerica

Chocó/Darién
Western Ecuador

Polynesia/
Micronesia

Tropical
Andes

Brazil's
Cerrado

West A
Forests

Brazil's Atlantic
Forest

Central Chile

Map adapted from Myers *et al.* (2000) with kind permission of the authors and *Nature*.
Data from Conservation International.

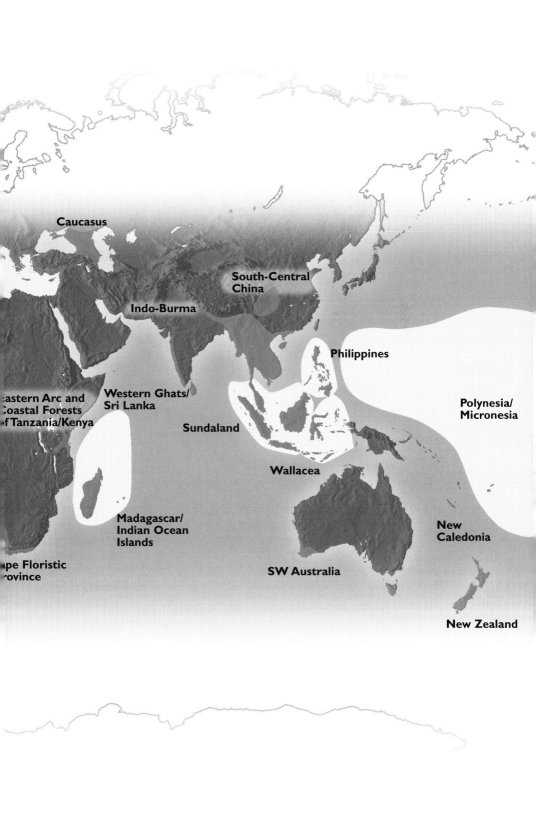

Caucasus

South-Central China

Indo-Burma

Philippines

Eastern Arc and Coastal Forests of Tanzania/Kenya

Western Ghats/ Sri Lanka

Polynesia/ Micronesia

Sundaland

Wallacea

Madagascar/ Indian Ocean Islands

New Caledonia

Cape Floristic Province

SW Australia

New Zealand

Forward

For those of us who learnt our botany in the temperate regions, the sheer diversity of plants in the tropics is both daunting and exhilarating. Yet, as in other parts of the world, this diversity, and the unique legacy and resource that it represents, is rapidly being lost through the destruction and degradation of natural habitats.

Within the framework provided by the Convention on Biological Diversity (CBD), this book promotes an integrated approach to plant conservation in the tropics and provides a variety of case studies from South America, Africa and Asia that illustrate practical responses to the loss of wild plant diversity. It communicates lessons from fieldwork, and from many different countries, about how scientists and resource managers are struggling to sustain plant diversity in the face of numerous and diverse threats. The essays illustrate both the problems and the opportunities confronted by plant conservationists in the tropics, and different chapters provide a variety of perspectives on the techniques and approaches needed to meet the challenges of habitat loss, species decline and erosion of culturally-important plant resources.

This book had its genesis in a three-year training programme for plant conservationists run in East Africa by the National Museums of Kenya and the Royal Botanic Gardens, Kew, with the support of the UK Darwin Initiative for the Survival of Species, the British Council and the UK Department for International Development (DFID). Building on case studies and experience in East Africa, further examples of conservation and practice from around the tropics have then been added so that the work now fills what was previously a gap in the conservation literature relating to the practical and in-country conservation of tropical plants.

It is a great pleasure for me to welcome this book into the literature of conservation biology and to congratulate the editors for their commitment and efforts that have brought this project to fruition. Conservation is something that is much easier to do in theory than in practice. This book provides illuminating and instructive case histories and many practical recommendations. I am sure that it will be an invaluable resource for those concerned with the conservation of tropical plant diversity well into the future.

Peter R. Crane FRS
Director
Royal Botanic Gardens, Kew, UK

Preface

Plant conservation relies on good science and policy, as well as both human and financial resources. It is essential to find the right person to do the right job, and this document is an excellent compilation of plant conservation projects which not only demonstrate the techniques needed to conserve the world's plants, but also show the devotion of the authors to the cause – a passion which is essential if a great deal of the world's biodiversity is not to be lost in the twenty first century.

Good science, as well as bringing together the human and financial resources needed to conserve biodiversity, underpins the activities of the Species Survival Commission (SSC), which is a Commission of volunteers dedicated to preventing biodiversity loss, operating under the banner of IUCN – the World Conservation Union. The SSC Plant Conservation Programme is working through its many Specialist Groups and collaborators to assess the status of plant diversity and to encourage and undertake practical responses to the erosion of plant diversity. Many authors in this volume are members of the SSC and have made key contributions in the area of plant conservation. It is my hope that this work will motivate others to increase efforts on plant conservation in their regions, as well as contribute to the work of the SSC, a global network of over 7,000 scientists.

I welcome this volume as a major contribution to increasing professional capacity in plant conservation, one of the key objectives of the Plant Conservation Programme of the SSC. I wish that I could have had such a compendium of experience to peruse when I worked on the island of Mauritius, as we tried different techniques to conserve a unique flora. While there is still room for more 'nuts and bolts' manuals to help the manager with specific plant conservation techniques, I think that this book is inspirational. The editors have pulled together a broad range of authors who all speak with experience from the field, and the true utility of this book will be measured, not by the number of copies that are mailed out, but by those falling apart through frequent use.

Through my work with the SSC, I am familiar with many of the projects described in this book, although I have learnt of a number of interesting projects of which I was previously unaware. Communication is essential to move the frontline of plant conservation forward, and not to re-invent the wheel. This volume also makes an important contribution to the SSC Plant Conservation Programme by providing access to both relevant SSC policy documents, as well as case studies, on the practical issues of plant conservation to build capacity. My felicitations to both the editors as well as the authors of this excellent compendium.

Dr Wendy Strahm, Plants Officer
IUCN/SSC, Gland, Switzerland

Acknowledgements

This book has a complex history. Whilst it is derived primarily from our work in East Africa, it has been guided by many people. Working with colleagues and students in East Africa and with students on the plant conservation courses at Kew we became aware that our discussions, workshops and field activities were taking us into arenas that are rarely found in conservation biology textbooks or journals. We hope that this volume will be a useful guide to the important challenges facing plant conservationists in the tropical regions. We would like to thank all those who participated in the Darwin courses, who challenged us to put conservation into a real world context and showed us wonderful friendship and hospitality.

This book is a direct product of the National Museums of Kenya – Darwin Plant Conservation Techniques Course for East Africa, that could not have taken place without the support of the UK Government's Darwin Initiative for the Survival of Species and also The British Council and UK Department for International Development (DFID). We are particularly grateful to the staff of the Darwin Initiative in London (Valerie Richardson, Jonathan Tillson, Maria Stevens and Sarah Collins), British Council in Nairobi (particularly Catherine Gitau) and DFID in Nairobi (Tim Pearce and Katherine Morris).

We owe a great deal to our project partner, the National Museums of Kenya (NMK), and, especially, the NMK-based project management team of Stella Simiyu and Perpetua Ipulet (course co-ordinator), as well as Patrick Muthoka, Quentin Luke, Dr Beatrice Kayotha, Benny Bytebier, William Wambugu, George Mugambi, Abel Atiti, Anthony Githitho, Geoffrey Mungai, Geoffrey Mwachala and Dr Joseph Mutangah. The course would have been more difficult without the support and enthusiasm of two extraordinary people: David Laur and Hamida Hassan of NMK. We are grateful for the support of the NMK Directors, Dr Mohammed Isahakia and Dr George Abungu, and Director of Science, Dr Rashid Aman. Special thanks to Dr Peggy Olwell of the United States National Park Service for her commitment and unique contribution to teaching whilst joining the courses in 1996 and 1998.

During our work in East Africa, we have been fortunate to work with a number of close collaborators who gave generously of time and guidance, including Quentin and Trish Luke, Diani, Kenya; Dr Mark Stanley Price and Micky Pritpal S. Soorae of the African Wildlife Foundation; Ann Robertson, Malindi, Kenya; Dr John Mulamba Wasswa, Entebbe Botanic Gardens, Uganda; Dr William Mziray, National Herbarium of Tanzania, Arusha; Dr Leonard Mwasumbi, Department of Botany, University of Dar es Salaam, Tanzania; Professor Len Newton, Botany Department, Kenyatta University, Kenya; Dr Geoff Howard, IUCN Regional Office, Nairobi, Kenya; Henry Kamau, Dan Kiambi, Dr Ehsan Dulloo, Dr Luigi Guarino, International Plant Genetic Resources Institute;

Anne Birnie, Nairobi Arboretum, Kenya; Dr John Mugabe, African Centre for Technology Studies (ACTS), Kenya; Dr John Donaldson, National Botanical Institute, RSA; Dr Robert Hoeft, UNESCO; Dr Hannah Jaenicke, International Centre for Research in Agroforestry, Nairobi, Kenya; Dr Stig Johansson, Dr F. Kilahama, Shedrack Mashauri of the East Usambara Catchment Forest Programme (EUCFP) and Catharine Muir, Mwaluganje Elephant Sanctuary. Special thanks to the staff of the Hotel Boulevard, Nairobi for providing home base in Nairobi. Dr Kaj Vollesen of RBG Kew co-ordinated the provision of the RBG Kew Landrover in Nairobi.

The following organisations provided funding for the East African courses: UK Government funding from The Darwin Initiative for the Survival of Species, The British Council, Nairobi and DFID; International Plant Genetic Resources Institute, Royal Netherlands Embassy, Nairobi, UNESCO, FINNIDA-EUCFP, African Wildlife Foundation and SIDA/Regional Soil Conservation Unit. Andrew McRobb, official photographer from RBG Kew, joined the 1999 course and took many of the photographs used in this book. Other photographs were taken by Mike Maunder, Colin Clubbe and chapter authors (the latter are credited individually).

During the formulation and editing of this book, we are very grateful for the support of the following who read and commented upon draft manuscripts: Dr Alan Hamilton, World Wide Fund for Nature (WWF); Dr Mike Gillman, The Open University; Dr Hew Prendergast, Dr Henk Beentje, Kerry ten Kate, Barbara di Giovanni, Dr John Dransfield, Stewart Henchie, and Noel McGough of RBG Kew; Dr Tony Kendle, University of Reading; Professor Ole Hamann, Copenhagen Botanic Garden, Denmark; Dr Wendy Strahm and Professor David Given, IUCN Species Survival Commission; and Dr Peter Wyse Jackson and Fiona Dennis of Botanic Gardens Conservation International (BGCI).

On a personal note we would like to thank our respective partners for their patience and support whilst we worked abroad and on this book. Special thanks to Madeleine Groves who joined the team while Clare Hankamer took maternity leave; without Mad's involvement, this book would not have been completed on time. We would like to thank Dr Erica Bower for casting a fresh pair of eyes over the manuscript at the final proof stage.

Mike Maunder
Colin Clubbe
Clare Hankamer
Madeleine Groves

List of Contributors

José Maria Assumpçã
Rio de Janeiro Botanic Garden Research
   Institute
Forest Nursery
Rua Pacheco Leão, 920
22.460-030
Rio de Janeiro
Brazil
Fax: +55 21 274 9799
e-mail: jmaria@jbrj.gov.br

Paul Blackmore
Limbe Botanic Garden
PO Box 437
Limbe
S.W. Province
Cameroon
e-mail: PCBlackmore@aol.com

Remigius Bukenya-Ziraba
Botany Department
Makerere University
Box 7062
Kampala
Uganda
e-mail: botany@swiftuganda.com

Benny Bytebier
University of Stellenbosch
Department of Biochemistry
Private Bag X 1 7602
Matieland
South Africa
e-mail: bytebier@maties.sun.ac.za

Lena Chan
National Parks Board
Singapore Botanic Gardens
1 Cluny Road
Singapore 259569
e-mail: Lena_CHAN@NPARKS.GOV.SG

Lynn Clayton
Renewable Resources Assessment Group
T.H. Huxley School of Environment, Earth
   Sciences and Engineering
Imperial College London
8, Princes Gardens
London SW7 1NA
UK
Fax: +44 20 7589 5319
e-mail: dr_lynn@manado.wasantara.net.id

Colin Clubbe
Royal Botanic Gardens, Kew
Richmond
Surrey TW9 3AB
UK
Fax: +44 20 8332 5640
e-mail: c.clubbe@rbgkew.org.uk

Ian Dawson
International Centre for Research in
   Agroforestry (ICRAF)
PO Box 30677
Nairobi
Kenya
Fax: +254 2 521001
e-mail: i.dawson@cgiar.org

M. Ehsan Dulloo
IPGRI-SSA
c/o ICRAF
P.O.Box 30677
Nairobi
Kenya
e-mail: e.dulloo@cgiar.org

Frauke Fleisher-Dogley
Botanic Garden
PO Box 445
Mahe
Republic of the Seychelles
Fax: +248 22 45 00
e-mail: natpark@seychelles.net

Anthony N. Githitho
Coastal Forest Conservation Unit
Box 596
Kilifi
Kenya
e-mail: cfcu.kilifi@swiftmombasa

David Given
Chair of the IUCN/SSC Plant Conservation
   Committee
Lincoln University
Canterbury
New Zealand
e-mail: givend@attglobal.net

Madeleine Groves
Conventions & Policy Section (CAPS)
Royal Botanic Gardens, Kew
Richmond
Surrey TW9 3AB
UK
Fax: +44 20 8332 5757
e-mail: m.groves@rbgkew.org.uk

Clare Hankamer
Royal Botanic Gardens, Kew
Richmond
Surrey TW9 3AB
UK
Fax: +44 20 8332 5640
e-mail: c.hankamer@rbgkew.org.uk

Héctor M. Hernández Macías
Director
Instituto de Biología, UNAM
Apartado Postal 70-233
Ciudad Universitaria
Deleg. Coyoacán
04510 Mexico D.F.
Mexico
Fax: +52 616 2326
e-mail: hmhm@servidor.unam.mx

Carlos Gómez-Hinostrosa
Departamento de Botánica
Instituto de Biología, UNAM
Apartado Postal 70-233
Ciudad Universitaria
04510 Mexico, D.F.
Mexico
e-mail: cgh@ibiologia.unam.mx

Carlos Iglesias
Instituto de Ecologia, A.C.
Sistematica Vegetal
Km. 2.5 carretera antigua a Coatepec A.P. 63
Xalapa
Veracruz 91000
Mexico
e-mail: iglesiac@ecologia.edu.mx

Perpetua Ipulet
East African Herbarium
PO Box 45166
National Museums of Kenya
Nairobi
Kenya
Fax: +254 2 741424
e-mail: plants@africaonline.co.ke

Hannah Jaenicke
International Centre for Research in
    Agroforestry (ICRAF)
PO Box 30677
Nairobi
Kenya
Fax: +254 2 521001
e-mail: icraf@cgiar.org

Seuram Jhilmit
Assistant Conservator of Forests
Forest Resource Inventory and Management
    Section (FRIM)
Forestry Division
Ministry of Agriculture, Land and Marine
    Resources
Long Circular Road
Port of Spain
Trinidad and Tobago
Fax: +1 809 654 4288

Stig Johansson
Regional Director,
Natural Heritage Services
Southern Finland
Metsähallitus – Forest and Park Service
PL 94, FIN-01301 Vantaa
Finland
e-mail: stig.johansson@metsa.fi

Maud Kamatenesi-Mugisha
Botany Department
Makerere University
Box 7062
Kampala
Uganda
e-mail: botany@swiftuganda.com

Massaba I.L. Katigula
Forest Project Officer
Tanga Region Catchment Forest Office
PO Box 1449
Tanga
Tanzania
Fax: +255 27 2643820
e-mail: usambara@twiga.com

Tony Kendle
Department of Horticulture and Landscape
The University of Reading
Whiteknights
Reading RG6 6AS
UK
Fax: +44 118 975 0630
e-mail: a.d.kendle@reading.ac.uk

Cintia Luchiari
Tree Phenology and Seeds Collection –
    Revegetation Project – ARFRP
Reserva Biológica de Poço das Antas
Caixa Postal 109.981
28860-970 – Casimiro de Abreu
Rio de Janeiro
Brazil
Fax: + 55 24 778 1540
e-mail: reveget@lagosnet.com.br

Shedrack Mashauri
Tourism and Conservation Education Officer
Tanga Region Catchment Forest Office
PO Box 1449
Tanga
Tanzania
Fax: +255 27 2643820
e-mail: usambara@twiga.com

Mike Maunder
Director of Conservation and Curator of
    Collections
The National Tropical Botanical Garden
3530 Papalina Road
Kalaheo
Kauai
Hawaii 96741
USA
Fax: +1 808 332 9765
e-mail: m.maunder@ntbg.org

Jeanette Mill
National Co-ordinator Australian Network for
    Plant Conservation
Chair, IUCN Species Survival Commission
    Australasian Plant Specialist Group
GPO Box 1777
Canberra, ACT, 2601
Australia
Fax: +61 2 62 509 528
e-mail: jeanette.mill@ea.gov.au

E.J. Milner-Gulland,
Renewable Resources Assessment Group
T.H. Huxley School of Environment, Earth
    Sciences and Engineering
Imperial College London
8, Princes Gardens
London SW7 1NA
UK
Fax: +44 20 7589 5319
e-mail: e.j.milner-gulland@ic.ac.uk

Ahmed Mndolwa
I/c Amani Botanic Garden
Tanzania Forestry Research Institute
PO Box 1
Amani
Tanzania
Fax: +255 27 2643820
e-mail: usambara@twiga.com

Luiz Fernando Duarte de Moraes
Coordinator for the Revegetation Project –
    Atlantic Rain Forest Research Program
Reserva Biológica de Poço das Antas
Caixa Postal 109.981
28860-970, Casimiro de Abreu
Rio de Janeiro
Brazil
Fax: +55 24 778 1540
e-mail: reveget@lagosnet.com.br

John Mugabe
Executive Director
African Centre for Technology Studies
    (ACTS)
PO Box 45917
Nairobi
Kenya
Fax: +254 2 524001/522987
e-mail: acts@cgiar.org

Moses Munjuga
Species Monograph Officer
International Centre for Research in
    Agroforestry (ICRAF)
PO Box 30677
Nairobi
Kenya
Fax: +254 2 524001
e-mail: m.munjuga@cgiar.org

Nouhou Ndam
Limbe Botanic Garden
PO Box 437
Limbe
SW Province
Cameroon

Mathias Ngonyo
P.O.Box 5067
Malindi
Kenya

Joseph Nkfor
Limbe Botanic Garden
PO Box 437
Limbe
SW Province
Cameroon

Tim Pearce
International Projects Co-ordinator
Seed Conservation Department
Royal Botanic Gardens Kew, Wakehurst Place
Ardingly, Haywards Heath
West Sussex RH17 6TN
UK
Fax: +44 1444 894110
e-mail: t.pearce@rbgkew.org.uk

Tânia Sampaio Pereira
Conservation Program Co-ordinator
Rio de Janeiro Botanic Garden Research
    Institute
Seed Laboratory
Rua Pacheco Leão, 915
22.460-030 - Rio de Janeiro
Brazil
Fax: +55 21 294 8696
e-mail: tpereira@jbrj.gov.br

Miguel Angel Pérez-Farrera
Escuela de Biologia, Universidad de Ciencias
    y Artes del Estado de Chiapas (UNICACH)
Calzada Samuel Leon Brindis 151
Tuxtla Gutierrez
Chiapas
Mexico 29000
Fax: +961 2 19 57
e-mail: miguelangel@chiapas.net

Ghillean T. Prance
Eden Project
Watering Lane Nursery
Pentewan, St Austell
Cornwall PL26 6BE
UK
e-mail: gprance@edenproject.com

Rafael Puglia-Neto
Reserva Biológica de Poço das Antas
Caixa Postal 109.981
28860-970 – Casimiro de Abreu
Rio de Janeiro
Brazil
Fax: +55 24 778 1540

Ann Robertson
PO Box 162
Malindi
Kenya

Agung Priyo Sarjono
Asosiasi Pengusaha Hutan Indonesia
Manggala Wana Bakti Block 4–9th Floor, Jl.
    Jenderal Gatot Subroto – Senayan
Jakarta
Indonesia
Fax: +62 21 573 2564

Uwe Schippmann
Head
CITES Scientific Authority Plants
Chair IUCN/SSC Medicinal Plant Specialist
    Group
Bundesamt fur Naturschutz
Konstantinstrasse 110
D-53179 Bonn
Germany
Fax: +49 228 8491 119
e-mail: schippmu@bfn.de

Pritpal S. Soorae
Senior Conservation Officer
IUCN/SSC Re-introduction Specialist Group
    (RSG)
Environmental Research & Wildlife
    Development Agency (ERWDA)
P.O. Box 45553
Abu Dhabi
United Arab Emirates (UAE)
Fax: +971 2 693 4628
e-mail: PSoorae@erwda.gov.ae

Mark Stanley-Price
Director, Durrell Wildlife Conservation Trust
Les Augrès Manor
Trinity
Jersey JE3 5BP
UK
Fax: +44 1534 860001
e-mail: jerseyzoo@durrell.org

Terry Sunderland
African Rattan Research Programme
Herbarium
Royal Botanic Gardens, Kew
Richmond
Surrey TW9 3AB
UK
e-mail: afrirattan@aol.com

Zacharie Tchoundjeu
Project Leader
ICRAF Cameroon
PO Box 2067
Yaounde
Cameroon
Fax: +237 23 74 40
e-mail: z.tchoundjeu@camnet.cm

Marivo Vázquez Torres
Instituto de Investigaciones Biologicas
Universidad Veracruzana
A.P. 294
Xalapa
Veracruz 91000
Mexico
e-mail: mvazquez@bugs.invest.uv.mx

Alan Tye
Head
Dept of Plant & Invertebrate Sciences
Charles Darwin Research Station (CDRS)
Isla Santa Cruz
Casilla 17-01-3891
Galápagos
Ecuador
Fax: +593 4 564636
e-mail: atye@fcdarwin.org.ec

Andrew Vovides
Instituto de Ecologia, A.C.
Sistematica Vegetal
Km. 2.5 carretera antigua a Coatepec A.P. 63
Xalapa
Veracruz 91000
Mexico
Fax: +52 28 187809
e-mail: vovidesa@ecologia.edu.mx

James Were
Research Officer
Genetic Resources of Agroforestry Trees Unit
International Centre for Research in
    Agroforestry (ICRAF)
United Nations Avenue
PO Box 30677
Nairobi
Kenya
Fax: +254 2 524 001
e-mail: j.were@cgiar.org

Scope of the Book

Plant conservation teams working in the tropics are facing huge technical and scientific challenges. These 'front-line' teams are developing practical responses to such challenges, and their activities are of immediate relevance to others around the world. This book has been written to help to share those experiences with all who need effective and practical responses to tropical plant conservation.

The publication of a paper in the scientific journal *Nature* on Biodiversity Hotspots for Conservation Priorities (Myers *et al.*, 2000) identified 25 terrestrial hotspots where 'exceptional concentrations of endemic species are undergoing exceptional loss of habitat' (Table 1 and inside book cover). To qualify as a hotspot at least 0.5% of the plant species within the hotspot must be endemic and it should have lost 70% or more of its primary vegetation, thus encapsulating species endemism and the degree of threat based on habitat loss. Not surprisingly, fifteen of these hotspots are in the tropics. In a world where conservationists are not able to assist all species under threat, this paper and the associated book (Mittermeier *et al.*, 1999) identified priority areas where 'we can support the most species at the least cost'.

When this paper was published we had just come to the end of our Darwin Initiative project, the fieldwork for which took place within the eighth 'hottest' hotspot, according to Myers *et al.* (2000) – the Eastern Arc and Coastal Forests of Tanzania/Kenya. The paper stimulated us to invite a number of colleagues working in other tropical hotspots and similar areas to contribute their experiences in order to provide some much-needed plant-focused case studies of current conservation activities.

This book has been designed to deliver the following:

- Case studies from the tropics outlining actual experience
- Review of practical issues facing tropical plant conservationists
- Review of current policy and guidelines of relevance to tropical plant conservationists.

The case studies review a diversity of topics including the implications of the Convention on Biological Diversity, habitat management and restoration, botanic gardens, medicinal plant management and threatened species recovery. These case studies encompass sets of experience that are not often published in the scientific literature but that have great practical relevance for conservation practitioners. We hope this book will be a useful addition to the small but growing library of books for the tropical plant conservationist, that it will generate debate, and above all, encourage practical responses to the loss of tropical plant diversity.

Table I The distribution of biodiversity hotspots (Myers et al., 2000)						
Biodiversity Hotspot	% of original primary vegetation surviving	Area protected (% of hotspot)	No. of plant species	No. of endemic plants	Endemic plants as % of global total	Relevant case studies in this book
Tropical Andes	25	25	45,000	20,000	6.7	
Sundaland	8	72	25,000	15,000	5.0	
Mediterranean Basin	5	40	25,000	13,000	4.3	
Madagascar and Mascarenes	10	20	12,000	9,704	3.2	Dulloo, Chapter 13; Fleisher-Dogley and Kendle, Chapter 19
Brazil's Atlantic Forest	8	35	20,000	8,000	2.7	de Moraes et al., Chapter 9
Caribbean	11	100	12,000	7,000	2.3	Clubbe and Jhilmit, Chapter 11
Indo-Burma	5	100	13,500	7,000	2.3	
Cape Floristic Province	24	78	8,200	5,682	1.9	
Philippines	3	43	7,620	5,832	1.9	
MesoAmerica	20	60	24,000	5,000	1.7	Hernández and Gómez-Hinostrosa, Chapter 18; Vovides et al., Chapter 22
Brazil's Cerrado	20	6	10,000	4,400	1.5	Prance, Chapter 12
SW Australia	11	100	5,470	4,331	1.4	Mill, Chapter 6
South-Central China	8	25	12,000	3,500	1.2	

Region						Reference
Polynesia and Micronesia	22	49	6,560	3,334	1.1	
New Caledonia	28	10	3,330	2,551	0.9	
Chocó/Darien/Western Ecuador	24	26	9,000	2,250	0.8	
Western African Forests	10	16	9,000	2,250	0.8	Sunderland et al., Chapter 21; Jaenicke et al., Chapter 20
California Floristic Province	25	40	4,400	2,125	0.7	
Western Ghats and Sri Lanka	7	100	4,780	2,180	0.7	
Succulent Karroo	27	8	4,850	1,940	0.6	
New Zealand	22	88	2,300	1,865	0.6	
Central Chile	30	10	3,400	1,605	0.5	
Eastern Arc and Coastal Forests of East Africa	7	100	4,000	1,500	0.5	Pearce and Bytebier, Chapter 4; Johansson et al., Chapter 5; Maunder et al., Chapter 7; Githitho, Chapter 8; Robertson et al., Chapter 14; Hankamer et al., Chapter 15
Caucasus	10	28	6,300	1,600	0.5	
Wallacea	15	40	10,000	1,500	0.5	Clayton et al., Chapter 23

References

Mittermeier, R.A., Myers, N. and Mittermeier, C.G. (1999). *Hotspots: Earth's Biologically Richest and Most Endangered Terrestrial Ecoregions.* CEMEX S.A. and Conservation International, USA.

Myers, N., Mittermeier, R.A., Mittermeier, C.G., da Fonseca, G.A.B. and Kent, J. (2000). Biodiversity hotspots for conservation priorities. *Nature* **403**: 853–858.

Mike Maunder
Colin Clubbe
Clare Hankamer
Madeleine Groves

Editors' Biographies

Mike Maunder

Director of Conservation and Curator of Collections at the National Tropical Botanical Garden, Hawaii, USA. Formerly Head of the Conservation Projects Development Unit (CPDU) at RBG Kew, between 1993 and 2000. Deputy Chair of the Re-introduction Specialist Group, a member of the SSC Plant Conservation Committee and an Associate of the Conservation Breeding Specialist Group. He has a specialist interest in the conservation of tropical floras and the effective role of botanic gardens in supporting plant conservation.

Colin Clubbe

Co-ordinator of Graduate Studies at RBG Kew where he has responsibility for managing the postgraduate education and international training programme in addition to collaborating on a range of conservation projects in the tropics. An ecologist, with a particular interest in survey and monitoring techniques for plant communities and for threatened species; prior to joining RBG Kew, he was a lecturer in Plant Ecology at the University of the West Indies in Trinidad. A member of the IUCN/SSC Re-introduction Specialist Group.

Clare Hankamer

Co-ordinator for international projects in the Conservation Projects Development Unit, at RBG Kew. Her responsibilities include setting up and managing conservation projects to provide in-country support through training and technical advice in plant conservation techniques. Her current research work is on demographic modelling and species recovery of rare and threatened plants primarily in Tanzania and the British Virgin Islands. She is a member of the IUCN/SSC Re-introduction Specialist Group.

Madeleine Groves

Madeleine Groves is the CITES Implementation Officer in the Conventions and Policy Section at RBG Kew. Since graduating from the RBG Kew Diploma in 1991, she has specialised in issues relating to the international trade in plants, and the relevant policy and legislation. Formerly the Conservation Programme Co-ordinator at the Atlanta Botanical Garden, Georgia, USA, she has a particular interest in the conservation of carnivorous plants and the restoration of bog habitats, and is Secretary for the IUCN/SSC Carnivorous Plant Specialist Group.

Collectively, the Editors established and have managed the International Diploma course in Plant Conservation Techniques at RBG Kew since 1993.

Chapter 1

Global and Local Action for Plant Diversity

David R. Given
Chair of the IUCN Species Survival Commission
Plant Conservation Programme

As a recent import to the world of university academia, it has been a challenging exercise for me to become embroiled in student concerns. Students, at times, become concerned about many issues. One of the most pressing concerns is being felt not only in New Zealand, but also in other parts of the world. It is the rising cost of tertiary education. However, it is not just students who are concerned about cost of living. Cost of living is something that hits every person. Yet, one of the most important areas of cost goes unnoticed by most people. Few realise that we have just left perhaps the most expensive decade of the most expensive century in history. The high cost paid is the unique biological diversity of this planet. We often live as though there were an inexhaustible stock of biological wealth, even though current trends of extinction are estimated as 3–4 orders of magnitude higher than the background rates measured from the fossil record (Josephson, 2000; May *et al.*, 1995).

One of the more unusual plants illustrated in New Zealand botanist Audrey Eagle's magnificent volume on New Zealand trees and shrubs is *Trilepidea adamsii* (Cheeseman) Tieghem (Loranthaceae) (Eagle, 1986); it is a mistletoe, a very primitive one, and is the sole member of its genus. She painted it from a specimen collected in 1954. But little did she know at the time that she would be the last person on earth to see this unique plant alive (Given, 1980). Today, it is extinct – over-collected, over-browsed, and its habitat over-disturbed. Such a story can be replicated through time and space. It is the story of the Philip Island glory pea *(Streblorrhiza speciosa* Endl. (Leguminosae)), of the passenger pigeon *(Ectopistes migratorius)*, and of a growing number of other species that have failed to make it into the new millennium. It is little wonder that the extinction crisis is acknowledged as a global problem and there are strident calls for the rate of loss of species to be decreased. To quote the Gran Canaria Declaration (BGCI, 2000):

> 'As many as two-thirds of the world's plant species are in danger of extinction in nature during the course of the 21st century, threatened by population growth, deforestation, habitat loss, destructive development, over-consumption of resources, the spread of alien invasive species and agricultural expansion. Further loss of plant diversity is predicted through genetic erosion and narrowing of the genetic basis of many species. The disappearance of such vital and massive amounts of biodiversity provides one of the greatest challenges faced by the world community: to halt the destruction of resources that are so essential for present and future needs.'

(Text reproduced with the kind permission of BGCI).

Catastrophic or localised extinctions can occur without human intervention, but in many parts of the world, and for vulnerable ecosystems generally, humans are primarily responsible for current extinctions. Figures estimating such current extinction rates often do not take into account the fact that the extinction of an obvious, large organism such as a forest tree probably results in the loss of at least fifteen organisms dependent on or confined to that single species (Hawksworth, 1998). Such figures are being corroborated by other analyses that also paint bleak scenarios (Given, 2000). Pimm and Raven (2000) suggest that unless markedly increased steps are taken to protect remaining species-rich regions and habitats there may be impending catastrophic loss of biodiversity. Using a model that examines the relationships among species in specific areas, the rate of fragmentation of habitat, and estimates of survival rates, they forecast an accelerating rate of extinction peaking about half-way through the twenty-first century at nearly 50,000 species per million species of animals and plants each decade. Only near the end of the twenty-second century, assuming that the rate of increase of human population stabilises or declines, is this rapid rate of extinction expected to decrease.

The Gran Canaria Declaration, which has now been accepted into the work programme of the Convention on Biological Conservation (CBD), (Appendix 1), is a response to the recognition that the loss of plant diversity is a matter of major concern for the world's scientific and, importantly, botanic garden community. Accordingly, in August, 1999 the botanists of the world, convened at the XVI International Botanical Congress in St Louis, Missouri, USA, resolved to address the problem of plant diversity loss. A key element of the Declaration is its call for co-operation and partnerships at all scales from global to local:

> '*Such a Strategy should develop effective mechanisms to enhance collaboration and networking, which will strengthen and support plant conservation locally, regionally, and internationally; to formally and informally link different partners such as government ministries, institutions, NGOs, and local community initiatives. This Strategy should link existing efforts of the many significant international and national programmes that are already active in this area. It should draw on and extend the experience and resources of bodies already active in implementing global conservation programmes, such as the FAO Global Plan of Action for the Conservation and sustainable use of Plant Genetic Resources for Food and Agriculture, supported by over 150 countries, the UNESCO Man and Biosphere Program, DIVERSITAS, the Millennium Assessment of the World's Ecosystems, the International Agenda for Botanic Gardens in Conservation and the IUCN Species Survival Commission's Plant Programme. An element of this Strategy should be to integrate efforts in different disciplines (social, economic, and biological) towards plant conservation so that all appropriate and available resources, technologies, techniques and sectors are brought together in support of plant conservation (BGCI, 2000).*'

(Text reproduced with the kind permission of BGCI).

I have had a close association with the development of one of the potential partner programmes noted in the Declaration. This is the new Plant Conservation Programme of the World Conservation Union's (IUCN) Species

Survival Commission (SSC). This is just one of several key strategic approaches being developed. The SSC celebrated fifty years of conservation action in 1999. It has been an appropriate time to take stock and to chart new directions, because we simply cannot afford another decade, let alone century or millennium, as expensive in terms of biological loss as that which has just come to a close. The SSC programme had its genesis in the outstanding work of the IUCN and the World Wide Fund for Nature (WWF) Plant Advisory Group during the 1980s, under the leadership of Peter Raven and later, Arturo Gomez Pompa. The SSC Plant Conservation Programme has five major objectives:

Objective 1
Sound interdisciplinary scientific information underpins decisions and policies affecting plant diversity.

Objective 2
Collaboration and strategic alliances are increasingly used within the plant conservation community to achieve plant conservation success, focused especially on local, national, and international organisations outside the SSC.

Objective 3
Modes of production and consumption that result in the conservation of native plant diversity are adopted by users of plant resources.

Objective 4
SSC's plants policy recommendations, guidelines, and advice are valued, adopted, and implemented by relevant audiences.

Objective 5
Capacity to provide long-lasting, practical solutions to plant conservation problems is markedly increased.

Basic to this strategic plan is that sound interdisciplinary scientific information must underpin decisions and policies affecting plant diversity, coupled with a focus on those parts of the world that are biological hotspots. Hotspots, or for us, the Centres of Plant Diversity, are places with particular concentrations of species richness and those places where there is an extraordinary stock of the world's botanical diversity at the ecosystem or genetic level (Mittermeier *et al.*, 1999; Myers *et al.*, 2000; WWF and IUCN, 1994–1997). Such sites are often under pressure from human influences. However, it is also crucial to recognize that there are other sites of importance for plant conservation that may not qualify in terms of number of species or endemism. Such areas include, for instance, the Scottish highlands and the Australian alpine region where very small changes in global climate may have profound effects on local biota. Hence, the SSC Plant Conservation Committee has moved towards use of the term 'Sites of Plant Conservation Importance'.

However, most important of all is that there are parts of the world for which there can be no compromise. They include regions such as the South African fynbos, and further north the arid lands of Namaqualand and Namibia, ancient continental islands such as Madagascar, younger actively evolving islands such as the Hawaii archipelago, the limestone mountains of South East Asia, subantarctic islands, ancient ultra-basic rock sites, Mediterranean shrublands and the cloud forests of South America. High ranking in such a foundational list must be given to the tropics, especially the moist tropical forests, tropical islands, and vegetation associated with limestone and ultra-basic rocks.

There is need for development of more rigorous criteria not only to identify, classify and prioritise the sites that are plant diversity centres, but also to ensure that conservation actions in such regions are appropriate, of high quality, and well-resourced. Through partnerships with conservation, research and development organisations, the IUCN SSC Plants Programme (along with other global and regional programmes) wants to see robust criteria and methods adopted so that best use is made of scarce resources and, equally important, to ensure that the right resources are available. Along with this, communication channels need to be improved to ensure that people can learn from the mistakes and successes elsewhere. Information on its own is of little use unless it leads to intervention.

Information, used appropriately, is a vital resource. A key action must be to create and co-ordinate information gathering programmes, working with, alongside and through local people, in those regions of botanical importance, especially where there are rich resources of economically valuable plants, wild crop relatives and subsistence crops. Much more attention needs to be focused on 'forgotten crops' that, although neglected, may be the salvation for future generations especially in climatically stressed regions (Vietmeyer, 1986). A second important action for which better information and concerted action is needed is the neutralisation of invasive plant and animal species; these being an increasing source of species depletion and loss (Appendix 5). A third action, where information exchange is vitally important, is to promote the sustainable use of plant resources. There is a major challenge here to ensure that the spirit and letter of the CBD (Appendix 1) becomes reality.

Consideration of sustainability (ecological, economic and social), encompasses far more than just a place for conservation and another for production. In conserving biodiversity, a major battleground of the twenty-first century will be the interface between wilderness and production lands. This interface is a hard landscape zone to manage and too often is neglected. In the past, the boundaries between production land and wildlands were often sharply drawn; these boundaries need to be blurred through meaningful dialogue and a genuine understanding of the natural heterogeneity and dynamism of most landscapes (see Figure 1.1). Further, sustainability involves participation of all stakeholders. This includes nature itself, which has no obvious voice but needs its advocates.

Figure 1.1   The conservation of plant resources in the tropics will be dependent upon an increasing understanding of ecology and the social context of conservation particularly in the interface between wilderness and production lands. Here, John Donaldson of the National Botanical Institute, Republic of South Africa, and the Cycad Specialist Group of the IUCN SSC works with Denis Byabashaija of the Ugandan Forestry Research Institute assessing regeneration of *Encephalartos hildebrandtii* Braun & Bouché (Zamiaceae) within a community wildlife reserve in Kenya.

A fundamental mark of civilisation, distinguishing it from brute force, is the ability to co-operate. The new millennium needs a paradigm in which people work together, rather than against one another for whatever reason. Critics of this view claim that this brings everything down to a lowest common denominator and stifles innovation and intelligence. This does not need to be the case if the world also sees a new breed of explorers who are not afraid to strike out into the deep, just as European, Polynesian and Asian navigators set out centuries ago to conquer the unknown. Such a leadership is essential to countering the biodiversity crisis. The world is hungry for credible, dynamic and visionary leadership – as much in conservation as in any other human endeavour.

The acronym 'ACTS' (Editors' note: to avoid confusion the African Centre for Technology Studies (Chapter 2) possesses the same acronym) comes to mind in thinking about the essentials for ensuring that effective and long-lasting conservation of plant diversity is achieved, especially for the tropics. The acronym stands for Application, Co-operation, Training, and Success.

The first of these is *Application*. Application means acting and not just talking. Theory, even a good robust strategy, is of little consequence unless it results in positive actions on the ground. Action – the real test of a strategy – needs to be on the ground, with tribal people, farmers, business people, conservationists, students and others feeling a sense of ownership. We all need to get dirt under our fingernails by not just writing about conservation issues but by being out there alongside local people as solutions to conservation issues are sought.

A very practical test can be suggested here. When meeting in someone's office, the test is to ask what is really being done about a conservation issue. One is usually shown the rows of tidy reports on the bookshelf – the reports on what should be done. How much more satisfying it is to be taken to a vehicle and driven out to a field site, and shown what is happening away from the office – to be shown the action in the field.

Application is followed by *Co-operation*. No one organisation can achieve the goals of a multi-disciplinary, holistic crisis-driven programme such as biodiversity conservation. No one organisation can achieve all that is required in isolation. Interaction, co-operative collaboration and strategic alliances are essential to achieve plant conservation success. There is urgent need to seek and to nurture strategic alliances with appropriate international, national and local organisations, as part of an expanding global network that speaks a coherent and dynamic message.

Third is *Training*. People, organisations, trusts and community groups need tools and need to know how to use them. Hence, the SSC intention to encourage the promotion of 'best reasonable practice' in plant conservation. It is not possible to be globally prescriptive; that is we cannot have global formulae that will do everything. It is possible to have basic principles at global level, but prescriptions are a local matter; at each site and for every situation, one has to take into account the unique biological, cultural and socio-economic context.

A practical outcome of this should be the development of a range of training and facilitating products ranging from formal courses to inexpensive 'best practice' manuals on conservation management for key biodiversity areas. Alongside one can envisage websites with good, proven case studies and contact networks, widely accessible to conservation managers.

An important outcome of global co-operative strategies, translated into local 'on the ground' action, is the practice of integrated conservation. Integrated conservation is not complicated. It simply stresses the combined values of on-site conservation, off-site conservation through botanic gardens and gene banking, conservation biology research and education. It asks people to think through the most appropriate mix of approaches and techniques for any particular conservation situation.

The last letter of ACTS stands for *Success*. There will be outcomes that are not successful. These should not be buried, but are valuable indicators of what should have been done, and in life we probably learn more from mistakes than triumphs. But having a 'success attitude' means believing from the start that what we or anyone else does, can have a desirable end result. Negativity can readily destroy the successful outcome of a project before it even starts. We must also believe that solutions to the extinction crisis are achievable. This means steps of faith, trusting people and institutions, and being prepared to risk putting stewardship in the hands of people who are close to the action. And above all it means hope. Without that hope the war against biodiversity erosion will be lost.

There are some very significant conservation initiatives happening. There are some bold prospects. Many are very small, and local, sometimes driven by those who would be thought least likely to act decisively. It is a major challenge to learn about and facilitate what individuals and communities are doing in this way. Such situations ought to be a major focus of the ACTS approach. Although it is easy to be critical of small-scale and often poorly resourced local initiatives, they often have a powerful lesson for the wider world – that actions ultimately speak louder than words.

Finally, involvement in conservation means assuming a sense of both kinship and stewardship with nature. Stewardship itself, both global and local, means an assumption of responsibility, corporate and individual, as guardians of the world's plant diversity. I am tempted to say that we should be guardians of the world's plant diversity – and in the new millennium we must be. But it needs to go further and we need to ensure that all humans become guardians. We cannot become guardians by prescription and regulation. So I do not believe that we can 'save' the plants and the overall biodiversity of the world by coercion. People assume kinship and become guardians because they care, love and cherish, so we ought not enter lightly into assuming such a relationship with nature.

I believe that such a relationship with nature will have many dividends. People too readily underestimate the role of plants in their lives. A tragedy for the increasingly urban orientation of people worldwide is dislocation from their roots in nature. For too many of us, the world of nature starts at the supermarket shelf and we have lost vital connections (material, inspirational and spiritual) with the real world outside our human constructs. We need to put less stress on what the world of nature and its biodiversity owes us, and more effort into showing care and compassion for a world that is in pain. We need to change ourselves rather than force nature to make all the changes. To pledge to ensure the future and security of biodiversity on Earth might well be one of the best millennial resolutions we can make as citizens of planet Earth.

References

BGCI (2000). *The Gran Canaria Declaration.* Botanic Gardens Conservation International, London, UK.

Eagle, A. (1986). *Eagle's Trees and Shrubs of New Zealand.* Collins, Auckland, New Zealand.

Given, D.R. (1980). *Rare and Endangered Plants of New Zealand.* Reed, Wellington, New Zealand.

Given, D.R. (2000). Vanishing Act. *Forum for Applied Science and Research Policy,* Fall 2000: 11–15.

Hawksworth, D.L. (1998). The consequences of plant extinctions for their dependent biotas: an overlooked aspect of conservation science, pp. 1–16. In: C-I. Peng and P.P. Lowry II, (eds). *Rare, Threatened and Endangered Floras of Asia and the Pacific Rim.* Academica Sinica, Chinese Taipei.

Josephson, J. (2000). Going, going, gone. Plant species extinction in the 21st century. *Environmental Science and Technology, 1 March 2000*: 130A–135A.

May, R.M., Lawton, J.H. and Stork, N.E. (1995). Assessing extinction rates, pp. 1–24. In: J.H. Lawton and R.M. May (eds). *Extinction Rates.* Oxford University Press, Oxford, UK.

Mittermeier, R.A., Myers, N. and Mittermeier, C.G. (1999). *Hotspots: Earths Biologically Richest and Most Endangered Terrestrial Ecoregions.* CEMEX S.A. and Conservation International, USA.

Myers, N., Mittermeier, R.A., Mittermeier, C.G., da Fonseca, G.A.B. and Kent, J. (2000). Biodiversity hotspots for conservation priorities. *Nature* **403**: 853–858.

Pimm, S.L. and Raven, P.R. (2000). Extinction by numbers. *Nature* **403**: 843–845.

Vietmeyer, N.D. (1986). Lesser known plants of potential use in agriculture and forestry. *Science* **232**: 1379–1384.

WWF and IUCN (1994–1997). *Centres of Plant Diversity. A guide and strategy for their conservation.* 3 volumes. IUCN Publications Unit, Cambridge, UK.

Chapter 2

From Global Policy to National Action:

East Africa's opportunities and challenges in implementing the Convention on Biological Diversity

John Mugabe
African Centre for Technology Studies (ACTS),
Nairobi, Kenya

Introduction

The three countries of East Africa, Tanzania, Kenya and Uganda, have signed and ratified the 1992 Convention on Biological Diversity (CBD) (see Appendix 1), on the 8 March 1996, 26 July 1994 and 8 September 1993 respectively. Thus, they have incurred legal obligations to translate the Convention's provisions into national policy, programmes and actions. More specifically, the countries are expected to institute measures that promote the conservation of biological diversity, sustainable use of its components and the fair and equitable sharing of benefits arising from the utilisation of genetic resources (CBD, Article 1 (Objectives); Appendix 1, and Glowka *et al.*, (1994)). This is a formidable challenge for these countries given their poor economic conditions and a growing political instability in some of them. However, they are making major efforts to translate the provisions of the CBD into national policies, laws, programmes and actions (Figure 2.1).

Figure 2.1 The vegetation of tropical Africa has often been viewed as a mere backdrop to the issues of wildlife conservation and development, whereas it is in fact fundamental to the future of biodiversity and sustainable development.

This chapter argues that due to an absence of sound scientific and technological infrastructures – and biodiversity policies that aim at upgrading scientific and technological research institutions – the countries of East Africa will neither effectively manage biodiversity nor benefit from international trade in their genetic resources. It is observed that the mere formulation of national strategies, action plans and policies for biodiversity, will not realise conservation and its benefits. The primary challenge is to build a scientific and technological base for conservation and ensure that science and technology are harnessed and applied at the lowest levels of governance. Indeed, it is the extent to which local people and conservation agencies at the ground level will apply new knowledge and information for conservation that forms an indicator of national implementation of the CBD. This chapter calls for science-led biodiversity planning.

The first part of this chapter identifies and analyses a number of provisions (articles) of the CBD and the extent to which East African countries are implementing them. It is observed that the countries have largely taken a reactionary approach, that is, governments are largely reacting to decisions of the Conference of the Parties to the CBD. They are therefore in the process of developing and adopting national strategies and action plans on biodiversity and all of them have funding to do so.

The second part of this chapter focuses on the questions of access to genetic resources (Article 15) and technology transfer (Article 16) of the CBD. It observes that bioprospecting enterprises should be created and used as channels of technology transfer to the countries of East Africa. In negotiating for benefits from access to their genetic material, these countries should give emphasis to those arrangements that will enable them to participate in scientific research on the use of the material.

The last section of the chapter suggests a number of policy and institutional reforms that would strengthen the capacities of the countries to meet the objectives of the CBD. Such measures include the creation of economic and legal incentives for private sector engagement in conservation research, and the review and revision of conservation policies and programmes in order to integrate science and technology oriented considerations.

The Convention on Biological Diversity

1. Technical and scientific aspects

The entry into force of the CBD represents a major effort by the world community to establish consensus about the problems of both biodiversity degradation and loss. It is also a manifestation of a certain measure of convergence in governmental and non-governmental bodies' views of the wide range of actions that should be undertaken by various stakeholders to halt this global trend. Box 2.1 summarises the objectives and scope of the CBD.

Box 2.1　The objectives and scope of the Convention on Biological Diversity

The objectives of the Convention on Biological Diversity (Article 1) are: 'conservation of biological diversity, sustainable use of its components, and a fair and equitable sharing of benefits arising from the use of genetic resources'. The Convention is the first international agreement to address comprehensively all aspects of biodiversity conservation and to establish possible ways of addressing associated problems.

The issues covered by the Convention include:

- biodiversity strategies
- action plans
- intellectual property rights
- indigenous and traditional knowledge
- innovations and rights
- access to genetic resources
- economically and socially sound incentives
- financial resources and mechanisms
- regional co-operation
- handling of biotechnology and distribution of its benefits
- transfer of biotechnology
- *in situ* and *ex situ* conservation measures
- research and training
- public awareness and education

The CBD has so far been ratified by more than 176 nation states and the European Union (March, 2000). Its effectiveness now depends on how well these states translate its provisions into national policies, laws, programmes and actions. It really depends on the manner in which it is implemented at national and local levels. Compliance with, and effective implementation of, the CBD can only be achieved if the parties of the nation states consistently and wholly adopt its three objectives. It is neither the mere ratification nor participation in international fora that constitutes CBD implementation. In common with other global environmental treaties or agreements, the CBD does not possess internal mechanisms to ensure its effective and long-term implementation, at least at national and local levels.

To effectively engage in its implementation, parties to the CBD must have an adequate understanding of its provisions and an ability to consistently interpret and translate them into national activities. This is not an easy task, since the CBD has more than 30 articles, all with a wide range of provisions that can be interpreted differently by different parties and other stakeholders. This paper is concerned with those articles that are aimed at building the scientific and technological basis for the implementation of the CBD.

Articles requiring scientific and technological investment for implementation

Article 7 focuses on identification and monitoring of biodiversity and its components. It requires each contracting party to identify and monitor (through sampling and other techniques) components of biodiversity requiring urgent attention in terms of conservation and those offering the greatest potential for sustainable use. To undertake these tasks, countries must possess a critical mass of scientists in such areas as taxonomy. They also need a scientific infrastructure suitable for identification and monitoring. For example, monitoring of biodiversity and its components will, amongst other approaches, require satellite technologies as well as local competency to use them.

In situ conservation measures are articulated in Article 8 of the CBD. Parties are called upon to establish systems of protected areas and to develop guidelines for the selection, establishment and management of protected areas (Article 8(a) and 8(b)). In addition, they are expected to 'rehabilitate and restore degraded ecosystems and promote the recovery of threatened species, *inter alia*, through the development and implementation of plans or other management strategies' (Article 8(f)). Article 8 contains several other provisions that contain explicit science and technology policies. For example, 8(g) is about the management of risks associated with the development, use and release of living modified organisms from biotechnology; and Article 8(h) focuses on the control of alien species that threaten ecosystems. To implement these provisions, the countries of East Africa must harness and apply science and technology – particularly new technologies such as biotechnology and information technology. For example, to manage risks associated with the release of living modified organisms, a country must have an infrastructure for biotechnology assessment and should be able to monitor the evolution of organisms when released into the environment. This is a science-intensive exercise that only those countries with modern biological sciences research facilities are likely to effectively engage in.

Other provisions, whose implementation requires direct scientific and technological investments, are contained in Articles 9, 14 and 19. Article 9 deals with *ex situ* conservation of the components of biodiversity, while Article 14 provides measures for environmental impact assessment and minimising adverse impacts of various activities on biodiversity. To implement Article 9, countries will need to establish and/or upgrade such facilities as gene banks, and zoological and botanical gardens. In the area of gene banking, they will need to invest in techniques such as tissue and culture cryopreservation. In addition, the collection and appropriate documentation of germplasm or accessions will need to be guided by or based on taxonomic studies.

Article 19 of the CBD addresses two interrelated issues: biosafety and the distribution of benefits from the development and application of biotechnology. Biosafety is a term generally used to refer to regulatory measures (policies and procedures) to ensure that the application of biotechnology does not cause negative environmental impacts. Articles 8 and 19 of the CBD contain biosafety provisions. Article 8(g) requires parties to 'establish or maintain means to regulate, manage or control the risks associated with the use and release of living modified organisms resulting from biotechnology which are likely to have adverse environmental impacts that could affect the conservation and sustainable use of biological diversity, taking also into account the risks to human health'. Countries are required to formulate specific policy and legal measures to regulate the application of biotechnology. Article 19(3) of the CBD calls for development of a specific protocol on biosafety.

National responses and challenges for East Africa

The CBD was negotiated while the three countries of East Africa were engaged in major political and economic reforms. These included those associated with managing the transition from a socialist form of governance to a capitalist one (in the case of East Africa), introduction of multi-party politics (in both East Africa and Kenya), and economic liberalisation (in all the countries). The countries were largely preoccupied with these reforms and their participation in the negotiation and formation of the CBD was very limited. Indeed, national political and economic problems and changes consumed most of the region's attention and energies; to the extent that it was not capable of mobilising political support and scientific talent to participate in the negotiations.

The limitations associated with their participation in the negotiation of the CBD not withstanding, the countries are now legally bound to implement its provisions. A host of factors will determine how well they do so. These include:

- whether there is an adequate understanding among the public and policy-makers of the provisions of the CBD and what specifically has to be done;
- the level of public support to change the trend in loss of biodiversity;
- the level of commitment amongst the political elites;
- the nature and strength of relationships among and between public agencies;
- the kinds of policies and laws (as well as organisations) that will be created to address new issues and problems of biodiversity; and more importantly,
- the extent to which the countries will harness and apply science and technology for conservation and sustainable use of biological diversity.

The implementation of the CBD is a new challenge for developing countries in general and East African ones in particular. It is new in a number of respects:

a) The nature of the instrument calls for holistic and systemic approaches. Biodiversity is about variety in and diversity of ecosystems, species, genes, and cultures. It is holistic. Reductionist and stand-alone organisational approaches cannot assist countries to implement the CBD. Meeting the goals set by the instrument requires institutional synergies – new ways of organising public agencies and programmes.

b) Unlike other biodiversity related conventions, such as the Convention on International Trade in Endangered Species of Wild Fauna and Flora (CITES), the CBD is erected on certain new policy contours (e.g. issues of biosafety, access to genetic resources, and technology transfer) where there are no countries with prior experience. The countries of East Africa have very few sources for accumulated experience. Issues such as access to genetic resources and benefit sharing have just entered the international policy arena. Experiences on how to institute policy and laws to regulate access to genetic resources and ensure benefit sharing are few and short. Opportunities for learning from others are therefore limited. This makes prospects for the region uncertain in implementing the access and benefit sharing provisions. However, it is possible to devise policy learning and renewal processes to address such limitations.

c) Environmental policy and planning are recent preoccupations of the governments of East Africa. They have fairly short histories of national environmental policy-making and planning. When considering biodiversity policy and planning which are even newer areas for them, it becomes clear why institutional capacity has to be upgraded. The emphasis however should be on the integration of science and technology oriented measures into such regimes, and then the implementation of such measures – rather than the formulation of general biodiversity regimes.

A review of national biodiversity planning and policy-making in the three countries demonstrates that science and technology have not formed a core aspect or concern in these efforts. In Kenya, for example, the draft National Biodiversity Strategy and Action Plan (NBSAP) does not contain specific strategies and actions that will be invested in to mobilise science and technology for biodiversity conservation and the sustainable use of its components. There is only passing reference to the provisions of Article 19 of the CBD. Another concern about the nature of biodiversity planning in the countries is their level of generality. The planning processes are largely preoccupied with generalities – their engagement in the core provisions of the CBD are very limited. They are also founded on a weak scientific base and are cast to reflect the general national concerns and modes, and often ignore ecosystem-specific issues. On the whole, there is a mismatch between scientific activities and policy-making.

2. Access to genetic resources and transfer of technology: the potential and limitations for East Africa

East Africa is richly endowed with genetic resources that form a significant portion of raw material for global industrial development in such sectors as pharmaceuticals, agriculture and chemicals. Its national economies are also heavily dependent on the use of genetic resources in such sectors as crop production, livestock and health. Furthermore, there is an increasing global interest in the region's genetic capital, particularly by pharmaceutical companies of the industrialised countries.

The last few decades have witnessed a revival of interest in East Africa's genetic diversity. Scientific advancements, especially in the field of biotechnology, have stimulated interest in medicinal plants and the related indigenous knowledge. As a result, academic and commercial institutions have increased their investment in the search for new biological and chemical compounds from East Africa's biodiversity. In Uganda, a private company is prospecting for *Cymbopogon* Spreng. (Gramineae) spp. (commonly referred to as 'mutete') for the production of toothpaste. *Aloe* L. (Aloaceae) spp. have been harvested from Kenya and extensively used in the USA for cancer research. Though there is scanty, and in many cases a complete lack of, information on the extent of bioprospecting and the specific plants and animals as well as micro-organisms targeted, it is clear that firms are prospecting in the region. In Tanzania, it is reported that one firm has worked out collaborative arrangements with the Institute of Traditional Medicine in bioprospecting for compounds in locally and traditionally used medicinal plants.

Recent evidence demonstrates that there is increasing private industry interest in *Prunus africana* (Hook. f.) Kalkman (Rosaceae) – a medicinal plant found in Kenya (see Jaenicke *et al.*, Chapter 20 and Sunderland *et al.*, Chapter 21). The bark of *P. africana* contains active compounds for the treatment of benign prostate hyperplasia. Major markets for the extraction of the plant are in Italy, Switzerland, France and Austria. Italian companies import bark extract from Kenya and Uganda. Capsules and tablets are marketed under various trade names including: 'Tadenan' and 'Pygenil'. The total market value of trade in *P. africana* was estimated to be US$150 million in the early 1990s (Cunningham and Mbenkum, 1993).

Regulation and legislation

On the whole, there is considerable potential for existing and future bioprospecting in East Africa. Modest estimates would put the market value of bioprospecting enterprises in the region to US$1.5 billion. Despite the considerable potential and increasing interest in the region's biodiversity, governments of these countries have not devised systemic regimes to regulate access to genetic resources and to ensure that they share (in a fair and equitable manner) benefits arising from bioprospecting. They have not instituted measures to implement Article 15 of the CBD. However, these countries have policies, laws and administrative measures to promote conservation of biological diversity. These measures contain provisions to regulate trade in biological resources, with emphasis on wildlife, fisheries and forests. They are not obtained in any one policy regime or body of legislation. They are found in sectoral policies and legislation in such specific instruments as national wildlife acts, national wildlife policies, fisheries acts and forest legislation. Existing wildlife and forestry laws contain provisions on ownership of and access to the resources. For example, Kenya has the Wildlife Conservation and Management Act (amended in 1989) to control illegal access to and exploitation of wildlife resources. Individuals and/or institutions have no right to extract wildlife or parts thereof protected by law without authority of the national agency in charge of wildlife. The Act requires that any person and/or institution seeking access to wildlife or parts thereof shall obtain prior consent of the relevant authorities; currently the Minister of Natural Resources and the Kenya Wildlife Service. The Act does not, however, contain provisions requiring the sharing of benefits arising from access to and utilisation of wildlife resources. It is also silent on participation of local people in determining access to wildlife, particularly that found on private lands.

Apart from wildlife laws, another national instrument that could be used to regulate access to genetic resources is forest legislation. The countries have formulated and instituted forest legislation that aims at regulating the exploitation of forests and forest resources. In Uganda, the Forest Act stipulates that individuals are not to enter a forest reserve for purposes of extracting any forest resource or undertaking activities that may cause damage to the forest ecosystem. Permission to enter forest reserves is granted by the Chief Forest Protection Officer. This law, as is the case in most other countries, was enacted to preserve forests. It does not specifically cover benefit sharing. However, the law could be amended to integrate various provisions for regulating access to genetic resources. Box 2.2 highlights the processes and approach Uganda has taken to develop draft regulations. Other instruments that could be applied by the countries to implement the provisions of Article 15 of the CBD include: quarantine regulations, seed and agricultural trade policies and laws, research permits, plant variety protection laws, and such administrative tools as presidential and ministerial decrees.

The application of existing national legislation and sectoral oriented instruments, to regulate access to genetic resources and promote the sharing of benefits from the utilisation of those resources, suffers from a number of limitations. These include their very sectoral nature, as well as the fact that they were not established to address concerns of benefit sharing. A review of existing natural resources laws and policies shows that these instruments are inadequate in addressing benefit sharing.

Building the effectiveness of legislation

A number of key points can be made about current efforts to build effective legislation:

a) There is little evidence to show that the countries are giving adequate, if any, attention to the potential of new technology, particularly biotechnology, in promoting illegal bioprospecting as well as in monitoring this activity. While the countries invest in formulating and adopting national policies and legislation to regulate access to their genetic resources, it will be crucial for them to focus on the urgency of establishing a technological infrastructure to monitor bioprospecting in order to quantify illicit trade in national resources. In fact, they could integrate specific provisions on technological capacity building for monitoring into their legislation. Part of the benefits that they would derive from access arrangements would be in the form of both soft and hard monitoring techniques. Furthermore, the countries will need to upgrade infrastructure and capacities of institutions such as customs departments and the police to enforce laws aimed at regulating illegal collections. They will also need to improve recording-keeping systems and create links between those institutions charged with monitoring and enforcement and those responsible for granting prior informed consent and permits to prospect for genetic material.

b) The extent to which the countries will derive maximum benefits from bioprospecting largely depends on their abilities to add value to the genetic resources. Bioprospectors are now targeting countries in which taxonomic studies have been conducted, unique genetic material has been identified, and where local and traditional knowledge is organised in such ways as to provide easy technical leads in the collection and screening processes. In the absence of organised local knowledge of the material and indigenous capacity to sample genetic material for prospecting purposes, the countries are unlikely to attract and benefit from companies. The East African countries should, therefore, give attention to the need to create local and national enterprises that add value and trade in genetic resources. Specific measures that enlarge the legal personality of local communities, promote their knowledge of plants and build their technological basis for engaging in bioprospecting should be integrated into national access and benefit sharing legislation.

Opportunities for Implementing the CBD

The countries of East Africa have numerous opportunities to implement the CBD. These opportunities have been created through:

1. **Significant donor interest and support to the countries' policy and planning processes.** Over the past decade or so, new donors have targeted project and programme support to East Africa's biodiversity activities. Such donors as the Global Environment Facility (GEF), the World Bank, the Norwegian Agency for Development Co-operation (NORAD) (www.environment. norad.no/), the Swedish International Development Agency (SIDA) (www.sida.se), and the Canadian International Development Agency (CIDA) (www.w3.acdi-cida.gc.ca/index.htm) are providing significant financial and technical resources to support biodiversity activities in East Africa.

2. **Consensus among policy-makers on the importance of biodiversity management**. There is also recognition of the need to undertake major reforms in national policies and institutions to achieve the goals of the CBD. Both policy-makers (at least within environmental agencies) and politicians (some participated in the negotiation of the CBD) are increasingly demonstrating their commitment to reform.

3. **The preparation of national biodiversity country studies for all three East African countries supported by United Nations Environmental Programme (UNEP)**. The studies acknowledge that there are major limitations in policy and legislation pertaining to conservation. While not as comprehensive and up-to-date as desired, these studies are a stepping-stone to biodiversity policy and planning.

4. **Open discussion on issues relating to environmental reforms between East Africa's policy-makers and other stakeholders (related to point 2 above)**. The level of openness among policy-makers has made it easy for other NGO policy advisors, such as ACTS, to start making contributions to policy processes. It offers the countries an opportunity to debate issues and experiment with certain policies. It also facilitates the flow of information to the public domain.

5. **Recognition of the need to decentralise environmental planning and management responsibilities to local levels**. This recognition is reflected in new policies, such as the wildlife and forest policies, that make reference to strengthening the capacities of communities to manage the resources. It offers government and NGOs an opportunity to devolve policy-making, as well as biodiversity management responsibilities, to local levels and could create the necessary space for establishing and experimenting with ecosystem-specific policies.

6. **The engagement of a number of non-governmental organisations (NGOs) (such as the wildlife societies, the World Conservation Union (IUCN), World Vision, African Wildlife Foundation (AWF)) in conservation concerned with advocacy in certain legal issues**. These create the institutional diversity required to develop and test new policy ideas. It also reduces the burden on governments. NGOs are beginning to provide proactive support to government. Some of them are becoming strong watchdogs for biodiversity and the environment in general.

Box 2.2: Uganda – process and approach to the development of draft regulations

Kenya and Uganda have initiated national processes to prepare specific legislation on access and benefit sharing. While Kenya has taken the route of preparing a subsidiary legislation – essentially a gazette notice under the current Science and Technology Act – Uganda (with the assistance of the African Centre for Technology Studies (ACTS)) has prepared detailed regulations on access and benefit-sharing.

Draft Regulations on Access to Biological Resources and Benefit Sharing in Uganda were discussed within national government agencies, non-governmental organisations and the private sector before they were revised and adopted at a national workshop.

There are a number of innovative provisions in Uganda's draft regulations:

1. The draft contains specific provisions on technology transfer and the building of technological capabilities in such new areas as biotechnology. Technology partnerships are perceived as mechanisms of technology transfer.

2. Uganda treats issues of access to genetic resources as largely technological, and it requires oversight by the National Council for Science and Technology. The Council is vested with, in the current draft regulations, the administrative and overall supervisory responsibilities of implementing the access and benefit sharing requirements of the legislation (Republic of Uganda, 1999).

There are however a number of limitations with Uganda's process and approach in formulating a regime to regulate access to genetic resources and promote benefit sharing. These are primarily as a result of the mismatch (at least the absence of explicit convergence) between the process and overall national biodiversity planning activities being undertaken to implement Article 6 of the Convention on Biological Diversity (CBD).

An effective regime on access and benefit sharing should draw from and be guided by (or even build upon) a National Biodiversity Strategy and Action Plan (NBSAP). One effective way to do this is to prepare a component of the NBSAP specifically on access and benefit sharing. This could assess the biodiversity and market opportunities in the country and the human, institutional and financial resources available to allow sustainable use of genetic resources by adding value to them within the country. It could also identify priorities for conservation and development that could be used to negotiate benefits in return for access that would be most useful to the country. Uganda, for example, is still preparing its NBSAP. The extent to which the proposed access and benefit sharing regulations will conform to the broader national agenda to implement the CBD is therefore not clear.

Conclusions: Building National Capacity for the CBD

The above sections have provided an indication of the challenges and opportunities that countries in East Africa have in implementing the CBD. To confront the challenges by tapping the opportunities, the countries may consider:

1. Establishing standing committees on the CBD, whose overall task would be to provide an inter-agency mechanism for co-ordinated biodiversity planning and policy-making. Such a committee could oversee the development of a national biodiversity strategy and action plans.

2. Investing in training some of the policy-makers in new issues, particularly those pertaining to access to genetic resources, technology transfer and benefit sharing (e.g. East African Biodiversity and Law Training, Nairobi, July–August, 1999).

3. Exploring, through case studies and pilot projects, the feasibility of formulating ecosystem-specific policies. Institutions, such as ACTS, could undertake studies to assess what impacts national policies have on the management and sustainability of local ecosystems and whether and how local and ecosystem policies can be established without generating conflict with national ones.

4. Raising public awareness and understanding of the core issues of biodiversity, national obligations under the CBD, specific national policies and laws on biodiversity, etc. Programmes that target politicians, youth, women, and the business community to inform and educate them, on CBD issues that directly affect and benefit them, would enlarge the national constituency for implementation.

This paper has provided an overview of policy issues from the CBD and the challenges, as well as the opportunities, that East Africa has to implement this global agreement. It has been noted that three areas of focus are crucial: establishing and strengthening institutional arrangements; devolving biodiversity policy and planning to make policies sensitive to ecosystem management needs as well as local people's aspirations and capacities; and capacity building to enhance understanding of policy-makers and other stakeholders in biodiversity.

References

Cunningham, A.B. and Mbenkum, F.T. (1993). *Sustainability of harvesting* Prunus africana *bark in Cameroon: a medicinal plant in international trade.* UNESCO, Paris, France.

Glowka, L., Burhenne-Guilman, F., Synge, H., McNeely, J.A. and Gündling, L. (1994). *A Guide to the Convention on Biological Diversity.* Environmental Policy and Law paper No. 30. IUCN, Gland, Switzerland.

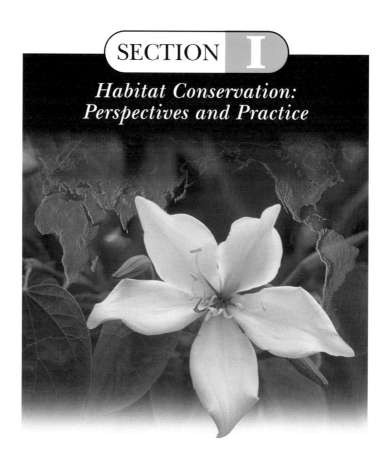

SECTION I

Habitat Conservation:
Perspectives and Practice

Chapter 3

Section Overview: Conserving Tropical Botanical Diversity in the Real World

Mike Maunder[1]
Royal Botanic Gardens, Kew, UK

Colin Clubbe
Royal Botanic Gardens, Kew, UK

[1] Now at: The National Tropical Botanical Garden, Kauai, Hawaii, USA

Introduction

"We are the generation for whom the only message for a tropical biologist is: *Set aside your random research and devote your life to activities that will bring the world to understand that tropical nature is an integral part of human life.* If our generation does not do it, it won't be there for the next."

Dan Janzen, 1986.

Botanical diversity provides vital life support services including the fundamentals of oxygen production, the formation of soils and the regulation of watersheds. For human society, wild and cultivated plants provide food, fibre, medicinal products, dyes, aromatics, ornamentals and more. This relationship is vividly illustrated in the tropics, where a large proportion of the world's population is directly dependent upon fragile soils, seasonal rains and local plant resources. The documentation, management, use and retention of botanical diversity are subjects of urgent debate, driven by a realisation that plant genetic resources continue to be eroded. These plant resources, valued in financial, cultural, ethical and emotional terms, form the very fabric of human life (Box 3.1).

Box 3.1 **The value of ecosystem services and natural capital**	
Ecosystem Services	**Value (trillion US$)**
Soil formation	17.1
Recreation	3.0
Nutrient recycling	2.3
Water regulation and supply	2.3
Climate regulation (temperature and precipitation)	1.8
Habitat	1.4
Flood and storm protection	1.1
Food and raw materials production	0.8
Genetic resources	0.8
Atmospheric gas balance	0.7
Pollination	0.4
Other services	1.6
Total value of ecosystem services	**33.3**
Source: Costanza *et al.*, (1997).	

The Convention on Biological Diversity (CBD) is the major legislative influence on the conservation of biodiversity directly influencing national activities through the requirement to produce and implement Biodiversity Action Plans (Box 3.2, Appendix 1 and Glowka *et al.*, 1994).

Box 3.2 The Convention on Biological Diversity

The Convention on Biological Diversity makes specific recommendations on species level conservation:

Article 8. *In situ* Conservation

(d) Promote the protection of ecosystems, natural habitats and the maintenance of viable populations of species in natural surrounding;

(f) Rehabilitate and restore degraded ecosystems and promote the recovery of threatened species, *inter alia*, through the development and implementation of plans or other management strategies;

(h) Prevent the introduction of, control or eradicate those alien species which threaten ecosystems, habitats or species;

(k) Develop or maintain necessary legislation and/or other regulatory provisions for the protection of threatened species and populations;

Article 9. *Ex situ* Conservation

(a) Adopt measures for the *ex situ* conservation of components of biological diversity, preferably in the country of origin of such components;

(b) Establish and maintain facilities for *ex situ* conservation of and research on plants..., preferably in the country of origin of genetic resources;

(c) Adopt measures for the recovery and rehabilitation of threatened species and for their reintroduction into their natural habitats under appropriate conditions;

(d) Regulate and manage collection of biological resources from natural habitats for *ex situ* conservation purposes so as not to threaten ecosystems and *in situ* populations of species, except where special temporary *ex situ* measures are required under subparagraph (c) above.

Outlined under Article 6 (General Measures for Conservation and Sustainable Use) is the requirement for each Contracting Party to develop a national biodiversity strategy, action plan or programme. Can conservationists use the CBD as a lever to obtain resources for the conservation of plant diversity or will we face a split between conservation policy (the committee room) and conservation reality (the reserve boundary and other biodiversity front-lines)? (Given, Chapter 1; Mugabe, Chapter 2).

Conservation actions are being employed to halt or reverse the damaging impacts of human induced change upon biodiversity. These efforts vary in the scale of spatial, financial and scientific investment, and encompass a range of activities including single species management through to wilderness retention. Management actions aimed at the conservation of biodiversity will take place at various levels of the biodiversity hierarchy – a nested hierarchy of spatially defined units often with ill-defined boundaries (Soulé, 1991):

• Ecosystem
• Community
• Species
• Population
• Individual
• Gene

Conservation problems are manifest at different levels of this hierarchy, for instance, a change in flooding or burning regimes will require action at the ecosystem level, while a species specific problem, e.g. overharvesting or the decline of a pollinator, will require action at the species level. The ecosystem and community level are the most complex and least understood units of conservation management. Ecosystem or habitat conservation is recommended as the foundation for effective conservation planning by both the CBD and Agenda 21. It is generally acknowledged that extensive protected areas are an effective mechanism for retaining a large proportion of a region's biota (e.g. Figure 3.1). The focus of management is both ecological processes, e.g. nutrient flow, water systems etc., as well as composition, e.g. species.

Habitat destruction is widely recognised as the most important anthropogenic cause of biodiversity loss, with invasive species becoming increasingly important as agents of extinction. Human activities, both local and global, are modifying botanical diversity; with nearly 40% of the earth's potential terrestrial net primary productivity either used by mankind or lost as a result of land change (Vitousek, 1994). The extensive conversion of wild habitats has resulted in changes to both the ecological and taxonomic composition of these areas. In many parts of the tropics the change is almost complete. Taxonomically and ecologically diverse wildlands and agro-ecosystems have been converted into intensive agricultural landscapes dominated by a relatively small range of domesticated species, for instance, the loss of natural grasslands and their replacement by cereal monocultures of wheat, or the conversion of species rich forests to single species plantations (Figure 3.2). The process of agricultural expansion into wilderness areas and the modernisation of agriculture in previously traditional working landscapes have left only scattered fragments of the original habitats. These fragments, defined as residual and isolated habitat patches with a high edge to area ratio, are undergoing high levels of species loss (Gascon *et al.*, 2000; Turner, 1996) and modifications in patterns of genetic diversity (Young *et al.*, 1996) often with a reduction in the level of genetic diversity in persisting populations (Hall *et al.*, 1996). The extent of fragmentation is often massive; for example, the area of fragmented and disturbed forest in the Brazilian Amazon is estimated to be 150% larger than the area actually deforested (Skole and Tucker, 1993).

Figure 3.1   The future of tropical plant diversity will depend upon the establishment and retention of large protected areas that include samples of habitats and allow ecosystem and evolutionary processes to continue. Here undisturbed moist lowland rainforest cascades down to the beaches of the southern coast of Trinidad. This is an increasingly uncommon situation as coastal habitats come under increasing pressure from development.

As a result of human activities the rate of extinction has increased beyond natural or background levels; indeed, we may be accelerating the rate of extinction four orders of magnitude beyond the background rate recorded in the fossil archives (Lawton and May, 1995). The majority of known extinctions are the poorly documented artefacts of habitat loss and degradation. For the vast proportion of tropical regions we have no clear idea of species decline and extinction; only for a few charismatic and favoured groups, such as orchids, succulents and palms, do we have even a preliminary understanding of how many species are threatened. Whilst species extinction has preoccupied many conservationists, we are also seeing the increasing loss of populations and associated patterns of genetic diversity (Hughes *et al.*, 1997).

As a result of deliberate introductions and as artefacts of agriculture, international trade, forestry and horticulture an increasing number of plant species are growing in new territories and habitats (Lonsdale, 1999). Whilst such introductions may in the short term increase local biodiversity, in the long term they are having disastrous impacts on both species diversity and ecological processes (Pimental *et al.*, 2000). In many parts of the tropics exotic

Figure 3.2 Fragmentation of moist lowland forest is exemplified by the encroachment by plantations of teak (*Tectona grandis* L.f., Verbenaceae), lower left and right, in south Trinidad.

plants are now a permanent component of the landscape; for instance, the dry lowland forests of Mauritius, have been replaced by a weedy assemblage of African forage grasses, trees of the Asian *Albizia lebbeck* (L.) Benth. and thickets of neotropical *Furcraea* Vent. and *Lantana* L.

Human activities, notably the combustion of fossil fuels and the subsequent release of carbon dioxide, are altering the composition of the atmosphere. It is predicted that the historic concentration of carbon dioxide in the earth's atmosphere may double in the twenty-first century and modify global climate. This could result in an increase in mean global temperature of approximately 1.5–4.5°C (WRI/UNEP/UNDP/World Bank, 1996). Botanical diversity will change in spatial distribution as species react through both local extinctions and the colonisation of new areas (Morse *et al.*, 1995). Ecological processes will change as fire and hydrological systems alter. It is likely that these changes will cause the extinctions of plant species as populations die out and are unable to migrate out of isolated habitats. Changes in phenology and population growth patterns are also possible. In addition it is likely that invasive weed species will spread quickly through these new and unstable communities (Dukes and Mooney, 1999). The potential impact of climate change threatens to undermine existing protected areas and conservation investments; indeed, as a problem it both defies prediction and effective planning. The synergistic impacts of changing climate regimes, nitrogen deposition, invasives, elevated carbon dioxide levels and land use are difficult to predict, and each will have differing impacts according to the sensitivity of the different ecosystems; for instance, it is expected that freshwater ecosystems will suffer more particularly from invasives whilst grassland ecosystems will be heavily impacted by all the above listed factors (Sala *et al.*, 2000). A few early signs suggest future trends. In 1998, northern Borneo suffered serious droughts related to the El Niño-Southern Oscillation event of 1997–1998. This drought interrupted the flowering patterns of figs and has led to local extinctions of pollinating wasps (Harrison, 1999). It is already suspected that high altitude tropical communities, such as cloud forests, are being affected as mist and precipitation patterns change and influence bird and amphibian populations (Pounds *et al.*, 1999).

All of these damaging impacts are driven by one fundamental force, the increase in human population. The world population currently stands at more than 6 billion. Estimates for future totals, based on current demographic trends, suggest that by 2050 the world population will reach 9.4 billion. However, future trends will be influenced by changing fertility rates, life expectancies and the progress of economic transition for developing countries. Human populations are still suffering from poverty; this is driving millions of people into the unsustainable use of resources. Some 1.3 billion people live in absolute poverty, 840 million people are undernourished and about 1.4 billion do not have access to safe water. In addition, 900 million people are still illiterate (Box 3.3).

The retention of tropical plant resources and associated habitats cannot be separated from the fundamental needs of human society. The retention of functioning ecosystems will be essential in maintaining economic growth. The majority of jobs in the tropical world are in rural areas and directly linked to natural resources. These natural resources (crops, timber, animal and plant products) still make major contributions to national and global economies. However, current patterns of land use, crop production and marketing threaten agro-ecosystem resources such as water supplies, soil stability and fertility, and biodiversity. Many of the long-term responses to the biodiversity crisis will depend upon establishing stable economies and better lives for the poor that remove the plague of poverty; a shared challenge for conservationists, governments and business (Cincotta *et al.*, 2000; Daily and Walker, 2000; Thrupp, 2000). As economies increase in wealth so spending patterns will change and so attitudes to the consumption and conservation of biodiversity will change.

Box 3.3 Urban Growth and Biodiversity

The world's population continues to grow, but the distribution of that population is changing. Not only is it certain that the world's population will increase beyond its current total of 6 billion people, but also that the world will see a further growth in huge urban areas, the mega-city.

The world's urban populations are growing faster than rural populations; between 1990 and 2025 the number of people living in urban areas is projected to double to more than 5 billion, an estimated 90% of this growth will take place in developing countries. Sub-Saharan Africa has the highest urban growth rate of 5% per year in the world. In 1980, 27% of the region's population was urban, by 2020 it is expected to reach 49%. Over the same period Europe will experience a shift from 69% in 1980 to an expected 80% by 2020. The massive increase in urban populations is stretching the capacity of many cities to provide even basic living conditions. An estimated 25–50% of all urban inhabitants in developing countries live in impoverished slums and squatter communities with restricted or no access to safe water, sanitation or refuse collection. This trend towards an increasingly urban population will contribute to the growth of the mega-city, escalating the number of cities of 8 million inhabitants; in 1950 there were only two such cities, by 2015 the number is expected to grow to 36, with 23 of these being in Asia.

The growth of urban areas will present a number of challenges to the plant conservationist. Firstly, the physical conversion of farmland or wild habitat to urban structures will often impact on high diversity areas. For instance, the Brazilian mega-cities of Sao Paulo and Rio de Janeiro are in the midst of the highly threatened and exceptionally rich Atlantic rainforests. This urban growth will create a mosaic of housing, industry and habitat, where the conservation challenge will be the retention of the original biodiversity in highly fragmented landscapes. For example, the Nairobi National Park, Kenya, with its annual migration of wildebeest, is increasingly isolated from other natural areas by urban and industrial growth. Nairobi has grown from 87,000 in 1950 to 1.8 million in 1995. Cape Town, in the Republic of South Africa, is situated in the unique Cape Floristic hotspot, with expanding suburbs threatening the Cape Flats, an area rich in endemic plants. Not only are the orginal habitats destroyed or fragmented by urbanisation, but ecological parameters are influenced by urban expansion with natural fire regimes, flooding and grazing affected, and importantly, urban gardens can introduce new suites of potentially damaging invasive plants.

Apart from the physical destruction and fragmentation of original vegetation, the mega-cities will have other conservation impacts. Two starkly different trends are developing; firstly the comfortable portion of the population, often relatively affluent and politically powerful, become increasingly isolated from the rural environment. Urban areas act as trading centres for natural products, such as charcoal, medicinal plants or bush-meat. In contrast the poor, a relatively mobile population, may contribute to rural settlement.

These trends mean that urban biodiversity facilities have three main responsibilities:

1. The retention of surviving biodiversity in urban areas through retaining habitat fragments and restoring degraded areas.
2. Working with trade groups to modify consumption patterns for natural products.
3. Working with land use agencies to reduce the impact of the migrating urban poor on distant habitats.

Sources:
Urban and Rural Areas, 1950–2030 (1996 Revision); United Nations, 1997.
World Urbanization Prospects (1996 revision); United Nations, 1999.

Figure 3.3 As protected areas become increasing isolated the status of the retained botanical diversity within these enclaves will be increasingly linked to the management of large mammal populations. Increasing elephant populations within the Mwaluganje Elephant Sanctuary, Kenya have destroyed extensive areas of *Brachystegia* Benth. woodland.

Protected areas for protecting plant diversity

"In the end, the future population of this earth will inherit and guard no more of tropical species and wildlands than we set aside in our lifetime."

Dan Janzen, 1986.

The greatest proportion of the world's plant diversity will be retained through the 'coarse filter' approach (*sensu* Hunter *et al.*, 1988), whereby effective habitat conservation could potentially conserve all levels in the biodiversity hierarchy and their interactions. The surviving major wilderness areas, large relatively undisturbed natural areas, such as the Zaire basin, Western Amazon basin, New Guinea, offer the best opportunities for retaining ecosystem and evolutionary processes (Hannah *et al.*, 1994; Mittermeier *et al.*, 1998). In addition Myers *et al.* (2000) have identified 25 endemic rich enclaves, largely tropical or mediterranean, that are under immense pressure from habitat loss. These areas contain 44% of the world's vascular plants and provide a strategic opportunity to secure a large proportion of the world's plant diversity. However, current protected area portfolios are heavily influenced by historic decisions based on the conservation of scenic upland areas and spectacular savannah mammal populations. Accordingly, important habitats and ecosystems may not be represented. A recent study of protected areas in Costa Rica revealed that only nine of the 23 life zones in Costa Rica were represented in protected areas. The addition of about 400,000 hectares to the protected area system through expanding 12 existing protected areas and the creation of a new protected area would adequately protected 19 life zones (Powell *et al.*, 2000).

The effective development of a protected area network will be dependent upon identifying target areas that will represent, or sample, the full range of biodiversity, and then ensuring their long-term persistence or viability (derived from Margules and Pressey, 2000):

- Measure and map diversity. The first stage of effective planning will be based on the measurement and mapping of biodiversity, this can be done through mapping of habitats, sets of species or species assemblages (e.g. charismatic groups such as orchids or more utilitarian groups such as legumes or palms). Other surrogate data sets can include climate, geology, locations of threatened species as well as more traditional indicators, such as vascular plants, birds, reptiles and mammals (Du Puy and Moat, 1996; Jones *et al.*, 1997; Valverde and Montana, 1996) (see Pearce and Bytebier, Chapter 4 Hernandez and Gómez-Hinostrosa, Chapter 18).

- Identify conservation goals for the planning region. The conservation goals should deal with the needs of both species and landscape processes. The emphasis should be on retaining extensive wild landscapes that allow local and regional scale population dynamics, and take into account traditional land use patterns. The design of these reserves will be influenced by an understanding of ecological linkages, for instance mobile frugivorous birds as umbrella species for the fragmented Atlantic forests of Brazil (Cardosa da Silva and Tabarelli, 2000). Soulé and Sanjayan (1998) have stated that more land is needed beyond the conventional target of allocating 10–12% of total land area for biodiversity. They suggest about 50% of the total land area is required. However, there is strong, and often bitter, competition between allocating land for conservation and the demands for agriculture and industry. Musters *et al.* (2000) argue that this objective may be feasible in the long term, but will be dependent, at least in Africa, on serious and long-lasting investment in economic development to deal with the immediate conflicts between conservation and food scarcity (see Johansson *et al.*, Chapter 5).

- Review existing reserves. Use spatial information to support the process of gap analysis, an established tool in conservation planning (Scott *et al.*, 1993). As with the Costa Rica study, an understanding of habitat types and level of protection can guide the location of new and needed reserves complementing any existing reserves.

- Select additional reserves. Using data derived from the previous steps the identification of needed additional reserves allows gaps in the portfolio to be filled. This can be a complex process and increasingly algorithms are being utilised to aid decision-making. These calculations utilise distribution data to calculate the most efficient allocation of investments using the concepts of complementarity and irreplaceability (Freitag and Van Jaarsveld, 1998; Lombard *et al.*, 1997; Trinder Smith *et al.*, 1996).

- Implement conservation actions on the ground. The establishment of protected areas will be dependent upon careful and detailed negotiations with local populations and national agencies, indeed, traditional land management may favour the retention of valuable habitat areas (Fairhead and Leach, 1999).

- Management and monitoring of reserves. The borders of protected areas and wilderness areas are permeable to a range of threats including disease, invasive species, poaching, civil unrest and climate change. The management of protected areas will directly influence the status of retained (or restored) plant diversity. A study in Tanzania has shown that the status of the vegetation is linked to the management and policing of protected areas. Those areas with a high investment in policing (e.g. conventional national parks) showed better vegetative cover than areas allocated for multiple use (Pelkey *et al.*, 2000). Whilst this study points to the importance of on-site management and regulation, the long-term effectiveness of these measures are uncertain (see Johansson *et al.*, Chapter 5).

All of the above activities can be undermined if insufficient time is given to involving local communities in the planning and management of protected areas. Very few parts of the world are devoid of humans, and indeed, the continued presence and activities of local populations may be necessary to maintain desired levels of biodiversity. For the long term success and viability of any conservation project local support and understanding is vital (see Kamatenesi-Mugisha and Bukenya-Ziraba, Chapter 24).

Scientific challenges for managing plant diversity

Plant conservation, encompassing the needs for managing increasingly fragmented habitats and populations within human dominated landscapes, is facing a number of scientific challenges:

- Important plant habitats are subject to increasing levels of encroachment as pressure for land and plant resources forces people further into natural habitats and protected areas (e.g. Figure 3.4). These areas are becoming ecological islands, and many of their enclosed species' populations will increasingly face issues of viability; can we both retain these core areas for the retention of diversity and improve the biodiversity value of the adjacent modified habitats. The latter can play a fundamental role in maintaining biological connectivity between protected areas (see Robertson *et al.*, Chapter 14 and Maunder *et al.*, Chapter 7).

Figure 3.4   Plant resources are a fundamental economic resource for a large proportion of the rural population in the tropics. Here confiscated charcoal produced illegally from Tanzanian mangrove reserves illustrates the pressure on protected areas.

- Habitat conversion produces an 'extinction debt' – a pool of species destined for extinction unless the habitat is repaired or restored (*sensu* Tilman *et al.*, 1994). Can we identify these species in anticipation of habitat loss and can we co-ordinate short term *ex situ* custody with large-scale habitat restoration? There is an urgent need to develop tools for the cost effective restoration of large tropical landscapes, tools that effectively promote natural regeneration and serve the goals of biodiversity restoration whilst providing some economic service to resident communities (see de Moraes *et al.*, Chapter 9; Robertson *et al.*, Chapter 14).

- Plant communities will be subject to the continued impact of invasives (plants, animals and pathogens). There is an urgent need for cost-effective tools for predicting and preventing invasions. It is likely that some fragmented tropical reserves will need permanent supervision and management to prevent invasion (see Chan, Chapter 10; Dulloo, Chapter 13; Tye, Chapter 17).

Figure 3.5   The future of Nairobi National Park as part of a functioning ecosystem is increasingly threatened by urban encroachment. Many tropical cities are adjacent to important areas of tropical diversity.

- There is a need to develop effective and easy to use monitoring protocols and plant identification tools (particularly for use by non-specialists) to assess the priorities and effectiveness of plant conservation activities. This will be of particular value in ecologically isolated reserves subject to continuing degradation from large mammal populations or human recreational use (e.g. the transition from natural protected area towards heavily managed 'mega-zoos' where management is directed by the needs of visible fauna, e.g. Figure 3.5). Whilst the taxonomic impediment has been recognised as a problem retarding the documentation of plant diversity there is an urgent need to break the larger and less publicised 'management impediment' for the management of plant resources. There is an enormous gap between ecological research and its application to the practical management of threatened habitats and species by community groups, protected area managers and motivated volunteers (see Clubbe and Jhilmit, Chapter 11).

Policy and financial challenges

Plant conservation in the tropics needs to identify specific national and local incentives and sanctions that can be used as connectors between government institutions (the legal infrastructure) and the public and private sectors (the financial source). These activities must be built upon sound science, and particularly an understanding of how plant populations respond to change. The conservation challenge is to maintain plant populations as evolutionary lineages, as ecological components of functioning ecosystems and as economic resources within a changing ecological, social and political context. This needs effective institutions that can deal with this increasing range of scientific, social, political and economic challenges:

- It is unlikely that we will see a major investment in staffing for conservation agencies. With the need for more management, how can we bring more people into the management of plant resources? The increased need for species and habitat management will require greater continuity of information between successive management regimes and individuals (see Pearce and Bytbier, Chapter 4).

- If, as expected, we witness an increased number of extinctions as economic pressures and habitat loss acts in synergy with invasive species and climate change, how can national authorities most effectively utilise existing facilities and personnel to maintain plant diversity (see Johansson *et al.*, Chapter 5; Mill, Chapter 6; Maunder *et al.*, Chapter 7).

- Most of the financial support for tropical biodiversity conservation is through externally funded projects. However, projects, with a finite life, as tools for managing evolutionary entities, are inherently limited in their effectiveness. This requires other types of interventions, including modifying financial and other incentive structures that influence decisions on land and resource use, and promoting biodiversity-friendly practices (see Fleisher-Dogley and Kendle, Chapter 19; Jaenicke *et al.*, Chapter 20; Sunderland *et al.*, Chapter 21; Vovides *et al.*, Chapter 22; Clayton *et al.*, Chapter 23; Kamatenesi- Mugisha and Bukenya-Ziraba, Chapter 24) (Figure 3.6).

- The ever-growing hunger for land and natural resources with it's associated conflicts has promoted the view that, if biodiversity is to be conserved in the long term, it must 'pay for itself'. In most cases, however, these sustainable use initiatives target only particular species or biological products, so that the users' main economic interest is to sustain the supply of these products,

Figure 3.6   Conservation on a landscape scale will be imperative for maintaining biodiversity in the tropics. Here in the Rift Valley of Kenya cattle ranching and ecotourism have allowed the retention of extensive wild landscapes.

rather than to conserve the total biodiversity of area. This is still likely to result in a loss of biodiversity, however, the loss may well be less than under alternative land use options. Promoting the sustainable use of biological resources may be an appropriate strategy (depending on stable land tenure and administration) where it is the most feasible alternative to a more destructive activity. However, the conservation priority remains the provision of incentives for biodiversity conservation *per se* (see Johansson *et al.*, Chapter 5; Clayton *et al.*, Chapter 23, Kamatenesi-Mugisha and Bukenya-Ziraba, Chapter 24; Sunderland *et al.*, Chapter 21).

• The international community (government and non-governmental organisations) places a high existence value on tropical biodiversity, demonstrating a willingness to contribute towards its conservation despite the fact that they will never use and probably never even see it. In contrast for rural people in the tropics conserving biodiversity represents a low priority for them relative to alternative land uses and economic activities. How do external stakeholders give biodiversity and natural habitats a direct economic value that is competitive with prevailing and damaging patterns of land use (see Maunder *et al.*, Chapter 7; Sunderland *et al.*, Chapter 21).

- Strategic land use planning should be the core of any national or local plant conservation strategy. However, few tropical countries, are implementing effective, strategic land use planning. Instead, *ad hoc* land use changes are occurring, often with very negative environmental and economic impacts. A number of countries are reviewing their protected area portfolios to more effectively meet the national objectives, which may include biodiversity conservation, preservation of cultural heritage and tourism development. This may involve redrawing boundaries, de-gazetting areas or establishing new ones, or changing management designations. Botanical data need to be provided in an accessible form before any spatial analysis and strategic planning is undertaken (see Pearce and Bytebier, Chapter 4; Johansson *et al*, Chapter 5; Hernandez and Gómez-Hinostrosa, Chapter 18).

- Building an effective constituency for biodiversity amongst senior government officials is essential for implementing conservation. The conservation of plant diversity is promoted by concerned specialist agencies and individuals that often have limited political and economic influence. The traditional agenda and range of players needs expanding to reflect both the importance of biodiversity conservation and the need for broad-based collaboration (Czech *et al.*, 1998; Le Maitre, *et al.*, 1997). The conservation of plant diversity, especially since plant diversity may be perceived as a non-economic resource, can be regarded as a luxury pre-occupation of the northern developed nations. However, plant conservation is a fundamental subset of environmental management, which is in turn an element of sustainable development, which is acknowledged by most governments as an important objective. Accordingly, it may be more effective to build upon an existing constituency for sustainable development (for instance the implementation of Agenda 21) and to demonstrate the importance of biodiversity to that objective (including specific aspects such as maintaining soil fertility and water supply and quality). (Prance, Chapter 12; Vovides *et al.*, Chapter 22).

- Biodiversity planning needs both local bottom-up and participatory negotiations, and priority setting and planning at a regional level. The secure retention of biodiversity will ultimately be controlled by the local communities who use or live with those resources. In some cases traditional land use or harvesting regimes will be fundamental to conservation management. On the other hand, human societies (urban and rural) are undergoing profound changes where traditional approaches to biodiversity will be modified or lost (see Githitho, Chapter 8; Kamatenesi-Mugisha and Bukenya-Ziraba, Chapter 24);

- Policy reform and institutional change must be central elements of biodiversity planning. Biodiversity loss is largely driven by land use and this in turn is most effectively influenced through policy reform. Effective plans for dealing with the causative factors for biodiversity loss should be given priority. To deal with the dynamic nature and cross-sectoral nature of biodiversity management will require investment in institutional strengthening. The full range of conservation techniques and technologies must be considered in developing biodiversity conservation plans. No one technique or tool can effectively save biodiversity; the full range of national facilities and resources need to be employed to develop a shared understanding and commitment to conservation (see Hankamer *et al.*, Chapter 15).

- Just as many protected areas exist only as 'paper parks', too many national action plans and strategies only exist as non-implemented and dusty documents. Institutions and agencies responsible for biodiversity planning and management should evaluate their abilities to implement the plan. To assist implementation partnerships with other agencies and NGOs should be considered (see Dulloo, Chapter 13; Tye, Chapter 17; Vovides *et al.*, Chapter 22).

Fundamental to all of the above is the need to raise the professional and public profile of plant conservation – it is still a poor relation to animal/wildlife conservation. This may require a move from the species as a conservation goal towards the broader promotion of habitat and ecosystem conservation as components of sustainable development. It is only by recognising these challenges, developing practical tools and techniques to solve them, and working towards a more cross-disciplinary approach to plant conservation that we can ensure that we maintain a natural capital worthy of passing on to the next generation and thus ensure that Janzen's challenge is met. Time is running out.

References

Cardosa da Silva, J.M. and Tabarelli, M. (2000). Tree species impoverishment and the future flora of the Atlantic forests of northeast Brazil. *Nature* **404**: 72–74.

Cincotta, R.P., Wisnewski, J. and Engelman, R. (2000). Human population in the biodiversity hotspots. *Nature* **404**: 990–992.

Constanza, R. *et al.* (1997). The value of the world's ecosystem services and natural capital. *Nature* **387**: 254–257.

Czech, B., Krausman, P.R. and Borkhataria, R. (1998). Social construction, political power, and the allocation of benefits to endangered species. *Conservation Biology* **12**: 1103–1112.

Daily, G.C. and Walker, B.H. (2000). Seeking the great transition. *Nature* **403**: 243–245.

Dukes, J.S. and Mooney, H.A. (1999). Does global change increase the success of biological invaders? *Trends in Ecology and Evolution* **14**: 135–139.

Du Puy, D.J. and Moat, J. (1996). A refined classification of the primary vegetation of Madagascar based on the underlying geology; using GIS to map its distribution and to assess its conservation status, pp. 205–218. In: W.R. Lourenco (ed.). *Proceedings of the International Symposium on the Biogeography of Madagascar.* Editions de l'ORSTOM, Paris, France.

Fairhead, J. and Leach, M. (1999). *Misreading the African Landscape: Society and Ecology in a Forest-Savanna Mosaic.* Cambridge University Press, Cambridge, UK.

Freitag, S. and Van Jaarsveld, A.S. (1998). Sensitivity of selection procedures for priority conservation areas to survey extent, survey intensity and taxonomic knowledge. *Proceedings of the Royal Society of London - Series B Biological* **265**: 1475–1482.

Gascon, C., Bruce Williamson, G. and da Fonseca, G.A.B. (2000). Receding forest edges and vanishing reserves. *Science* **288**: 1356–1358.

Glowka, L., Burhenne-Guilmin, F., Synge, H., McNeely, J.A. and Gündling, L. (1994). *A Guide to the Convention on Biological Diversity.* Environmental Policy and Law Paper No. 30. IUCN, Gland, Switzerland.

Hall, P., Walker, S. and Bawa, K. (1996). Effect of forest fragmentation on genetic diversity and mating system in a tropical tree, *Pithecellobium elegans. Conservation Biology* **10**(3): 757–768.

Hannah, L., Lohse, D., Hutchinson, C., Carr, J.L. and Lankerani, A. (1994). A preliminary inventory of human disturbance of world ecosystems. *Ambio* **23**: 246–250.

Harrison, R.D. (1999). Repercussions of El Niño: drought causes extinction and the breakdown of mutualism in Borneo. *Proceedings of the Royal Society of London - Series B Biological* **267**: 911–915.

Hughes, J.B., Daily, G.C. and Ehrlich, P R. (1997). Population diversity: its extent and extinction. *Science* **278**: 689–692.

Hunter, M.L., Jacobsen, G.L. and Webb, T. (1988). Palaeoecology and the coarse filter approach to maintaining biological diversity. *Conservation Biology* **2**: 375–385.

Janzen, D.H. (1986). The future of tropical ecology. *Annual Review of Ecology and Systematics* **17**: 305–324.

Jones, P.G., Beebe, S.E., Tohme, J. and Galwey, N.W. (1997). The use of geographical information systems in biodiversity exploration and conservation. *Biodiversity and Conservation* **6**: 947–958.

Lawton, J.H. and May, R.M. (eds) (1995). *Extinction Rates.* Oxford University Press, Oxford, UK.

Le Maitre, D. *et al.* (1997). Communicating the value of fynbos: results of a survey of stakeholders. *Ecological Economics* **22**: 105–121.

Lombard, A.T., Cowling, R.M., Pressey, R.L. and Mustart, P.J. (1997). Reserve selection in a species-rich and fragmented landscape on the Agulhas Plain, South Africa. *Conservation Biology* **11**: 1101–1116.

Lonsdale, W.M. (1999). Global patterns of plant invasions and the concept of invasibility. *Ecology* **80**: 1522–1536.

Margules, C.R. & Pressey, R.L. (2000). Systematic conservation planning. *Nature* **405**: 243–253.

Mittermeier, R.A., Myers, N., Thomsen, J.B., da Fonseca, G.A.B. and Olivieri, S. (1998). Biodiversity hotspots and major tropical wilderness areas: Approaches to setting conservation priorities. *Conservation Biology* **12**: 516–519.

Morse, L.E., Kutner, L.S. and Kartesz, J.T. (1995). Potential impacts of climate change on North American flora. In: E.T. LaRoe, G.S. Farris, C.E. Plunkett, P.D. Doran, and M.J. Mac (eds). *Our Living Resources: A Report to the Nation on the Distribution, Abundance, and Health of US Plants, Animals and Ecosystems.* US Department of the Interior, National Biological Service, Washington DC, USA.

Musters, C.J.M., de Graaf, H.J. and ter Keurs, W.J. (2000). Can protected areas be expanded in Africa? *Science* **287**: 1759–1760.

Myers, N., Mittermeier, R.A., Mittermeier, C.G., da Fonseca, G.A.B. and Kent, J. (2000). Biodiversity hotspots for conservation priorities. *Nature* **403**: 853–858.

Pelkey, N.W., Stoner, C.J. and Caro, T.M. (2000). Vegetation in Tanzania: assessing long term trends and effects of protection using satellite imagery. *Biological Conservation* **94**: 297–309.

Pimental, D., Lach, L., Zuniga, R. and Morrison, D. (2000). Environmental and economic costs of nonindigenous species in the United States. *Bioscience* **50**: 53–65.

Pounds, J.A. *et al.* (1999). Biological responses to climate change on a tropical mountain. *Nature* **398**: 213.

Powell, G.V.N., Barborak, K. and Mario Rodriguez, S. (2000). Assessing representativeness of protected natural areas in Costa Rica for conserving biodiversity: a preliminary gap analysis. *Biological Conservation* **93**: 35–41.

Sala, O.E. *et al.* (2000). Global biodiversity scenarios for the year 2100. *Science* **287**: 1770–1774.

Scott, J.M. *et al.* (1993). Gap analysis: a geographic approach to the protection of biological diversity. *Wildlife Monographs* **123**: 1–41.

Skole, D. and Tucker, C. (1993). Tropical deforestation and habitat fragmentation in the Amazon: Satellite data from 1978 to 1988. *Science* **260**: 1905–1910.

Soulé, M.E. (1991). Conservation: Tactics for a Constant Crisis. *Science* **253**: 744–750.

Soulé, M.E. and Sanjayan, M.A. (1998). Conservation targets – do they help? *Science* **279**: 2060–2061.

Thrupp, L.A. (2000). Linking agricultural biodiversity and food security: the valuable role of agrobiodiversity for sustainable agriculture. *International Affairs* **76**: 265–281.

Tilman, D., May, R.M., Lehman, C.L. and Nowak, M.A. (1994). Habitat destruction and the extinction debt. *Nature* **371**: 65–66.

Trinder Smith, T.H., Lombard, A.T. and Picker, M.D. (1996). Reserve scenarios for the Cape Peninsula: high, middle and low road options for conserving the remaining biodiversity. *Biodiversity and Conservation* **5**: 649–669.

Turner, I.M. (1996). Species loss in fragments of topical rainforest: a review of the evidence. *Journal of Applied Ecology* **33**: 200–219.

Valverde, P.L. and Montana, C. (1996). A rapid methodology for vegetation survey in Mexican arid lands. *Journal of Arid Environments* **34**: 89–99.

Vitousek, P.M. (1994). Beyond global warming: ecology and global change. *Ecology* **75**: 1861–1876.

WRI/UNEP/UNDP/World Bank. (1996). *World Resources 1996–1997.* Oxford University Press, Oxford, UK.

Young, A., Boyle, T. and Brown, T. (1996). The population genetic consequences of habitat fragmentation for plants. *Trends in Ecology and Evolution* **11**: 413–418.

Chapter 4

The Role of an Herbarium and its Database in Supporting Plant Conservation

Tim Pearce
Royal Botanic Gardens, Kew, Wakehurst Place, UK

Benny Bytebier[1]
National Museums of Kenya, Nairobi, Kenya

[1] Now at: University of Stellenbosch, Matieland,
South Africa

Introduction

Herbaria have an established role in providing voucher specimens for plant identification, compilation of local and regional Floras and production of monographic works (Figures 4.1 and 4.2). However, the potential for the herbarium to render rather more complex information on distribution, phenology, indigenous uses, local names and additional information – frequently required by conservation projects – is often frustrated by the repetitive and tedious nature of manually consulting large numbers of sheets.

Computer-based database systems now afford herbaria with an excellent opportunity not only to manage the curatorial aspects of the work of the herbarium, but also to provide powerful tools to extract and analyse data in a way that is more difficult when handling specimens and recording their data manually.

Figure 4.1   Collection of herbarium samples by Mollel Neduvoto, National Herbarium, Arusha, Tanzania, during inventory work in the Mwaluganje Elephant Sanctuary, Kenya.

Figure 4.2   George Mugambi of the National Museums of Kenya making field collections during inventory work in the Mwaluganje Elephant Sanctuary, Kenya.

Herbarium Specimens and their Data

The specimen will provide three types of information:

1. those data provided by the collector at the time of collection;

2. a specimen identification giving full binomial with author and sometimes even the citation
   e.g. *Kalanchoe boranae* Raadts in *Willdenowia* **13**(2): 378 (1983)

3. additional data that can be inferred from the plant and/or the label by the researcher who is looking at the specimen e.g. phenology.

Figure 4.3   A poor example of an herbarium specimen. It is represented by a single leaf and there are no details of locality, habitat nor habit.
Photo: Benny Bytebier.

A specimen will in most cases, consist of a dried plant (or at least part of a plant) and a label, glued together on a sheet of mounting paper. Any researcher who has spent time browsing herbarium sheets will fully appreciate that they vary in quality from the worst, being a single leaf and a label '*Polystachya* sp.?' Ethiopia' (Figure 4.3), to the most useful with well preserved taxonomic characters, description, distribution, habit, habitat and economic data. This is the limit of the herbarium; its usefulness is directly related to the professional practice of the collectors whose material is deposited and the curatorial ability of the herbarium.

Figure 4.4 Jane Nyafuono of the Forest Department, Uganda, using a laminated cibachrome copy of a herbarium specimen from the Herbarium of the Royal Botanic Gardens, Kew, to aid identifications during a Conservation Assessment and Management Planning Workshop (CAMP) in the Amani Nature Reserve, Tanzania.

Collection data

A proper herbarium label will not only contain the plant's name, but also where and when it was collected. A correctly identified specimen is, of course, a useful asset for the herbarium, but on its own it has limited use. Once there is geographical and temporal information about the specimen, then this allows researchers to compile data collectively over a taxon or geographical area that can have tremendous value for conservationists (Figure 4.4).

A good collector will also include ecological information, like a description of the habitat and the associated species. Again, these can be compiled over a taxon to assess the habitat specificity of the species. In addition, such data is often useful in indicating habitat change or loss.

Ethnobotanic information can be held on herbarium specimen labels. This may be just the name of the plant in the local language, but may extend to include a description of uses by the local population and provide an insight into the value of the species within a community; another important factor when considering conservation scenarios.

Identification data

There will be a species name (a binomial and author abbreviation) either on the main label or on separate smaller labels. The specimen may have been identified in the field by the collector, in which case caution should be used when accepting the accuracy of the name. More often, the specimen will be re-identified ('determined') afterwards in the herbarium by the collector or herbarium curators with the aid of a published Flora or monograph or, better still, by an expert in the group consulting the herbarium collection during research on the production of a monograph. In addition to the original identification, other names may be found on the specimen in the form of *determinavit* labels (annotated label giving name of plant, botanist who identified it and date of identification (www.ibiblio.org/unc-biology/ herbarium/courses/chpt31.html)), again either provided by curators or specialists in the plant group. These may reflect original misidentifications or taxonomic changes made to the species. They should be dated and can therefore act as a record of such changes over the course of time.

Inferred data

Information inferred by the researcher studying the specimen is usually that of a taxonomic or of a phenological nature. Researchers interested in information on the life history of the plant that may prove important for seed collecting programmes will be able to plot peaks of fruiting in a particular area.

Specimen and species information

At this stage we should consider the differences between what can loosely be called 'species data' and 'specimen data'. As described above, specimen data are those that can be found directly on the sheet (of varying quality) or that can be inferred by a researcher (normally limited to taxonomic and life history information). The important issue is that these data can only be attached to the particular specimen from which it was derived or inferred; it is therefore specific to the individual collection. Species data, on the other hand, is that which can be attributed to a taxonomic name. It is important to remember that the majority of information one can find in literature about a particular species has its origins from herbarium specimens. Some species data may be collected from particular field surveys and some is purely anecdotal. It is important to note that, unless citations are made (either to herbarium specimens or specific reports), species data remains unverifiable.

Computerisation of Botanical Data

Database software requirements

Much progress has been made in the development of software capable of managing large quantities of data associated with specimens, species and geographical localities. Given the correct tools, there is much potential for the conservation biologist to extract and manipulate data originating from herbarium specimens, in order to observe patterns (distribution, habitat specificity etc.) that would have taken many long hours of manual browsing to discover.

There will always be arguments about the best available software to use for compiling herbarium and species data. It would probably be true to assume that all herbaria around the world with a reasonable piece of hardware would be able to benefit from a number of purpose-built herbarium databases. Some are freely available for downloading from the Internet, while others may require some form of agreement, perhaps as part of a conservation or capacity building project. The choice of software suitable for a particular institution or project will depend upon answers to the following questions:

- Who are your major collaborators?
- What system do they use?
- Is the chosen database compatible with theirs?
- Can I get help if I need it?

The last question – that of technical support is the most important. Good technical support can come in the form of easy access to the authors of the application software or other users. The experience of the authors is that any given popular piece of herbarium software on the market is as good as the next. Conservation questions require a compilation of both species and specimen data so ensure that you are not selecting a herbarium management tool which is excellent for managing herbaria but limited in scope for managing species data or *vice versa*. Your software will need to be able to compile both specimen and species data for your final analysis and reporting.

A word of caution is necessary at the outset of this discussion. It is a fact that many tropical herbaria are limited in the size of their collections. This limited number of specimens reduces the credibility of inferences on distribution and conservation status. Care should be taken that computerisation of herbarium data is not somehow seen as a method of improving the quality of those data. If the collection is well curated then computerisation is frequently a worthwhile exercise as this paper outlines. However, the computerisation of poorly-curated herbarium collections will always result in poor quality conservation inferences. The best advice would be to ensure that the computerisation of herbarium data, either for herbarium management or research, should be part of the standard curatorial process.

Computerisation projects at the East African Herbarium, Nairobi, Kenya

Over the past five years a number of projects have been implemented by staff at the East African Herbarium that have assisted species conservation efforts. We will use examples from two databasing projects. A large exercise was undertaken to computerise the herbarium collections of the Orchidaceae (6,396 specimens) and a smaller, but similar project databased the collection of the genus *Kalanchoe* Adans. (Crassulaceae) (839 specimens).

In addition to these specimen databasing projects, the List of East African Plants (LEAP) represents the first electronic list of taxa for the East African flora (Knox and Vanden Berghe, 1996). This list was compiled from the Flora of Tropical East Africa (FTEA) and, in the case of unpublished families, from the herbarium specimen collections. The altitude range and FTEA Floral Region distribution augment each species name.

Such reference data tables massively improve the speed and accuracy of data entry. Our specimen databasing projects also benefited from the early development of lists of collector series (Bytebier *et al.*, 1996) and access to electronic gazetteers providing grid references (and thence distribution maps) on the basis of the locality name (Pearce *et al.*, 1996). These reference data tables should act as dictionaries or 'pick-lists' so that repetitive typing of similar data is avoided. Therefore, once the species list is created and an acceptable quality of data achieved, there should be no need to type the name of the species ever again!

Gathering and Compiling Data Sets for Conservation

Taxonomic lists

Accurate taxonomic information is one of the cornerstones of species conservation work. An important first task is the compilation of a list of the 'units of conservation' (normally the species). From both literature and herbaria, a list of species names can be compiled to which can be attached basic species data. For conservation projects in the FTEA region, the LEAP database (Knox and Vanden Berghe, 1996) should be the first data set consulted. If such an electronic list of names is not available, it is worth investing in the time to compile one. Whilst not essential, a full list of synonyms is of great benefit.

For a small group of taxa in East Africa, such as the genus *Kalanchoe*, there are relatively few species (only 23 in this case) currently recognised in the FTEA. Compiling all synonyms for these 23 names resulted in a list of some 94 names, a task that required consultation of only five major publications (Table 4.1). Indeed, the relatively recently published FTEA volume for Crassulaceae provided most of the information. For this small genus, getting the taxonomy up-to-date was a relatively straightforward task. However, the parallel project for the Kenyan orchids proved more problematic and synonymy has not yet been compiled. At this stage, we are simply trying to create a list of species that gives us the basis on which to assess conservation priorities and take our conservation project forward.

Specimen data entry

The process of data entry should be used as an opportunity to review these data and the species themselves. It is unlikely that there will be a better opportunity to review all the herbarium specimens again. Our experience has been that it is best to get it right at data entry stage, rather than starting with the notion that it can always be corrected later on. The first author used the data entry process to physically inspect all the *Kalanchoe* specimens in the East African Herbarium. This resulted in a much better understanding of the taxonomic variation between and within recognised species. It also revealed other workers in the region that had previously inspected the specimens and therefore presumably had a good knowledge of the group. Having a vested interest in the quality of these data resulted in accurate data entry, which required little editing. Subsequent time was spent updating the taxonomic information on the specimens, which was consequently reflected in the electronic database.

The second author had a larger task with some 282 species names and nearly 6,500 herbarium specimens to log. Employment of staff to enter these data on to the computer was necessary. However since these staff did not have such a vested interest in the outputs, much time has been spent checking and editing these data. Consequently, less time has been available for improving the quality of these data. Data entry is not a 'secretarial' task; it is a thankless one and thus it is important that the process of data entry is seen as the first and best opportunity to become familiar with the species.

The data entry process and the database system should be 'smart' enough not to accept new data that are in conflict with those already stored on your database. If a new collector or gazetteer is entered then the database should be able to flag it as such. Database design should also make input of incorrect data (e.g. century 91 instead of 19) impossible.

The process of data entry is about ensuring the consistency and quality of data entered on to the database. As mentioned before, the use of standard dictionaries and employment of data entry staff with a vested interest in the output of the process is crucial to this. Once the compilation of the 'live' data set is begun, it should be considered a sanctuary for high quality data. Access should be read-only for all users except for the researchers on the project. Rather than fix problems later, it is easier to ensure that data entered onto the database at the start are of high quality.

Species data entry

The database should allow data entry (often textual) that can be attributed to the species. It may be as simple as a description of the species or as complex as data on specimen numbers, reproductive habit or seed production in monitored populations. As an example, the East African Herbarium has adapted the Botanical Research and Herbarium Management System BRAHMS – Copyright, Department of Plant Sciences, University of Oxford, UK (www.brahms.co.uk) to accept ten textual memo fields to compile the following species information:

1. Taxonomic/identification notes
2. Standard plant descriptions
3. Habitat preferences
4. Population status
5. Recorded uses
6. Cultivation and propagation notes
7. Coded distribution (FTEA Floral Regions)
8. Global distribution (by countries)
9. Miscellaneous project notes
10. Bibliography

As well as data fields for standard species data, such as the World Conservation Union (IUCN) Category of Threat (see Appendix 2), taxonomic status and habit, the ability to add as many types of numerical data as are required (such as numerical indices for flower colour, status of *ex situ* collections, recorded trade, etc.) is available through a linked species table. Essentially the amounts and the types of data that can be stored, and thereafter compiled, are limitless.

Much of the information is found in published material will have been compiled from herbarium specimens, but further anecdotal or distribution data from inventories, travel logs and the like, may be valuable for improving the understanding of the species and should be recorded as species data.

Table 4.1　Understanding the taxonomic units of conservation, such as this list of currently recognised *Kalanchoe* species from Kenya with their synonymy, is the first step to compiling the available data necessary for developing species conservation projects

***Kalanchoe aubrevillei* Cufod.**
- ⇒　　*Kalanchoe aubrevillei* Raym.-Hamet

***Kalanchoe ballyi* Cufod.**

***Kalanchoe bipartita* Chiov.**

***Kalanchoe boranae* Raadts**

***Kalanchoe citrina* Schweinf.**
- ⇒　　*Kalanchoe citrina* Schweinf. var. *erythraeae* Schweinf.
- ⇒　　*Kalanchoe citrina* Schweinf. var. *ballyi* Raym.-Hamet ex Wickens
- ⇒　　*Kalanchoe brachycalyx* A. Rich. var. *erlangeriana* Engl.
- ⇒　　*Kalanchoe citrina* Schweinf. var. *longipetiolata* Raym.-Hamet
- ⇒　　*Kalanchoe citrina* Schweinf. var. *ballyi* Raym.-Hamet

Kalanchoe crenata* (Andrews) Haw. ssp. *crenata

Kalanchoe densiflora* Rolfe var. *densiflora
- ⇒　　*Kalanchoe glaberrima* Volkens
- ⇒　　*Kalanchoe bequaertii* De Wild.
- ⇒　　*Kalanchoe petitiana sensu* auct.
- ⇒　　*Kalanchoe tayloris* Raym.-Hamet

***Kalanchoe densiflora* Rolfe var. *minor* Raadts**

***Kalanchoe fadeniorum* Raadts**

***Kalanchoe glaucescens* Britten**
- ⇒　　*Kalanchoe holstii* Engl.
- ⇒　　*Kalanchoe flammea* Stapf.

***Kalanchoe magnidens* N.E. Br.**
- ⇒　　*Kalanchoe marinelli* Pampan.
- ⇒　　*Kalanchoe beniensis* De Wild.
- ⇒　　*Kalanchoe crenata sensu* Br. & Mass.
- ⇒　　*Kalanchoe rotundifolia sensu* Raadts

***Kalanchoe laciniata* (L.) DC.**
- ⇒　　*Kalanchoe schweinfurthii* Penzig.
- ⇒　　*Kalanchoe rohlfsii* Engl.
- ⇒　　*Cotyledon laciniata* L.
- ⇒　　*Verea laciniata* (L.) Willd.
- ⇒　　*Kalanchoe* sp. *sensu* Bally

***Kalanchoe lanceolata* (Forssk.) Pers.**
- ⇒　　*Kalanchoe floribunda* Wight & Arn.
- ⇒　　*Kalanchoe glandulosa* Hochst. ex A. Rich.
- ⇒　　*Kalanchoe ritchieana* Dalz.
- ⇒　　*Meriostylis macrocalyx* Klotzsch
- ⇒　　*Kalanchoe modesta* Kotschy & Peyr.
- ⇒　　*Kalanchoe platysepala* Britten
- ⇒　　*Kalanchoe glandulosa* A. Rich. var. *benguellensis* Engl.
- ⇒　　*Kalanchoe pilosa* Baker
- ⇒　　*Kalanchoe crenata* (Andrews) Haw. var. *collina* Engl.
- ⇒　　*Kalanchoe pentheri* Schlechter

⇒   *Kalanchoe glandulosa* A. Rich. var. *rhodesica* Baker f.
⇒   *Kalanchoe glandulosa* A. Rich. var. *tomentosa* Keissler
⇒   *Kalanchoe goetzei* Engl.
⇒   *Kalanchoe diversa* N.E. Br.
⇒   *Kalanchoe ellacombei* N.E. Br.
⇒   *Kalanchoe homblei* De Wild.
⇒   *Kalanchoe homblei* De Wild. forma *reducta* De Wild.
⇒   *Kalanchoe laciniata* (L.) DC. var. *brachycalyx* Chiov.
⇒   *Kalanchoe lanceolata* (Forssk.) Pers. var. *glandulosa* (A. Rich.) Cufod.
⇒   *Kalanchoe brachycalyx* A. Rich.
⇒   *Cotyledon lanceolata* Forssk.
⇒   *Verea lanceolata* (Forssk.) Sprengel
⇒   *Kalanchoe sp. sensu* Britten

Kalanchoe lateritia Engl. var. *lateritia*
⇒   *Kalanchoe velutina sensu* auct.
⇒   *Kalanchoe zimbabwensis* Rendle
⇒   *Kalanchoe lateritia* Engl. var. *zimbabwensis* (Rendle) Brenan
⇒   *Kalanchoe integra* (Medic.) Kuntze var. *subsessilis* (Britten) Cufod.
⇒   *Kalanchoe sp. A sensu* UKWF
⇒   *Kalanchoe coccinea* Britten var. *subsessilis* Britten
⇒   *Kalanchoe cuisini* De. Wild. & Th. Dur.
⇒   *Kalanchoe kirkii* N.E. Br.
⇒   *Kalanchoe sp. sensu* Bally

Kalanchoe lateritia Engl. var. *prostrata* Raadts
Kalanchoe lateritia Engl. var. *pseudolateritia* Raadts
Kalanchoe marmorata Baker
⇒   *Kalanchoe marmorata* Baker *forma somaliensis* (Baker) Pampan.
⇒   *Kalanchoe somaliensis* Baker
⇒   *Kalanchoe kelleriana* Schinz
⇒   *Kalanchoe grandiflora* A. Rich. var. *angustipetala* Engl.
⇒   *Kalanchoe macrantha* Baker
⇒   *Kalanchoe marmorata* Baker var. *maculata* Senni
⇒   *Kalanchoe rutshuruensis* Lebrun & Toussaint
⇒   *Kalanchoe grandiflora* A. Rich.

Kalanchoe mitejea Leblanc & Raym.-Hamet
⇒   *Kalanchoe lugardii sensu* UKWF

Kalanchoe nyikae Engl. ssp. *auriculata* Raadts
⇒   *Kalanchoe lugardii sensu* UKWF

Kalanchoe nyikae Engl. ssp. *nyikae*
⇒   *Kalanchoe hemsleyana* Cufod.
⇒   *Kalanchoe lugardii sensu* UKWF

Kalanchoe obtusa Engl.
Kalanchoe prittwitzii Engl.
⇒   *Kalanchoe dielsii* Raym.-Hamet
⇒   *Kalanchoe lugardii* Bullock
⇒   *Kalanchoe secunda* Werderm.
⇒   *Kalanchoe robynsiana* Raym.-Hamet
⇒   *Kalanchoe germanae* Raym.-Hamet

Extracting Information for Conservation Projects

Analysing herbarium collection data

As mentioned earlier, access to computerised herbarium data provides a more powerful tool for observing patterns within specimen/species data. Put another way, having captured all the Orchidaceae data stored at the East African Herbarium, researchers are now aware of the patterns of collection as well as the species information collated from all the specimens. We know where most of the collecting has been done (and therefore where it *has not* been done). We can assess the amount of collecting over time and focus on a need for further collections from particular geographical locations. Distribution maps show 'hotspots' for priority site conservation.

Categories of Threat and candidate lists

The IUCN Categories of Threat (see Appendix 2) are a useful tool to set species conservation priorities. However, they demand a thorough understanding of the status of the species in the wild, especially attributes such as: extent of occurrence, area of occupancy and the trends in numbers of individuals. Obviously herbarium data alone cannot provide all this information, but they can offer a rapid approach to selecting the critical species that will demand such rigorous field research (see Clubbe and Jhilmit, Chapter 11).

The BRAHMS database has the facility to produce a summary of the collection per species in terms of: the numbers of specimens, the number of countries, administrative areas, floral regions and localities from which the species has been collected, the last year each species was collected, and its altitude range (Figure 4.5). If required, additional inferred attributes can be added such as habitat specificity and economic importance based on local ethnobotanic information or, in the case of orchids, potential usefulness in horticulture.

Such a table (Figure 4.5) can then be manipulated to display priority species lists based on these characters. These characters alone are insufficient to assess the status of a given species in the wild. However, an initial assumption can be made that the number of collection localities can be related to the number of subpopulations of given species. This is obviously a crude assumption as it relies on the fact that all localities where a species is found are known. Therefore, a table sorted on numbers of collection localities gives a good initial indication of the priority species. Such a sort can be represented as simple tables or exported to other software to produce charts (Figure 4.6). We have used a figure of five localities or less, relating to part of the data required for criteria 'B' of the IUCN Categories of Threat (see Appendix 2), in combination with the 'country' characters and 'major area', to use as a cut-off point for conservation priorities (Table 4.2). We have tended to coin the phrase 'Candidate Lists' for this first

SPECIES	TOTAL	COUNTRIES	MAJORS	LOCALITIES	COLLECTORS	LASTYEAR	REGTOTAL	REGIONS
Kalanchoe densiflora var. densiflora	122	4	33	87	76	1997	8	K1;K2;K3;K4;K6;K7;T2;T7
Kalanchoe lanceolata	152	12	63	120	106	1986	17	K1;K2;K3;K4;K5;K6;K7;T1;T2;T3;T5;T6;T7;T8;U1;U2;U4
Kalanchoe mitejea	22	4	14	21	16	1989	6	K2;K3;K4;K7;T2;U1
Kalanchoe prittwitzii	61	5	34	53	51	1978	15	K1;K2;K3;K4;K5;K8;T1;T2;T3;T5;T6;T7;U1;U2;U3
Kalanchoe citrina	48	3	19	41	32	1995	6	K1;K3;K4;K5;K8;U1
Kalanchoe nyikae ssp. nyikae	16	3	8	14	10	1990	4	K7;T2;T3;U4
Kalanchoe glaucescens	68	7	33	53	47	1980	12	K2;K3;K4;K5;K6;K7;T1;T2;T3;T6;U1;U2
Kalanchoe fadeniorum	1	1	1	1	1	1977	1	K7
Kalanchoe obtusa	16	2	6	15	10	1992	4	K4;K7;T3;T6
Kalanchoe lateritia var. lateritia	82	5	30	61	55	1991	11	K1;K4;K5;K6;K7;T1;T2;T3;T4;T6;T8
Kalanchoe lateritia var. prostrata	4	2	3	4	4	1990	2	K7;T3
Kalanchoe lateritia var. pseudolateritia	3	2	3	3	3	1963	3	K5;K7;T2
Kalanchoe bailyi	11	1	2	9	8	1989	1	K7
Kalanchoe nyikae ssp. auriculata	20	2	8	17	12	1981	5	K4;K6;K7;T1;T2
Kalanchoe densiflora var. minor	9	1	5	9	8	1973	3	K3;K5;K8
Kalanchoe laciniata	40	6	21	33	28	1988	10	K1;K3;K4;K5;K6;T1;T2;T5;T6;U2
Kalanchoe crenata ssp. crenata	57	8	29	48	44	1989	8	K3;K5;T3;T6;T7;T8;U2;Z
Kalanchoe marmorata	45	7	18	39	36	1979	7	K1;K4;K6;T2;U2;U3;U4
Kalanchoe bipartita	2	1	1	1	2	1984	1	K4
Kalanchoe rotundifolia x glauc	3	1	2	3	3	1985	1	K7
Kalanchoe	2	1	2	2	2	1977	2	K1;K7
Kalanchoe nyikae aur. x glaucescens	1	1	1	1	1	1972	1	K6
Kalanchoe aubrevillei	17	1	5	11	16	1994	3	K6;T1;T2
Kalanchoe densiflora x lanceolata	1	1	1	1	1	1973	1	K6
Kalanchoe boranae	2	1	1	1	2	1972	1	K1
Kalanchoe sp. B sensu FTEA	6	2	4	4	6	1993	4	K1;K7;T6;T8

Figure 4.5   An unsorted collection summary from a Botanical Research and Herbarium Management System (BRAHMS) database based on herbarium collections of the genus *Kalanchoe* Adans. (Crassulaceae) from East Africa.

For each species the summary shows (from left to right) in data columns: the total number of specimens, the number of countries the species is found in, major areas, localities from which the species has been collected, the number of collectors, the last year each species was collected, the number of 'Flora of Tropical East Africa' (FTEA) regions the species was collected in and the regions themselves.

attempt at species prioritisation, recognising that they will be the candidate species towards which resources should be channeled in order to assess Category of Threat following the recognised IUCN guidelines. The Candidate Species List needs to be treated as a first attempt to understand the patterns of species distribution and abundance available to the researcher through herbarium data alone. It must be considered as a method of guiding the conservation project. The Candidate List can be used to challenge other known experts in the plant group. Much factual and anecdotal information can be added to your knowledge at this stage by bringing other experts and amateur botanists on board (see Maunder and Clubbe, Chapter 16). The Conservation Breeding Specialist Group of the IUCN Species Survival Commission (SSC) has developed the Conservation Assessment and Management Planning (CAMP) process. This provides a very effective forum and process for collating conservation data from a candidate list. A case study is provided in Chapter 15 by Hankamer *et al.* and guidelines for holding a CAMP Workshop are located in Appendix 3.

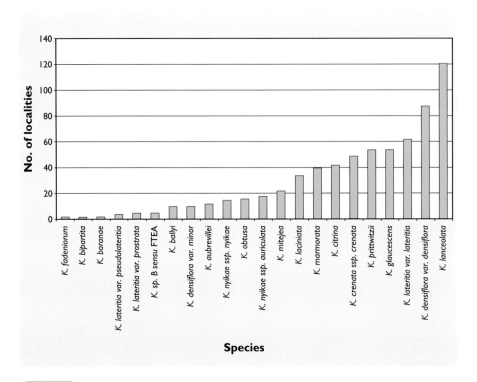

Bar chart of selected *Kalanchoe* Adans. (Crassulaceae) species from East Africa held as herbarium collections in the East African Herbarium and sorted by number of collection localities.

It is important to remember the inadequacies of herbarium collections. Any inference solely from herbarium data will be subject to the vagaries of collecting bias. Small plants are often overlooked and consequently will be under-collected. Rare plants may be over-collected and conversely other common plants may be under-collected when they are difficult for collectors to reach, such as epiphytes. Large flowered and conspicuous plants, such as *Ansellia africana* Lindl. (Orchidaceae) (leopard orchid), tend to stand out in the landscape and will often be over-represented in the collection. Areas along major roads and holiday sites will have good collections; a plot of the distribution of large numbers of Kenyan herbarium specimens will produce a fairly accurate road map of the country!

Table 4.2 **Collection review from specimens of Kenyan *Kalanchoe* Adans (Crassulaceae) sorted by number of collection localities**

Kalanchoe species	Countries	Major areas	Collection localities
K. bipartita Chiov.	I	I	I
K. boranae Raadts	I	I	I
K. fadeniorum Raadts	I	I	I
K. lateritia var. *pseudolateritia*	2	3	3
K. sp. B sensu FTEA	2	4	4
K. lateritia Engl. var. *prostrata* Raadts	2	3	4
K. ballyi Cufod.	I	2	9
K. densiflora Rolfe var. *minor* Raadts	I	5	9
K. aubrevillei Cufod.	2	5	II
K. nyikae Engl. ssp. *nyikae*	3	8	I4
K. obtusa Engl.	2	6	I5
K. nyikae Engl. ssp. *auriculata* Raadts	2	8	I7
K. mitejea Leblanc & Raym.-Hamet	4	I4	21
K. laciniata (L.) DC.	6	21	33
K. marmorata Baker	7	I8	39
K. citrina Schweinf.	3	I9	41
K. crenata (Andrews) Haw. ssp. *crenata*	8	29	48
K. glaucescens Britten	7	33	53
K. prittwitzii Engl.	5	34	53
K. lateritia Engl. var. *lateritia*	5	30	61
K. densiflora Rolfe var. *densiflora*	4	33	87
K. lanceolata (Forssk.) Pers.	I2	63	I20

Legend: ▢ species prioritised (with less than or equal to 5 collection localities) for conservation action.

Species reports

Information about individual species can now be compiled from the specimen and species data into reports designed appropriately for the conservation project in hand. At this stage, the value of a database that is capable of a fairly sophisticated compilation of data and output, directly to the report format that you desire, will become apparent. The alternative is to download all the available and compiled data into a word processor and create the report manually. This is a less efficient method and can be rather frustrating. Species and specimen data then edited on the database are not reflected in the word-processed document. Reports produced on each of the priority species as defined by the Candidate List.

Some Conclusions and Cautions

The creation of Candidate Lists and Species Reports place the plant conservation practitioner in a better position to design species conservation projects accordingly, especially effective fieldwork programmes. This will enable the practitioner to solicit the support of other workers who may be in a better position to reach remote areas or look out for particular species in particular habitats or sites. However, it is the implementation of the field research that will eventually lead to a refined Candidate List and a better understanding of distribution patterns and thus areas of high species diversity and/or threats, giving clearer objectives for conservation interventions.

Very often the inferences on conservation status that can be compiled from herbarium data will be the only available method of setting initial species conservation priorities. However, the kind of activities outlined in this discussion should not be seen as the end product of the conservation project. Data compiled from the herbarium are woefully inadequate to arrive at the best conservation practices to adopt. What they can and should do is act as a guide towards the necessary data gathering from field observations, assessments and monitoring programmes.

Kalanchoe boranae Raadts (Crassulaceae) could be considered Extinct in the Wild (*sensu* IUCN, 1994). It has been collected twice from the same type locality, but not been seen in the wild for 29 years. This IUCN Category of Threat can be adopted simply from analysis of the specimen data set. The appropriate action, however, will involve continued searches of the species in the original collection locality with the assistance of local communities. We have found that the Species Report or the Red Data List is frequently considered by many projects as the *product* of the conservation initiative. This is dangerous and shows a lack of strategic thinking and long-term planning. High impact conservation work will view them as stepping-stones to achieving *in situ* species survival and proliferation — the real products of species conservation projects.

References

Bytebier, B., Nduma, I. and Pearce, T. (1996). *Directory of Botanical Collectors Series for East Africa.* National Museums of Kenya. Centre for Biodiversity Research Reports; Biodiversity Database No. 7. National Museums of Kenya, Nairobi, Kenya.

Knox, E.B. and Vanden Berghe, E. (1996). The use of LEAP in herbarium management and plant biodiversity research. *Journal of East African Natural History* **85**: 65–79.

Pearce, T.R., Filer, D. and Vanden Berghe, E. (1996). The computerisation of the East African Herbarium, pp. 18–31. In: L.J.G. Maesen, *et al.* (eds). *The Biodiversity of African Plants.* Proceedings XIVth AETFAT Congress, Wageningen, The Netherlands.

Chapter 5

Conservation of Plant Diversity in Tanzania:

approaches and experiences from the East Usambara Conservation Area Management Programme, Tanzania

Stig G. Johansson
Metsähallitus Forest and Park Service
Vantaa, Finland

Massaba I.L. Katigula
Tanga Region Catchment Forest Office
Tanga, Tanzania

Shedrack Mashauri
Ministry of Natural Resources and Tourism
Amani Nature Reserve, Tanzania

Ahmed Mndolwa
Amani Botanic Garden
Tanzania Forestry Research Institute
Tanzania

Introduction

Tanzania is one of the biologically richest African countries: it is at the convergence of several major vegetation types or phytochoria; the semi-arid and arid Somali-Masai Regional Centre of Endemism (RCE) in the north, the savannah Sudanian RCE in the north-west, the wooded Zambezian RCE in the south, and the forested Guineo-Congolian RCE in the west (White, 1983). The range of landforms, including the old geologically stable block fault mountains and the recent volcanic mountains, combined with recurrent geological climate change, have generated a diverse and dynamic ecology. This has been further modified by successive patterns of human land use (Figure 5.1), including pastoralists and agro-pastoralists, agrarian communities and hunter-gatherers.

Figure 5.1   The main threat to the montane forests of East Africa is from felling of forest for small-scale agriculture. Here agricultural plots encroach on to montane forest, East Usambara Mountains, Tanzania.

Biodiversity, broadly defined, makes a significant contribution to the Tanzanian economy (Kaiza-Boshe *et al.*, 1998):

- Agriculture and tourism are directly derived from and dependent upon biodiversity. Agriculture accounts for almost 50% of the country's Gross Domestic Product (GDP) and 75% of foreign exchange earnings;

- Tourism is fundamentally based on biodiversity (albeit mammalian!);

- The Tanzanian Forestry Action Plan (TFAP) indicates that in 1988 the forestry sector contributed between 2 and 3% of the GDP;

- Wood fuel accounts for more than 90% of the national energy requirement;

- Forestry contributes 10% of registered exports and provides employment to 730,000 people per annum.

A wide range of linked influences are driving the degradation of biodiversity resources (Kaiza-Boshe *et al.*, 1998):

- Increasing human populations – in terms of both national totals and regional densities – generated by internal growth and refugee movements. Tanzania has a population of around 30 million people with a population growth of between 2.7–3.7% per annum;

- Inappropriate economic and land use policies which generate disincentives to conserve/maintain biodiversity and the sustainable use of its components;

- Limited national capabilities to acquire and utilise new conservation technologies; and

- Absence of a national policy regime and strategy on conservation and sustainable use of biodiversity.

Tanzania and international conservation values

In Sub-Saharan Africa, only the Republic of South Africa (23,000 species) and probably the Democratic Republic of Congo have larger numbers of flowering plants than Tanzania (about 10,000 species). In terms of numbers of endemics, Tanzania (11%) is regionally important. Only Ethiopia (>10%) and Somalia (16%) in the broad East African region have higher levels of endemism (WWF and IUCN, 1994–1997). Myers *et al.* (2000) identified 25 global hotspots that contain between them about 44% of all vascular plant species, one such hotspot being the Eastern Arc and Coastal Forests of Tanzania and Kenya. This hotspot contains about 4,000 plant species, of which 1,500 are endemic. In a review of global centres for plant diversity undertaken by the World Wide Fund for Nature (WWF) and the World Conservation Union (IUCN) (WWF and IUCN, 1994–1997), 12 coastal and upland forest sites were identified as Centres of Plant Diversity in Tanzania (Table 5.1).

Table 5.1 **Globally significant centres of plant diversity in Tanzania (WWF and IUCN 1994–1997)**

Phytochorion	Area name	Location
Zambezian RCE	Itigi thicket	Central Tanzania
Zambezian RCE	Mahale-Karobwa Hills	Kigoma, Lake Tanganyika
Indian Ocean Coastal Belt	Msumbugwe Forest Reserve	Coast, S of Pangani
Indian Ocean Coastal Belt	Pugu Hills Forest Reserve and Kazimzumbwi Forest Reserve	SSW of Dar es Salaam
Indian Ocean Coastal Belt	Rondo Plateau	Lindi District
Indian Ocean Coastal Belt	Middle Ruvuma River area	Tanzania/Mozambique border
Afromontane RCE	East Usambara Mountains	W of Tanga, NE coast
Afromontane RCE	Nguru Mountains	Morogoro District
Afromontane RCE	Uluguru Mountains	Morogoro District
Afromontane RCE	Udzungwa Mountains	Kilornbero District
Afromontane RCE	Kilimanjaro	Arusha Region
Afromontane RCE	Kitulo Plateau/Kipengere	Lake Nyasa Mountains

Note: RCE – Regional Centre of Endemism

National policies and strategies for biodiversity conservation

Biodiversity management in Tanzania, at an institutional level, is undertaken through:

- Government ministries: Wildlife Division, Forestry and Beekeeping Division (FBD), Fisheries Division, Division of the Environment, Crop Division and Livestock Division;

- Parastatal organisations or public corporations: Tanzanian National Parks (TANAPA), Tanzanian Forest Research Institute, National Environment Management Conservation (NEMC), and others; and

- Non-governmental organisations: Wildlife Conservation Society of Tanzania, the Tanzania Association of Foresters, the Tanzania Natural Development Company Ltd., World Wide Fund for Nature (WWF) national office, Agenda for the Environment, Roots and Shoots, Tanzania Environmental Society, etc.

Tanzania has prepared a forestry action plan (TFAP) (MTNRE, 1994*a*) which includes programmes on biodiversity conservation. A new forestry policy has been drafted (FBD, 1997) providing for the establishment of nature reserves with the objective to conserve biodiversity. It also makes provisions for several other mechanisms and instruments for better management of the forestry estate, including increasing local participation and ownership of both forests and forest products.

In the 1990s, several important policy documents were prepared which are relevant for biodiversity conservation. These include the National Conservation Strategy for Sustainable Development (NEMC, 1995), the National Environmental Policy (Division of Environment, 1994), the National Environmental Action Plan (MTNRE, 1994*b*), the Policy for Wildlife Conservation (Department of Wildlife, 1996), and the National Land Policy (MLHUD, 1995). In 1996, Tanzania signed the Convention on Biological Diversity (CBD) (Appendix 1). The National Environmental Policy for Tanzania (1997) recognises the need for an integrated approach to biodiversity conservation (cited in Kaiza-Boshe *et al.*, 1998); it states that 'conservation of biological diversity is to be attained, in the wild (*in situ* conservation) in protected areas, outside protected areas and on the farm, in gene banks and botanical gardens (*ex situ* conservation) and laboratories relevant to conservation of biological diversity'.

Tanzania and the National Protected Area Network

The national priority for biodiversity management has been given to the protected areas network. These are officially categorised as national parks, game reserves, multiple land use areas, game controlled areas and, more recently, nature reserves. Marine reserves are managed by the Fisheries Department. In addition to these areas are the forest reserves, subdivided into productive and protected forest reserves. In total, the 540 forest reserves account for 132,000 km², of the total woodland and forest area (440,000 km²), including Man and the Biosphere Reserves (Table 5.2). Productive forests account for 108,000 km², and protective forests account for 26,000 km². However, the status of these forests (legal, social and ecological) needs urgent review.

In Tanzania, about 14% of the land area is under some form of wildlife protection. However, only 4% is totally protected, while 10% is partially protected. The protected area portfolio consists of 12 national parks (41,000 km²), one conservation area (Ngorongoro Conservation Area), one nature reserve (Amani Nature Reserve – Figure 5.2), one marine park, 23 game reserves (104,000 km²), 56 game controlled areas, and 405 central government and 166 local authority forest reserves. These extensive areas contain one of the world's largest protected areas, the Selous Game Reserve. However, there is increasing concern that these reserves are too fragmented and isolated from each other. Whilst Eastern Africa, and Tanzania in particular, has a high number of protected areas, many are relatively small (between 100 and 1,000 km²) and the large distances between national parks

Table 5.2 World Heritage Sites and Man and Biosphere Reserves in Tanzania		
World Heritage Sites	**Area ($\times 10^3$ ha)**	**Year notified**
Mount Kilimanjaro National Park	76	1987
Ngorogoro Conservation Area	829	1979
Selous Game Reserve	5,000	1982
Serengeti National Park	1,476	1981
Total	**7,381**	
MAB Reserves		
Lake Manyara National Park	33	1981
Serengeti-Ngorongoro	2,305	1981
Total	**2,338**	

Figure 5.2   View of the Amani Nature Reserve showing the fragmented nature of the forest reserve resulting from the establishment of farm plots and tea and forestry plantations in the area. Conservation in this area will accordingly work across a variety of land uses, tenures and social expectations.

increasingly prevents biological linkages between them. This can only be addressed by regional conservation plans, encompassing protected areas and forest lands, which consider the matrix surrounding the parks and aim to maintain biological links and processes (Siegfried *et al.*, 1998).

In 1991, the state financing allocated to the protected area network in Tanzania was about US$3.5 million. In the same year the total earnings from the Wildlife Division, central government returns, TANAPA, and the Ngorongoro Conservation Area Authority (NCAA) were US$9–10 million. In 1995, gross foreign exchange earnings from international tourism to Tanzania amounted to about US$205 million spent by an estimated 280,000 visitors. The target for 2005 is a doubling of international visitors to 575,000, staying for just over 2.6 million 'bed nights' and spending US$570 million at 1995 rates. This equates for about 7.5% of GDP, nearly 25% of total export earnings and directly supports 25,700 jobs (MNRT, 1996).

Much of the protected area system in Tanzania and East Africa has evolved from the wildlife sector with the aim of protecting large mammals and their savannah habitats. Hence, the professional capabilities and skills have been primarily geared towards other disciplines than forest management and conservation. For instance, until recently only small areas of closed forests have been included in the national parks. The establishment of the Udzungwa National Park, which primarily is a forest park, signifies a positive change.

The forest reserve system covers some 132,000 km^2. The land is vested in the Director of the FBD and the management is co-ordinated by the FBD for the central government forest reserves and by the District Councils and District Forest Officers for the Local Authority forest reserves. The reality in most cases is that the forestry authorities do not have the resources, whether manpower or financial, to effectively enforce the law and protection of the forests. Largely the same applies to many of the Game Reserves and Game Controlled Areas, which also contribute some 175,000 km^2 to the protected area network. Illegal hunting and poaching is difficult to contain and, in most cases, control is geared towards the protection of the larger, politically valued mammals.

Most of the organisations, possibly with the exception of TANAPA and the NCAA, which generate considerable revenue from tourism, are struggling with decreasing resources. Their hands are often tied by limiting flexibility and innovation that so often characterise government bodies. Simultaneously, pressure from communities, increasingly dependent on the resources, is growing. It is probably this area of institutional-financial interaction, combined with increasing local pressures, that forms the greatest challenge for conservation of biodiversity in Tanzania (and probably East Africa as a whole).

The existing protected areas in Tanzania have often been created as a result of lobbying for specific areas rather than considering a comprehensive representation. An optimal protected area (PA) system should include representative samples of the biological regions, ecosystems, natural communities and species of the country. They should also be large enough to be viable in the foreseeable future. Tanzania has an extensive PA network covering 25% of the land surface (if forest reserves and game controlled areas are included). However, this PA network in primarily based on the protection of large mammals and savannah habitat and therefore there are several gaps in representation of habitat types and their associated biodiversity.

Practical Challenges for Forest Biodiversity Conservation

A case study from the Usambara Mountains

The East Usambaras form part of the Eastern Arc, a chain of isolated, forested mountains near the north-eastern coast of Tanzania. These forests are globally recognised as a biodiversity 'hotspot' (Myers *et al.*, 2000) and an IUCN Centre of Plant Diversity (WWF and IUCN 1994–1997). The diversity of flora and fauna is considerable and the forests support an exceptionally high degree of endemism. This is a result of the area's relatively stable environmental conditions, over more than 30 million years isolation from the larger central African forest blocks, and the proximity of a diverse number of phytogeographical elements (particularly the Afromontane and Somalia-Masai regional centres of endemism) (Iversen, 1991; Kingdon, 1990; Lovett and Wasser, 1993; Rodgers and Homewood, 1982). From a total of 2,855 plant species, about 25% are endemic to the Usambaras (Iversen, 1991). Perhaps the best known endemic plant species are the African violets (*Saintpaulia* Wendland spp. (Gesneriaceae)) (Eastwood *et al.*, 1998). The East Usambara forests are considered amongst the most valuable areas for conservation in Africa (e.g. Rodgers, 1993) and some authors have compared their biological importance to the Galápagos Islands (Rodgers and Homewood, 1982; Kingdon, 1990). Furthermore, the forests also form the catchment for the Sigi River and are crucial to the water supply of Tanga, the regional administrative centre.

The largest surviving blocks of lowland coastal forest in Tanzania are located on the Matumbi Massif to the south-east of Utete (*c.* 25 km^2), Zaraninge Plateau to the east of Sadaani (*c.* 20 km^2), Gendagenda to the west of Pangani

Figure 5.3 Tea plantations are an important part of the East Usambara landscape; they depend upon the forest as watersheds. The management of adjacent forest reserves is supported through collaborative land use agreements between the Forest Department and tea estates.

(*c.* 26 km²), Pugu Kazimzumbwi (*c.* 30 km²) and inland from Kilwa (up to 100 km²). The Tanzanian coastal forests are being destroyed by unsustainable human actions generally following the sequence:

(a) logging for timber and fuel;

(b) pole-cutting to build houses;

(c) wholesale burning for charcoal;

(d) wholesale conversion to agriculture (Burgess *et al.*, 1992; 1998).

The present land use patterns in the East Usambaras can be traced to the early German colonisation. The forests of these original German estates have developed into the present network of forest reserves, while the agricultural areas developed mainly into the present-day tea estates (Figure 5.3), which cover about 3,000 ha on the plateau. In a similar progression, the lowland foothill areas developed into sisal and other large-scale agricultural estates, forcing local small-holders to carve out their living along the slopes. Small-holder agriculture in the Usambaras is still on a semi-shifting cultivation basis, where pressures on land and productivity are released through the clearing of new forest for cultivation. Cash crops, such as cardamom, have played a major role in the expansion of agriculture into the forests.

The forests contain valuable timber species and commercial logging was started in the late nineteenth century, while large-scale forest clearing created the coffee and tea plantations, a process that continued up to independence. Simultaneously, small-scale agriculture has expanded, often at the expense of forests (Hamilton and Bensted-Smith 1989; Iversen 1991). Commercial timber harvesting continued until the mid-1980s, but was stopped in 1987 due to environmental concerns. However, pitsawing continued until the early 1990s. Rodgers and Homewood (1982) estimated that 50% of the public land forests in Amani disappeared between 1954 and 1978. The Amani Forest Inventory and Management Plan (AFIMP) (1988) estimated that just over 5,000 ha were intact by 1988. Clearing for agriculture has increased forest fragmentation, and some authors consider this a major threat to biological diversity in the East Usambaras (Newmark, 1993) (Figure 5.2). Presently, only 4–5 major blocks of forests remain and proposals for establishment of corridors between these have been made (EUCFP, 1995; Newmark, 1993).

The establishment of a nature reserve in Amani, which is the area with the highest frequency of endemic plants, was officially proposed in 1988 by AFIMP. Large parts of the area were formerly gazetted into six forest reserves, managed by the FBD. The proposal was to combine these reserves into a nature reserve, with an explicit biodiversity conservation objective. An important part of the Amani Nature Reserve is the Amani Botanical Garden. The Biologisch Landwirtschaftlische Institut Amani was initially established in 1902 by the Government of German East Africa. Between 1902 and 1914 about 200 ha, out of a total area of 300 ha, were developed as trial plantations, with around 1,000 exotic species planted in a facility parallel to the development of the Limbe Botanical Garden in Cameroon (see Sunderland *et al.*, Chapter 21). These plantations now form what is known as the Amani Botanical Garden, which still is one of the largest botanical gardens in Africa.

i) The general approach of the project

The East Usambara Conservation Area Management Programme (EUCAMP)[1] aims to:

- establish the Amani Nature Reserve (ANR)
- protect water sources
- establish and protect forest reserves
- sustain villagers' benefits from the forest
- rehabilitate the Amani Botanical Garden

The project is implemented by the FBD of the Ministry of Natural Resources and Tourism with financial support from the Government of Finland, and implementation support from the Finnish Forest and Park Service. The total financial support for the first two phases (1991–1998) amounts to a total of US$6.1 million.

[1] Formerly known as: The East Usambara Catchment Forest Project (EUCFP).

The assumption is that ultimately conservation will not be possible unless the Tanzanian government has both the political will and sufficient available resources. This will only be possible if the East Usambaras are sufficiently well known – in an ideal situation they would be as internationally known as the Ngorongoro or Serengeti. Fundamental to this process will be both local and international commercial and economic support. Potentially, this would generate revenue and foreign exchange for Tanzania. For the project, this means a focus on publicity and public relations work, so that the East Usambaras stand out from the mass of national parks and conservation areas and asserts an identity of its own. The focus on visitors, many of whom would come out of a specific interest in flora and fauna, also means that the area would be staffed permanently, thus allowing the control and monitoring of developments which are not in line with the conservation objectives. One approach may be to privatise, through franchise or other similar arrangements, parts of or the whole of the protected area service activities.

ii) New protected forests

In 1993, mapping of land use patterns was undertaken, covering a total area of about 83,601 ha, including all submontane areas and most of the lowland forest areas except north of the main East Usambara range. In this exercise, about 50% of the area was classified as forest, with about 19,713 ha of the natural forest (roughly 50% of both the sub-montane and lowland forest) within the legally protected forest reserves (Hyytiainen, 1995). Despite problems associated with the protection of forests in forest reserves, it appeared obvious that without a legal basis for protection, the remaining forests in the unreserved public lands would disappear from the East Usambaras.

Between 1992 and 1996, EUCAMP surveyed more than 12,000 ha of forests as part of extensions to old reserves or the establishment of new reserves. A proportion of this is, however, still in the process of being gazetted. The inclusion of some of the northern areas that were not included in the AFIMP aerial survey in 1986, brought the forest area up to 45,137 ha. Presently about 32,000–35,000 ha or approximately 75% of the total forest area is or will be within the ANR or the gazetted forest reserves (Table 5.3). Presently the ANR and 13 of the 17 forest reserves (four are teak plantations) in the East Usambaras are managed with support from EUCAMP, covering a total area of 29,351 ha.

iii) Mapping biodiversity

Baseline biodiversity surveys were initiated by EUCAMP in July 1995, executed by the UK charity, Frontier Tanzania, under contract to EUCAMP. The aim of the surveys was to provide baseline information on the biological values of different forests as a basis for management planning and long-term

Table 5.3 **Distribution of the East Usambara forests into forest types, density classes, and between reserved (forest reserves) and non-reserved forest lands including tile surveyed and tile proposed forest reserve enlargements, but excluding the proposed forest corridors**

Forest class/sub-class	Total (ha)	Outside Forest Reserve (ha)	% of total	Inside Forest Reserve (ha)	% of total
Sub-montane rainforest:					
Dense forest	6,940	839	12	6,101	88
Poorly stocked forest	471	454	96	17	4
Cultivation under forest	5,506	3,270	59	2,236	41
Sub-total	12,917	4,563	35	8,354	65
Lowland forest:					
Dense forest	15,180	349	2	14,831	98
Poorly stocked forest	8,757	2,583	30	6,174	70
Cultivation under forest	5,561	3,612	65	1,950	35
Sub-total	29,498	6,544	22	22,955	78
Plantations:					
Eucalyptus spp.	493	493	100	0	0
Maesopsis eminii Engl.	529	21	4	509	96
Tectona grandis L.f.	1,683	11	1	1,671	99
Other species	19	19	100	0	0
Sub-total	2,724	544	20	2,180	80
Total	45,139	11,651	26	33,489	74

monitoring, as well as training in the use of biological inventory techniques. The programme has been carried out over four phases of ten weeks. Altogether, this will provide systematic inventory information on the flora and fauna of eight forest reserves or more than 9,000 ha in the East Usambaras. Earlier work (during 1994/95) by Frontier Tanzania, using a slightly different method, provided additional information on the biodiversity values of other forests. The field work involves short-term expatriate volunteer Research Assistants, as well as permanent EUCAMP, Frontier and University of Dar es Salaam staff, and an international network of taxonomists and other experts. The work has become progressively more systematic and quantitative. The project has already resulted in the discovery of several taxa previously unknown to science, and this will undoubtedly raise awareness of the unique biodiversity values of the East Usambaras (Figure 5.4). The surveys have proved to be very useful as they provide information on new species and range discoveries, which are used for publicity for the conservation of the East Usambaras. Secondly, they are providing information for the zoning and

Figure 5.4   Opening ceremony of a Conservation Assessment and Management Plan workshop (CAMP) hosted by the East Usambara Catchment Forest Project in association with the Darwin Initiative training programme. An example of collaborative training for conservation (see Hankamer *et al.*, Chapter 15), where a variety of stakeholders worked together to assess conservation risks to the endemic flora of the East Usambara Mountains.

management of the individual forest reserves so that areas of priority conservation value can be identified and protected. Thirdly, the exposure and training that the forestry staff received while participating in the ten-week field surveys has considerably changed their perspective on biodiversity and conservation issues.

iv) Containing pitsawing and agricultural encroachment

A major effort has been made to contain pitsawing, agricultural encroachment, mining and other illegal activities, which were rampant within the protected areas until the early 1990s. This has been tackled through the re-establishment of reserve borders, patrolling, awareness and extension activities with the local villages. This has been successful and reported incidents have been reduced. Despite problems, the forest reserve system must be credited for having maintained much of the forests in the East Usambaras (Hamilton and Bensted-Smith, 1989). This may be as a result of having several generations who have lived with the forest reserves and are aware of the restrictions. For example, about 95% of respondents in 14 villages were aware of the forest reserves. Some of the most frequently mentioned benefits were water supply, building poles and fuelwood, while the problems were related to availability of land and supply of forest products.

A recent study in 1997, on the various approaches adopted by EUCAMP on the management of forest reserves, indicates that the consultative approaches used in villages where pilot village forest management has been tested are more successful than the more traditional enforcement approach. However, from the experience in the East Usambaras, it seems that without the initial enforcement with court cases and fines for illegalities, the consultative approach may be too slow in containing a situation where control has been lacking for several years.

v) Joint forest management of government forest reserves

A dialogue and interaction with the communities is particularly important in the areas where new forest reserves have been established or old ones have been enlarged. EUCAMP has made efforts to strengthen villagers' rights to manage their own forests, while a model for joint forest management of government reserves is drafted but yet to be tested (Johansson, 1997; Johansson and Kijazi, 1997; Johansson and Sandy 1996). These proposals are based on the following hypotheses:

- Local community management of non-reserved forest will not succeed unless provided with a sufficiently strong legal and institutional basis for administration and management that also recognises the tenure of forest resources.

- Joint management of government forest areas is only possible if the roles and aspirations of the partners are defined and understood in a practical and concrete manner, where the authority and legal mandate of the forest service is accommodated, and where any concerns of the local communities are dealt with.

- Ownership is a pre-requisite for good resource management, but land ownership and tenure can be substituted by rights to concrete economic benefits from well defined incentives linked to the forest, especially if those concrete benefits are directly retained at the village level.

- There will be neither sound forest management nor sustained agroforestry unless there is a price or value on the forest resources, especially for formerly free commodities, such as fuelwood, polewood, etc.

- A price on the forest resources is an incentive to tree planting, forest management, and agroforestry-based land husbandry by local, communities adjacent to the forest.

The above approach would call for discussions with the villages, village authorities and forest products user groups in the villages, where criteria would be defined for the collection of specific forest products. A method to determine village collection ranges would be developed and tested, and the ranges would be determined. A village institution would be identified, which would have the authority to issue permits to villagers, and the responsibility to control and collect revenue from utilisation. Prices would then be defined for different products, such as fuelwood, polewood, etc. These would need to take into account the ability of villagers and users to pay, and the availability and cost of alternatives. Moreover, the village should determine a pricing and control system for forests outside the forest reserves, but within the village lands. The revenue collected from the utilisation of the forest reserves would be split between the villagers, while the revenue from the 'public land forests' would be retained by the village. An arrangement for reporting and accounting should be drafted between the proposed village and the FBD, and the forest officer in charge of the forest reserve should review the reporting and accounts together with the village authority, and discuss problems and issues requiring attention.

This arrangement could have several advantages. It would give the authority and responsibility of utilisation and control to the village. It would reduce the distance between the foresters, the forest and the villagers, and it would decriminalise villagers entering forest reserves. The joint forest management model would also develop a price for previously free commodities and hence create an incentive for people to cater for their own needs e.g. through tree planting. It would create an incentive for the village to control and use the resource in an effective manner as it would generate revenue for village development. It has the potential to considerably reduce the amount of time and funds paid to patrol and enforce forest reserves.

There are also several risks. People may refuse to pay even nominal fees. The revenue may not be seen as an incentive because it primarily circulates and re-distributes funds within the village rather than bringing new economic resources to the village disposal. The FBD or the Districts may not accept a split of the revenue, or they may propose a split that is unattractive to the villages. There are always risks that the revenue will be mismanaged both at the village level and in the arrangements between the village and the forest authorities. Selected groups in villages may become marginalised.

vi) Village forest management and farmer-to-farmer extension

Traditionally, farmers have always exchanged knowledge. The project found it difficult to work with villagers using externally generated professional forestry knowledge, partly because of limited staff numbers and partly because foresters have not been trained to deal with farmers' problems, which are mostly concerned with growing crops or raising livestock. The use of local knowledge, using farmer-to-farmer extension, seemed a more promising approach, both in village forest management and farm forestry. The approach is based on the notion that a farmer will only readily appreciate information and experience from someone who is faced by the same practical constraints and experience, and with experience of issues with the potential to increase household income. In this context, farmer-to-farmer extension means that the project's role is primarily in facilitating communication, discussion and analysis between farmers in order for them to develop their own methods to manage land use problems and opportunities. The project also tries to challenge farmers with new information and new situations by taking them on study visits to other areas.

The farmer-to-farmer approach has been used to conserve village forests. Two villages, Vuga and Hemsambia, share a 30 ha forest known as the Mpanga forest. The Mpanga village forest belonged to the 'Wakilindi' clan, and was used for worshipping and ritual purposes. In 1993, the villagers petitioned EUCAMP to gazette the area as a central government forest reserve so that it could be rescued from over-exploitation. Instead, the project decided to make Mpanga into a pilot project to develop a model for local management of village forests. The approach was to facilitate discussions in the villages about the forest and how it should be managed, and to provide technical advice and inputs if required. In a joint effort with the villagers, the Mpanga forest was surveyed, the boundary was demarcated (and planted with *Grevillea robusta* A. Cunn. ex R. Br. (Proteaceae)), and a boundary map was drawn. The two villages drafted rules to guide the management of the forest and these were approved as a bye-law by the District Council.

Farmers from Vuga and Hemsambia visited other village forests and presented what they had done in their own forests. Pilot groups were used to facilitate the process in other villages. EUCAMP has also started to work with environmental education in primary schools by involving elders and 'forest specialists' in the villages.

Ideally in the farmer-to-farmer approach, decisions made at the household and farm level are products of local people's ideas, views, needs, arrangements and agreements. Farmers have appreciated opportunities to visit other areas and talk to other farmers; implementation is through the people themselves, and the EUCAMP role is through facilitating the process, providing ideas and technical support, monitoring and evaluation of the programme. Another advantage of this approach has been to involve both farmers and field staff in farmer-to-farmer activities to create a more productive relationship and working environment between professional extension staff and the farmers.

Why has this not taken place before, and why have local management initiatives not developed without the facilitation of the project? The project functions as a mediator; it brings ideas, new perspectives and innovations into the communities. Probably the process, such as the development of village forest management, also helps the communities to regain some strength to solve their own problems and establish a legal framework for management and control of their own resources.

Discussions and conclusions

Experience in the Usambara Mountains suggests that a clear legal authority is a prerequisite for proper land management and conservation. Unless such a basis is developed, most participatory or joint management arrangements may remain temporary social contracts. Equally as important is the need to ensure transparency at all levels whilst developing broad-based community arrangements. Also, there is a need to look for strong incentives and linkages between individuals or groups and the asset or resource that they manage; whether through direct privatisation or other commercial arrangements.

The pressure on natural resources is mounting. Many of the remaining large protected areas exist because they are mainly savannah areas, where relatively weak competing pressures have enabled these areas to be created. Many of the important areas for biodiversity conservation, such as the Eastern Arc and the Coastal Forests, are in areas where there are strong pressures for both traditional and estate agriculture because of favourable climate and soils. Along the coast, there are pressures from urban and tourism development. An example is the struggle to save the Pugu forest south of Dar es Salaam. Other areas, such as wetlands, are also attractive for agriculture. Many areas, which are important for conserving biodiversity, are so far away from major routes that their potential for generating resources (e.g. tourism traffic) is low, while the costs of conservation are high and the logistics are difficult.

Presently, regardless of minor problems with poaching, protection in the Tanzanian national parks is reasonably effective. This is because their management is geared towards tourism, which brings in considerable returns, thus linking the activity to one of the major economic and foreign exchange earning activities, and promoting the country's international image. Many other areas, such as the forest reserves, game reserves and game controlled areas, have much greater problems. Numerous questions still remain:

- What about all the other 'not so attractive', but still important, biodiversity areas?
- Should all the protected areas be brought under the National Parks?
- What about the benefits, and dependence of the local communities on the resources and assets within the proposed protected areas?
- Can these benefits be secured without compromising the conservation objectives?
- Can they be secured under increasingly resource restricted conditions?

- There are already great problems with the proper management of the protected area network. What will be the result if the area is considerably increased, while the expectations for the standard of protection will also increase?
- Will there be possibilities to maintain revenue per area protected?

Is the option, for example, to give the natural forests currently managed by the FBD to the national park authorities? The present trend in East Africa seems to be that the 'goodies', i.e. those natural forests which are both valuable biodiversity resources and tourist assets, are given to the national parks authorities. This has been partly done in Uganda, with discussions held in Kenya and in Tanzania. Can this solve the problem? The risk is that an increasing amount of low-yielding estates are loaded on to the national parks authorities which in the worst case may lead to a weakened institutional capacity as resources are spread thinner and the yield per area managed decreases. Moreover, this would still leave large areas of the forested and wooded environment, the 'general landscape and environment' with managing institutions stripped of their revenue generating assets and under different institutional cultures and procedures (Table 5.4).

Generating revenue from nature reserves or other protected forest areas through tourism may offer one strategy for forestry conservation. In fact, there are few equally attractive options that offer both positive economic incentives along with positive effects on the public image of the forestry sector at both the national and international level. A stronger and dynamic emphasis on forest conservation and tourism within the forestry sector would also be associated with a positive contribution to the national economic development. This could be through establishing a protected forest area network of 'nature reserves' or 'forest parks'.

However, such an approach requires a considerable change in institutional and organisational scope. In order to avoid damaging institutional 'tug of wars', the conservation and tourism assets of the forestry sector could be utilised through joint management arrangements, including revenue sharing, with established conservation area management institutions, such as national parks authorities (e.g. TANAPA). Such arrangements, through a memorandum of under-standing and joint management, have, for example, been tried in Kenya between the Forestry Department and Kenya Wildlife Service.

There may also be a need to create completely different institutions to manage these resources. Should the natural resources (be they savannah, natural forests or plantations for protection, production or services) be managed by one institution? Could one alternative be a Tanzania Forest and Park Service? New policies have been drafted during the last few years in many areas. This is the time for reformulating those institutions that will be implementing Tanzania's response to new economic, political and ecological challenges.

Table 5.4 Differences in working practice between wildlife and forestry sectors in East Africa (derived from Rodgers, 1995)

Parameter	Forestry	Wildlife
Biodiversity network	Biodiversity low priority in past network development	*Ad hoc* development, more recently with a scientific input
Size	Typically small and fragmented, no corridors	Many very large, some smaller, functional corridors exist in some cases
Status	Forest reserve = low status	National park or wildlife reserve = higher status
Biodiversity focus	Poor, increasing slowly	Strong focus on large mammals
Zonation	Complex – involving many small compartments that can be assigned to one of several functional roles	Less complex – some use of buffer and wilderness zones
Managerial effectiveness	Utilisation: Timber – fair, Conservation – poor	Utilisation: Tourism – good, Conservation – fair
Staffing	Low staff-to-area ratios, especially at level of field-based guards	High staff-to-area ratio in some cases, especially at level of field-based guards
Monitoring infrastructure	Little of biodiversity values	Adequate for larger species
Investment/ha	Low in field	High in field
People interactions, e.g. through Integrated Conservation and Development Projects	Starting-focus on timber and related products	Starting-focus on game products

References

Amani Forest Inventory and Management Plan Project (1988). *Amani Forest Inventory and Management Plan.* East Usambara Catchment Forest Project, Forest Division of Tanzania and Joint Venture Finnmap Silvestria, Dar es Salaam, Tanzania and Helsinki, Finland.

Burgess, N.D, Clarke, G.P. and Rodgers, W.A. (1998). Coastal forests of eastern Africa: status, endemism patterns and their potential causes. *Biological Journal of the Linnean Society* **64**: 337–367.

Burgess, N.D., Mwasumbi, L.B., Hawthorne, W.J., Dickinson, A. and Doggett, R.A. (1992). A preliminary assessment of the distribution, status and biological importance of coastal forests in Tanzania. *Biological Conservation* **62**: 205–218.

Department of Wildlife (1996). *Policy for Wildlife Conservation.* MTNRE – Ministry of Tourism, Natural Resources and Environment, United Republic of Tanzania, Dar es Salaam (Final Draft).

Division of Environment (1994). *The National Environment Policy.* United Republic of Tanzania, Dar es Salaam (Draft 1994).

Eastwood, A. *et al.* (1998). The conservation status of *Saintpaulia. Curtis's Botanical Magazine* **15**: 49–61.

EUCFP (1995). *East Usambara Catchment Forest Project – Project Document Phase II 1995–1998.* The Ministry of Finance and Department for International Development Co-operation, Dar es Salaam, Tanzania and Helsinki, Finland.

FBD (1997). *Tanzania Forest Policy.* Forestry and Beekeeping Division, Ministry of Natural Resources and Tourism, United Republic of Tanzania, Dar es Salaam (Final Draft).

Hamilton, A.C. and Bensted-Smith, R. (eds) (1989). *Forest Conservation in the East Usambara Mountains, Tanzania.* IUCN, Gland, Switzerland.

Hyytiainen, K. (1995). *Land use classification and mapping for the East Usambara Mountains.* East Usambara Catchment Forest Project Technical Paper no. 12. Forest and Beekeeping Division and Finnish Forests and Park Service, Dar es Salaam, Tanzania and Vantaa, Finland.

Iversen, S.T. (1991). The Usambara Mountains, NE Tanzania: Phytogeography of the vascular plant flora. *Acta Universitatis Upsaliensis Symbolae Botanicae Upsaliensis* **XXIX**: 1–234.

Johansson, S.G. (1997). Joint Forest Management in the EUCFP. *Misitu ni Mali (Forestry is wealth)* **3**: 32–34.

Johansson, S. and Kijazi, M. (1997). Farmer to farmer extension. Village forest management in the East Usambara Catchment Forest Project. *Misitu ni Mali (Forestry is wealth)* **3**: 7–9.

Johansson, S. and Sandy, R. (1996). *Protected areas and public lands. Land use in the East Usambaras.* East Usambara Catchment Forest Project Technical Paper no. 28. Forest and Beekeeping Division and Finnish Forest and Park Service, Dar es Salaam, Tanzania and Vantaa, Finland.

Kaiza-Boshe, T., Kamara, B. and Mugabe, J. (1998). Biodiversity management in Tanzania, pp. 121–152. In: J. Mugabe and N. Clark (eds). *Managing Biodiversity: National Systems of Conservation and Innovation in Africa.* African Centre for Technology Studies (ACTS), Nairobi, Kenya.

Kingdon, J. (1990). *Island Africa.* Collins, London, UK.

Lovett, J.C. and Wasser, S.K. (eds) (1993). *Biogeography and Ecology of the Rain Forests of Eastern Africa.* Cambridge University Press, Cambridge, UK.

MLHUD (1995). *The United Republic of Tanzania National Land Policy.* Ministry of Lands, Housing and Urban Development, The United Republic of Tanzania, Dar es Salaam.

MNRT (1996). *Integrated Tourism Master Plan Vol 1.* Ministry of Natural Resources and Tourism, The United Republic of Tanzania with CHL Consulting Group, Irish Tourist Board and Inter-Tourism Ltd.

MTNRE (1994*a*). *Tanzania Forestry Action Plan 1989/902007/08.* Update 1994. MTNRE, United Republic of Tanzania, Dar es Salaam.

MTNRE (1994*b*). *Tanzania National Environment Action Plan, a first step.* Ministry of Tourism, Natural Resources and Environment, United Republic of Tanzania, Dar es Salaam.

Myers, N., Mittermeier, R.A., Mittermeier, C.G., Da Fonseca, G.A.B. and Kent, J. (2000). Biodiversity hotspots for conservation priorities. *Nature* **403**: 853–858.

NEMC (1995). *Tanzania National Conservation Strategy for Sustainable Development.* National Environment Management Council, United Republic of Tanzania, Dar es Salaam (revised proposal).

Newmark, W.D. (1993). The role and design of wildlife corridors with examples from Tanzania. *Ambio* **22**: 500–504.

Rodgers, W.A. (1993). The conservation of the forest resources of eastern Africa: past influences, present practices and future needs, pp. 283–330. In: J.C. Lovett and S.K. Wasser (eds). *Biogeography and Ecology of the Rain Forests of Eastern Africa.* Cambridge University Press, Cambridge, UK.

Rodgers, W.A. (1995). The conservation of biodiversity in East Africa: the approaches of forestry and wildlife sectors compared, pp. 217–230. In: L.A. Bennun, R.A. Aman and S.A. Crafter (eds). *Conservation of Biodiversity in Africa: local initiatives and institutional roles.* National Museums of Kenya, Nairobi, Kenya.

Rodgers, W.A. and Homewood, K.N. (1982). Species richness and endemism in the Usambara mountain forests, Tanzania. *Biological Journal of the Linnean Society* **18**: 197–242.

Siegfried, W.R., Benn, G.A. and Gelderblom, C.M. (1998). Regional assessment and conservation implications of landscape characteristics of African national parks. *Biological Conservation* **84**: 131–140.

White, F. (1983). *The vegetation of Africa: a descriptive memoir to accompany the UNESCO/AETFAT/UNSO Vegetation Map of Africa.* Natural Resources Research, UNESCO, Paris, France.

WWF and IUCN (1994–1997). *Centres of plant diversity: A guide and strategy for their conservation.* 3 volumes. IUCN Publications Unit, Cambridge, UK.

Chapter 6

The Australian Network for Plant Conservation

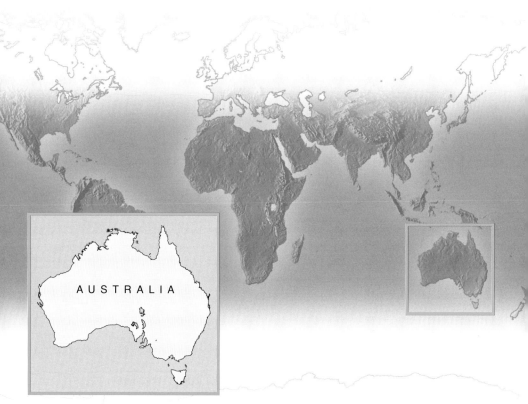

Jeanette Mill
ANPC National Co-ordinator, Canberra, Australia

Introduction

The Australian Network for Plant Conservation (ANPC) model of integrated conservation has received considerable international recognition. Other countries, such as Indonesia and Canada, are adopting this model successfully. This international recognition has also resulted in the ANPC becoming the World Conservation Union (IUCN) Species Survival Commission Australasian Plant Specialist Group.

Overview of plant conservation issues in Australia

Australia is one of the world's centres of plant diversity, with 11 sites listed in *Centres of Plant Diversity* (WWF and IUCN, 1994–1997). Australia has been nationally documenting rare or threatened vascular plants for less than two decades. The number of described vascular plant species is around 22,000, 85% of which are endemic. Of these, 10.5% are considered to be globally rare, threatened or extinct (Walter and Gillett, 1998). Official estimates are that 1,236 taxa (5.6%) are nationally extinct or threatened. Over 4,000 taxa (18.1%) are nationally rare or poorly known. National legislation to list and protect threatened species has only been introduced in the last decade.

Knowledge of the conservation status of non-vascular plants is extremely limited, and basic work on describing and documenting species still has a long way to go. Fewer than 5% of the estimated 250,000 species of fungi in Australia are known. Described species of non-marine algae total over 2,800, and it is thought that the number of undescribed species may equal that figure (Scott, *et al.*, 1997). Approximately 3,000 lichen species have been described, however the total number of species could be as high as 3,500. Further bryophyte species yet to be described will add to the 1,500 described species (McCarthy, pers. com.)

Background to and rationale for a plant conservation network

In 1987, a survey of *ex situ* collections of threatened Australian plant species was conducted by the Australian National Botanic Gardens (ANBG). It was found that many collections were being held by a wide range of groups and individuals. However, there was little co-ordination of collections. Following on from this survey, a conference was held to discuss and promote co-ordination of plant conservation efforts in Australia.

The conference was attended by a broad range of plant conservation stakeholders in Australia, as it was apparent that co-ordination of plant conservation was most effective when *ex situ* and *in situ* efforts were integrated. International models of plant conservation co-ordination, including those promoted by Botanic Gardens Conservation International (BGCI) and the US Center for Plant Conservation (CPC), were discussed. The notion of integrated conservation, as discussed by Don Falk, the then Director of the CPC, was adopted. The follow-on from the conference was a *Proposal for an Australian Network for Plant Conservation* in which the best and most relevant aspects and values of BGCI and CPC were combined to form a broadly based network. The foreword to the proposal expresses this combination of ideas in terms of 'hybrid vigour'!

A network was felt to be appropriate and timely in Australia due to the fact that many people and organisations were carrying out plant conservation work, and national co-ordination was required to reduce duplication of effort, set priorities and maintain maximum relevance with minimal resources. The proposal was submitted to the Endangered Species Program, a programme of the federal Environment Department established under the federal Endangered Species Protection Act, and seed funding for a network was provided. The ANBG also supported the formation of a network by hosting the national office and providing some staffing resources.

The process by which the ANPC was formed is now considered to be the basis for the success and relevance of the Network. Any region considering its future strategy for plant conservation should consider what is relevant and appropriate for the region before embarking on the costly and time-consuming process of establishing a new organisation or strategy. Studying other models is a productive step, and brief notes on some regional models can be found under the section 'Other networks' in this chapter.

The evolution of the ANPC paralleled the process that led to the Convention on Biological Diversity (CBD). The CBD and resultant strategies emphasise a multi-faceted approach to conservation (Glowka *et al.*, 1994; Appendix 1). The *National Strategy for the Conservation of Australia's Biological Diversity* (Commonwealth of Australia, 1996) includes bioregional planning and management, conservation outside protected areas, *ex situ* conservation, integrating biological diversity conservation and natural resource management, improving our knowledge, education and training, information exchange, technical and scientific co-operation, involvement of the community and Australia's international role in addressing key areas to achieve effective global conservation. The development of the ANPC is a prime example of this integrated approach, and was included in *Australia's National Report to the Conference of the Parties to the Convention on Biological Diversity* (Commonwealth of Australia, 1998) (www.biodiv.org/convention/cops.asp). This reflects the effectiveness of actively promoting the involvement of all stakeholders in the development of the Network.

Network Objectives and Activities

1. Mission and objectives

The Mission Statement of the ANPC is:

> *To promote and develop plant conservation in Australia.*

The objectives of the Network are to:

- develop and maintain a national Network of organisations and individuals that includes the range of stakeholders in plant conservation, including land managers;
- develop and maintain regional groups;
- promote co-operation and information exchange;
- encourage the participation of all relevant groups and individuals – community, industry and government – in the Network, and in plant conservation generally;
- contribute to the recovery and long-term survival of threatened plant populations, species, communities and ecosystems;
- co-ordinate and link the efforts of members and others towards plant conservation;
- encourage and assist in the filling of information gaps on the conservation status of Australia's flora;
- formulate or recommend guidelines and codes of best practice for plant conservation activities;
- solicit, collate and disseminate information relevant to plant conservation;
- support and promote education, training strategies and programs;
- maintain regular communication between members by means of publications, meetings and other appropriate media.

2. A diverse membership

It has often been said that we are all part of the problem, therefore we must all be part of the solution. One of the most vital components of successful conservation is to bring into the picture all those who have an impact on the survival of biodiversity. Good information needs to be available to all who have an interest in or an impact on biodiversity, and broad-reaching awareness programmes are vital. The ANPC's integrated approach therefore encourages all stakeholders to be part of the Network, across government, industry and community. The membership of 400 organisations and

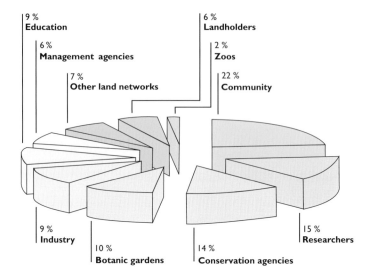

9 %
Education

6 %
Management agencies

7 %
Other land networks

6 %
Landholders

2 %
Zoos

22 %
Community

9 %
Industry

10 %
Botanic gardens

14 %
Conservation agencies

15 %
Researchers

Figure 6.1   Australian Network for Plant Conservation (ANPC) membership breakdown. (Figures for individual memberships are incorporated into the above categories as appropriate).

individuals is the backbone of the Network, and membership is open to all those who support the objectives of the ANPC. Figure 6.1 illustrates the current make up of the membership. It is clear from this figure that the ANPC model is very successful in bringing together a range of stakeholders in plant conservation in a co-operative manner. An annual membership fee is charged which is based on a sliding scale, with individuals paying a nominal charge, non-profit organisations a slightly higher amount, and corporations and government agencies paying the highest fee.

3. Conferences

Conferences are important in ensuring the Network remains relevant, and focused on the most pressing issues in plant conservation. Priorities for the Network are set by members at the conferences, and subsequent funding applications and biennial work programmes are based on conference recommendations. As all members can participate, conferences distil out the priorities for the ANPC from national co-ordination and international linkages through to on-the-ground action. One of the keys to the success of the ANPC is that it was generated from such a conference. Conferences are now held biennially, with the location rotating around the country.

The conference structure consists of:
- papers
- posters
- workshops
- practical demonstrations
- field trips
- social events

All aspects of the conference are contributed to by members from across the spectrum of stakeholders. Conferences are often commented on as being a major networking exercise, bringing members together at a personal level. Conference venues, activities and accommodation options are chosen with care to promote informal networking, and to encourage maximum participation.

4. Newsletter

The ANPC newsletter, *Danthonia* (Box 6.1 and Figure 6.2), named after a grass genus threatened with extinction, is the major means by which members communicate. Articles are encouraged from all members, and range from ANPC news, local projects, practical management techniques, strategies and legislation, snippets on events, publications and Internet addresses for international developments. The newsletter is co-ordinated by the National Office and the ANPC Management Committee. Articles are received and solicited with a view to presenting a balance of information and interest to the range of readers. As a newsletter rather than a refereed journal, *Danthonia* is important as an interface between science and practice.

Box 6.1 Aims of the Australian Network for Plant Conservation (ANPC) newsletter, *Danthonia*

The ANPC newsletter serves many purposes. It:
- documents information about threatened species often stored only in people's heads;
- is a crucial interface between science and practice;
- documents a range of information from stakeholders not available through other sources;
- is a promotional tool for members;
- has a circulation of 1,000 – reaching a wide and international audience;
- is a promotional tool for the ANPC to increase memberships and funds;
- is a ready digest of information about courses, conferences, publications, Internet addresses, etc.

Danthonia

Volume 8 Number 4
March 2000

NEWSLETTER OF THE AUSTRALIAN NETWORK FOR PLANT CONSERVATION INC.

Mundulla Yellows: A growing concern

David C. Paton[1] & Joanne Cutten[2],
[1]Dept of Environmental Biology, University of Adelaide, Adelaide SA 5005
[2]Biodiversity Branch, DEHAA, GPO Box 1047, Adelaide, SA 5001

In the late 1970s Geoff Cotton, an observant beekeeper, noticed unusual yellowing of the foliage on a few mature red gums (*Eucalyptus camaldulensis*) at Buckingham (near Mundulla) in the south-east of South Australia. By 1990 some of the trees showing the yellow foliage were dead and adjacent trees were showing the symptoms. Other eucalypts in the region were also showing symptoms, particularly along roads in and around the towns of Keith and Bordertown.

A 1992-93 survey of the health of hundreds of eucalypts in the Keith-Bordertown area (Paton & Eldridge 1994; Paton *et al.* 1999) showed that trees showing symptoms of Mundulla Yellows (patches of yellow foliage on one or more branches or sub-branches with inter-veinal chlorosis of the leaves) were largely found along roadways or waterways.

A survey of sections of roadside vegetation between 1994 and

1999 showed that of 477 eucalypts examined, none recovered over the five-year period. While some were about the same, most had deteriorated and a reasonable proportion had died (Table 1). The conclusion from these figures is that once a plant shows symptoms of Mundulla Yellows its condition deteriorates and death is inevitable, although it may take 10 or more years. It is likely that all of these trees will be dead within 10-20 years. Furthermore, the only eucalypt seedling to establish along these roadsides over the last five years also contracted Mundulla Yellows, severely dampening any thoughts of ever being able to re-tree these areas.

A range of possible treatments was tried, involving injecting infected trees with either phosphorous acid, tetracycline, rogor or aquasol nutrient solution. No obvious recovery was found, although the ability to record a response was limited

(Continued on page 4)

1

Figure 6.2   The Australian Network for Plant Conservation (ANPC) newsletter, *Danthonia*.

To assist with minimising time spent on preparing and editing information for the newsletter, articles are obtained from the source of the information, e.g. the scientist working on a particular technique for promoting seed germination, or a local group running a conservation project. Contact details are provided to enable readers to follow up if they require more information. *Danthonia* is produced quarterly, and sent to all members as part of their membership fee. Numbers of copies received are scaled according to the fee paid, i.e. corporate members receive five copies, non-profit organisations two, and individuals one.

5. Best practice guidelines

If plant conservation is to be successful, activities must be carried out following current best practice techniques. The ANPC set up working groups, comprising members who are specialists in their respective areas, to collaborate in producing two sets of best practice technical guidelines for plant conservation practitioners. Expertise was drawn from conservation agencies, community greening groups, botanic gardens, primary industries and research organisations. The guidelines are used to improve management practices, and in funding, policy and legislation implementation. For example, funding agencies now require applicants submitting proposals for translocations to use the guidelines. Endorsement of the guidelines has been sought and obtained from relevant government ministerial bodies. In the future, endorsement will also be sought from international conservation organisations, such as the World Conservation Union (IUCN).

i) Germplasm and seed banking guidelines

Germplasm Conservation Guidelines for Australia (ANPC Germplasm Working Group, 1997*a*) (Figure 6.3) have been developed in response to the need for:

- greater co-ordination of effort by genebanks in relation to threatened native flora;
- long-term storage of germplasm to be developed as a strategy for threatened plant conservation in Australia;
- best-practice technical guidelines for plant conservation practitioners, consolidating the currently available information on collection, storage, viability testing and germination.

The guidelines are subtitled: *An Introduction to the Principles and Practices for Seed and Germplasm Banking of Australian Species*, and include such topics as:

- germplasm collection
- seed storage
- vegetative propagule storage
- documentation and databasing of collections
- cryostorage
- Seed orcharding

Figure 6.3   The Australian Network for Plant Conservation (ANPC) Germplasm and
Translocation Guidelines.

ii) Translocation guidelines

Guidelines for the Translocation of Threatened Plants in Australia (ANPC
Translocation Working Group, 1997*b*) (Figure. 6.3) have also been produced.
These guidelines were developed in response to the need to provide best
practice technical guidance to those undertaking reintroductions or
introductions of threatened plants (see Figure 6.4). Translocation can be
defined as the movement of plants at the individual, propagule or population
level. The destination determines the more specific cases of introduction and
reintroduction 'from one place to another, regardless of source or destination'
(Berg, 1996). Translocation is a complex conservation measure, requiring
long-term monitoring (Howald, 1996; Sutter, 1996) to ensure:

- the viability of the translocated population;
- the quality of habitat in which the translocated population resides.

If the translocation is unsuccessful it may be impossible to retrieve the habitat
depending on the ecological niche occupied by the translocated taxon. The
ecology of such species is often unknown and translocation measures may
therefore involve educated guesswork (Howald, 1996).

Figure 6.4 Translocated plants of *Eucalyptus pulverulenta* Link (Myrtaceae) on a rural property in the Australian Newwork for Plant Conservation (ANPC) New South Wales Central West region. During a field trip here the owner explained her efforts to conserve the population on her land, and to propagate plants for use by the local Landcare group in revegetation work. The field trip was also an important opportunity to promote the ANPC Translocation Guidelines to a targeted group of practitioners. Photo: Jeanette Mill.

Translocations are often unsuccessful. Some reasons for the lack of success are:

- the original threats are not adequately controlled;
- the biological and ecological requirements for the taxon have not been adequately considered;
- the assessment of genetic variability is often neglected;
- ongoing commitment of resources to monitoring and follow-up maintenance is often lacking.

The translocation guidelines focus on techniques and issues once a decision to translocate has been made, and cover:

- the translocation process from project development through to monitoring;
- biological, ecological, *ex situ* and logistical considerations;
- case studies;
- community participation.

iii) *In situ* guidelines

The underlying premise stated in all ANPC guidelines is that *ex situ* measures should never be considered as an alternative to *in situ* conservation. Therefore, the next step, as concluded by the 1997 ANPC conference in Coffs Harbour, is to produce guidelines for *in situ* conservation. These guidelines are to focus on ecosystem management and will include management techniques, survey methods, fire research applications, case studies and guidelines for community consultation and involvement.

Regional Groups: networking at the grass roots

As conferences provide a meeting opportunity for members only once every two years, regional groups were proposed as a means by which members can meet and network more regularly. Volunteers co-ordinate regional groups locally. This enables the group to be relevant to the region, and better able to identify and deal with on-the-ground issues. Voluntary Regional Co-ordinators are often provided by member organisations, minimising the resources required from the National Office. Individual ANPC members, who take on voluntary Regional Co-ordinator roles, are assisted with costs for postage, etc, by the National Office.

Regional activities include:

- establishment of volunteer training and placement programmes for local projects;
- threatened plant surveys;
- field trips to view and discuss local projects (see Box 6.2 and Figure 6.4);
- seminars with guest speakers;
- promoting the need to conserve through field days and workshops;
- local networking to assist with community action for poorly conserved species (see case study below);
- hosting ANPC national activities, such as the conference and Plant Conservation Techniques Course; and
- assisting with ANPC core activities, such as incorporation and promotion.

Box 6.2 Case study: local networking in the New South Wales South West slopes – *Caladenia concolor* Fitzg. (Orchidaceae) (crimson spider orchid)

This region encompasses a rural area, straddling the border between the states of New South Wales and Victoria. The region is one of the most extensively altered landscapes in the country. Fragments of vegetation are poorly protected, yet of high biodiversity value (Johnson, pers. com.).

The Australian Network for Plant Conservation (ANPC) Regional Co-ordinator has been using local networking, notably with landholders, to conserve *C. concolor*, which is listed as nationally threatened. The concept of conservation can be interpreted as both intensely political and antagonistic, however the networking approach is often successful in bringing together stakeholders to the benefit of the species in question. Visits to sites of conservation interest have been arranged, where local people can gain an appreciation of the local flora, on public and private land. Co-operative management strategies have been formulated, whereby grazing regimes have been altered to minimise effects on *C. concolor*. Areas at high risk from illegal activities have been fenced and signage erected with co-operation from local agencies. The Regional Co-ordinator is on the ANPC Management Committee, and hosted the 1999 ANPC conference. The orchid was the mascot species for the conference, and conference activities and publicity raised its profile nationally and internationally as a species in need of conservation.

Education – Plant Conservation Techniques Course

The expertise within the ANPC membership presents a unique opportunity to effectively assemble and transmit information to practitioners, and one of the ANPC's major objectives is training and education. The ANPC Plant Conservation Techniques Course was inspired by the International Diploma Course in Plant Conservation Techniques run by Royal Botanic Gardens, Kew (see also Hankamer *et al.*, Chapter 15). The curriculum was designed using available literature, recommendations from ANPC conferences regarding the types of training practitioners required, and input from consultants.

The course runs for eight days and is aimed at plant conservation practitioners from government, industry and community groups (Figure 6.5). The range of topics covered includes:

- causes of rarity;
- accessing existing information;
- gathering new information, including survey techniques;
- conservation management techniques including:
  - integrated conservation
  - habitat management
  - threat abatement
  - germplasm collection and storage

Figure 6.5 ANPC Plant Conservation Techniques Course. John Neldner demonstrates survey techniques. Photo: Jeanette Mill.

  - translocation
  - smoke germination (Figure 6.6)
  - research
- conservation genetics;
- field trips to plant conservation projects;
- community awareness and involvement including:
  - philosophies of partnership
  - methods of engagement
- conservation instruments and initiatives;
- strategic planning;
- funding opportunities;
- determining funding priorities.

Course co-ordinators and lecturers include national and international specialists, and local expertise is used where possible. Written notes from

Royal Tasmanian Botanical Gardens staff demonstrate how to make smoked water, used in stimulating germination. Photo: Jeanette Mill.

presenters are photocopied for students to put in a folder and take home as a reference manual. In addition to the technical information students receive, crucial working relationships are forged with specialists. This latter aspect is most often quoted as one of the most valuable outcomes of the course. Course fees are designed to reflect the attendees' ability to meet the cost. Members of local community action organisations pay a lower fee than those from better-funded organisations. The text for the course is *Principles and Practice of Plant Conservation* by Dr David Given (1994). The ANPC plans to produce a manual to enable members to run the course in their regions.

Electronic networking

Electronic networking and use of the Internet are cost effective ways of communicating with members, transmitting information and promoting the Network. However, many members still do not have adequate access, and ANPC continues to use printed publications as its basic tool for communication.

The Internet is a rapidly growing source of information, and increasingly of information not published elsewhere. A common problem with the Internet is that valuable information including documents, which were until recently available in hard copy, are no longer printed, and their presence on the

Internet is inadequately publicised. Networks can serve a useful purpose in keeping a watchful eye on Internet developments, and informing members of newly available Internet resources. The ANPC newsletter devotes a section to Internet developments.

To assist members and other web users with access to plant conservation information, the ANPC website includes an Internet Directory of Plant Conservation Resources (Porigneaux, 1998). Compiled by a volunteer, it is an indexed set of links to relevant sites. One of the benefits of using the Internet as an information reference in this way is that the information is updated by the compiler at no cost to the user. It is planned to expand the Directory to become a directory of ANPC members.

The ANPC website address is www.anbg.gov.au/anpc, and the Internet Directory can be found at www.anbg.gov.au/anpc/web.html.

E-mail

E-mail mailing lists enable rapid communication between members. Committee functions can be largely carried out using e-mail, especially when distances between members hinder the use of meetings. Members use e-mail to contribute to ANPC activities, such as the newsletter, and to call for volunteers to assist with surveys and other ground work.

Network Structure and Funding

The ANPC is an independent association incorporated under state legislation. This enables the ANPC to control its own finances, and limits the liability of members. It also promotes members' participation, as incorporation requires membership, and a management committee elected by the membership. This independent, non-government, non-profit status enables ANPC to attract donations and tax deductibility for donations.

The ANPC has a constitution, copies of which can be obtained by contacting the National Office. It can also be viewed on the ANPC's website. The structure has evolved over time as the organisation has matured. The original structure, prior to the ANPC becoming an incorporated association, was as per Figure 6.7a, and has evolved to the structure in Figure 6.7b.

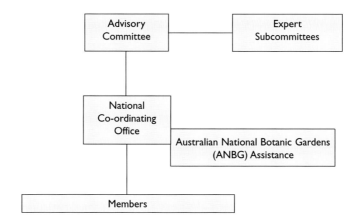

Figure 6.7a   Original Australian Network for Plant Conservation (ANPC) structure.

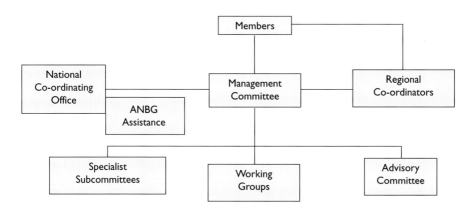

Figure 6.7b   Current Australian Network for Plant Conservation (ANPC) Inc. structure.

Members

The membership is enshrined in the constitution as the ultimate decision-making body. The ANPC Inc. Constitution was voted in by the membership as the ANPC's governing document. The Management Committee is elected by the membership biennially. Annual general meetings, incorporating the conference, are the regular decision-making meetings for members. Special resolution passed by the membership is the only mechanism for changing the Constitution. Postal ballots may be used to enable all members to vote on issues.

Management Committee

The Management Committee is elected by the membership, and has responsibility for the management of the Network, including recruitment of staff for the National Office, and establishment of subcommittees. The Committee is comprised of a representative cross-section of the general membership.

National Office

The National Office is hosted by the Australian National Botanic Gardens (ANBG), providing valuable infrastructure support, which would otherwise be costly for the Network to obtain. The paid staff of the Network are based in the National Office, and volunteers assist with office tasks and network co-ordination.

Regional Co-ordinators (see also Regional Groups)

Regional Co-ordinators are informally selected at the local level. Generally people willing to volunteer their time are enthusiastically welcomed by the local region to take on the role of co-ordinator. Regional Coordinators are encouraged to run for Management Committee positions, to streamline the implementation of ANPC national objectives at the local level.

Specialist subcommittees

As the ANPC membership encompasses a wide range of specialist skills, sub-committees and working groups are formed for specific purposes such as organising conferences and the production of guidelines and other publications. The use of national committees under the ANPC banner is also an effective method of overcoming political barriers.

Advisory Committee

The Advisory Committee was the principle steering committee for the Network between conferences during the formative years of the Network, prior to the ANPC becoming an Incorporated Association. The future role of the Advisory Committee will be reviewed by the Management Committee.

Resourcing

Money is often readily obtained for new projects, but can be problematic when keeping projects up and running. ANPC members generally donate their time, expertise and travel costs, and this in-kind support, along with membership fees

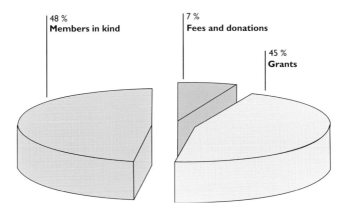

48 %
Members in kind

7 %
Fees and donations

45 %
Grants

Figure 6.8   Australian Network for Plant Conservation (ANPC) resources: financial and in-kind support.

and donations, makes up over 50% of the ANPC's resources (Figure 6.8). In preparing budgets for funding bodies, demonstration of 'dollar-for-dollar' matching of funds is often critical. All financial and in-kind contributions are costed into budgets. Funding agencies sometimes have a formula for valuing in-kind contributions.

Budgeting

Budgeting for non-profit non-government organisations, like the ANPC, with very limited financial resources is challenging. A balance needs to be struck between focussing on a limited set of the many actions members consider as priority, in order to complete these with a degree of quality, and not being overly restricted by lack of financial resources. For example, courses and conferences can often be costed to break even financially, particularly if participants can obtain funding. Members are often willing to donate time to assist with organisation, and provide services as presenters at no cost to the Network. Volunteers are a valuable resource. Flexibility is necessary in order to provide willing volunteers with meaningful tasks that use their existing skills at the time the volunteer offers their services.

Much of ANPC's committee and subcommittee work is achieved using e-mail and telephone conferencing, with face-to-face meetings held only as finances permit. Keeping track of the committee members' travel arrangements enables meetings to be arranged when committee members are travelling to locations for other purposes.

Sponsorship

Sponsorship is an increasingly necessary source of resources, and is obtained through financial and in-kind sponsorship. The latter is often easier to obtain, and is very valuable to the organisation. Staff and infrastructure resources are very costly, and whether provided in a small or in a large part by a sponsor, this support is of great benefit to the Network.

It has been said that the most important thing to do with sponsors is to thank them, thank them again, and then thank them again. This need not be a costly exercise, and ANPC currently achieves this with a simple ceremony, where a high profile patron of the Network presents a certificate. The event is publicised through the Network's newsletter. Everyone who contributes to the Network is considered a type of sponsor, including hosting organisations and Regional Co-ordinators. Conferences and courses are used to attract financial sponsorship as they are high profile events, particularly if media attention is gained. This is achieved by inviting high profile people to open the conference and launch publications, recovery programmes, etc. Member agencies, with dedicated media staff, assist with arranging media releases and co-ordinating enquiries.

Fund-raising

Living in a world increasingly oriented towards business practices and fund-raising, ANPC places emphasis on developing those activities that potentially generate income. The three major activities outside of membership fees are courses, conferences and publications. However, seed money is vital to fund the organisation of courses and conferences, and to produce publications, and these are also action items for securing grant money.

Promotion

The membership of the Network is its greatest asset in raising the profile of the Network. Targeted and high profile events, such as conferences, are very important opportunities for promoting both plant conservation and the Network simultaneously. The products of the Network are important promotional tools. Many people need only see a copy of the Newsletter or the Guidelines to be convinced to join. Free copies of such items distributed strategically are very effective promotion. ANPC's philosophy is to include a package of ANPC information, including a free newsletter in all mail sent out. ANPC volunteers prepare a supply of ready-made packages for promotional purposes.

Publicity and media

High profile events such as conferences offer great opportunities for publicity and mass media coverage. National and international focus is drawn to particular species in the region, and activities of the Network and its members. ANPC members contribute by writing articles for local papers, and publications to which they subscribe.

Other Networks

Many networks have been established over the years, which perform functions relevant to their own region, and these would be worth studying in considering an appropriate model.

- The Southern African Botanical Diversity Network (SABONET) focuses on capacity building in the ten southern countries of Africa. SABONET is a network of herbaria and botanic gardens with the objective of developing local botanical expertise. Website: www.sabonet.org

- Indonesia has established a network, the Indonesian Network for Plant Conservation (INetPC), based on the ANPC model, which has been adapted for the needs of the region. INetPC aims to facilitate communication and cooperation between conservation organisations, groups, institutions and individuals and their international counterparts working on the Indonesian flora. Website: www.bogor.indo.net.id/inetpc

The ANPC has Memoranda of Understanding with both these networks, outlining an agreement to pass information freely between the networks, and collaborate on projects of mutual benefit. The Wellington Plant Conservation Network in New Zealand focuses on a region within its own country. A comprehensive plant conservation strategy has been drawn up for the region. For contact details and information on these and other regional networks, contact the ANPC, BGCI, or the IUCN Species Survival Commission.

Contacts

> Australian Network for Plant Conservation
> GPO Box 1777
> Canberra, ACT, Australia, 2601.
> Tel: +61 2 62 509 509, Fax: +61 2 62 509 528
> E-mail: anpc@anbg.gov.au
> Website: www.anbg.gov.au/anpc

Botanic Gardens Conservation International (BGCI)
Descanso House
199 Kew Road, Richmond, Surrey TW9 3BW, UK
Tel: +44 (0) 20 8332 5953, Fax: +44 (0) 20 8332 5956
E-mail: bgci@rbgkew.org.uk
Website: www.bgci.org.uk

IUCN Plants Officer
Species Survival Programme
Rue Mauverney, 28
CH-1196 Gland, Switzerland
Tel: +41 22 999 0157, Fax: +41 22 999 0015
E-mail: was@hq.iucn.org
Website: www.iucn.org/themes/ssc/index.htm

Acknowledgements

I would like to thank the following colleagues for their help:

Patrick McCarthy, Australian Biological Resources Study.

Bob Makinson, Curator, Australian National Herbarium, Centre for Plant Biodiversity Research.

Mark Richardson, Alice Springs Desert Park.

References and Bibliography

Note: additional sources have been added to the list of references cited for further readimg

ANPC (1993). *The National Endangered Flora Collection: A Conservation Resource.* Australian Network for Plant Conservation, Canberra, ACT, Australia.

ANPC Germplasm Working Group (1997*a*). *Germplasm Conservation Guidelines for Australia.* Australian Network for Plant Conservation, Canberra, ACT, Australia.

ANPC Translocation Working Group (1997*b*). *Guidelines for the Translocation of Threatened Plants in Australia.* Australian Network for Plant Conservation, Canberra, ACT, Australia.

Australian and New Zealand Environment and Conservation Council Endangered Flora Network and Threatened Species and Communities Section, Environment Australia (1999). *Threatened Australian Flora.* Australian and New Zealand Environment and Conservation Council, Canberra, ACT, Australia.

Berg, K.S. (1996). Rare plant mitigation: a policy perspective, pp. 279–292. In: D.A. Falk, C.I. Millar and M. Olwell (eds). *Restoring Diversity: Strategies for Reintroduction of Endangered Plants*. Island Press, Washington DC, USA.

Briggs, J.D. (1996). *Rare or Threatened Australian Plants*. CSIRO Publishing, Collingwood, Victoria, Australia.

Commonwealth of Australia (1996). *The National Strategy for the Conservation of Australia's Biological Diversity*. Department of Environment, Sport and Territories, Canberra, ACT, Australia.

Commonwealth of Australia (1998). *Australia's National Report to the Fourth Conference of the Parties to the Convention on Biological Diversity*. Environment Australia, Canberra, ACT, Australia.

Given, D.R. (1994). *Principles and Practice of Plant Conservation*. Timber Press, Portland, Oregon, USA.

Glowka, L., Burhenne-Guilmin, F., Synge, H., McNeely, J.A. and Gündling, L. (1994). *A Guide to the Convention on Biological Diversity*. Environmental Policy and Law Paper No. 30. IUCN, Gland, Switzerland.

Howald, A.M. (1996). Translocation as a mitigation strategy: lessons from Calfornia, pp. 293–330. In: D.A. Falk, C.I. Millar and M. Olwell (eds). *Restoring Diversity: strategies for reintroduction of endangered plants*. Island Press. Washington DC, USA.

Meredith, L.D. (ed.) (1995). *Integrated Plant Conservation in Australia. Proceedings of the Conference 'Cultivating Conservation: Integrated Plant Conservation for Australia' Hobart, December 1993*. Australian Network for Plant Conservation, Canberra, ACT, Australia.

Morse, J., Whitfield, S. and Gunn, B. (1992). *Australian Seed Banks and Cryogenic Storage of Rare and Threatened Flora*. Australian Tree Seed Centre, CSIRO Division of Forestry. Report to the Endangered Species Unit, Australian Nature Conservation Agency, Canberra, ACT, Australia.

Porigneaux, J.M. (1998). *Australian Network for Plant Conservation Internet Directory: a guide for web sites on threatened plant species*. Australian Network for Plant Conservation, Canberra, ACT, Australia.

Scott, G., Entwisle, T., May, T. and Stevens, N. (1997). *A Conservation Overview of Australian Non-marine Lichens, Bryophytes, Algae and Fungi*. Endangered Species Program, Environment Australia, Canberra, ACT, Australia.

Sutter, R.D. (1996). Monitoring, pp. 235–264. In: D.A. Falk, C.I. Millar and M. Olwell (eds). *Restoring Diversity: strategies for reintroduction of endangered plants*. Island Press, Washington DC, USA.

Walter, K.S. and Gillett, H.J. (eds.) (1998). *1997 IUCN Red List of Threatened Plants*. Compiled by the World Conservation Monitoring Centre. IUCN – The World Conservation Union, Gland, Switzerland and Cambridge, UK.

WWF and IUCN (1994–1997). *Centres of plant diversity. A guide and strategy for their conservation*. 3 volumes. IUCN Publications Unit, Cambridge, UK.

Chapter 7

The Role of Tropical Botanic Gardens in Supporting Species and Habitat Recovery:

East African opportunities

Mike Maunder[1]
Royal Botanic Gardens, Kew, UK

Mark Stanley-Price[2]
IUCN/SSC, Reintroduction Specialist Group
Nairobi, Kenya

Pritpal Soorae
IUCN/SSC Reintroduction Specialist Group,
Nairobi, Kenya[3]

Shedrack Mashuari
East Usambara Conservation Area Management
Programme (EUCAMP), Tanga, Tanzania

[1] Now at: The National Tropical Botanical Garden Kauai, Hawaii, USA
[2] Now at: The Durrell Wildlife Conservation Trust, Jersey, UK
[3] Note: The IUCN/SSC Reintroduction Specialist Group Secretariat is now based at Environmental
   Research and Wildlife Development Agency (ERWDA), Abu Dhabi, United Arab Emirates

Introduction

This paper examines the need for tropical botanic garden facilities and the potential role they can play in supporting conservation at the landscape level and linking directly with human development issues. We present experience from East Africa as a case study. East Africa supports a unique and spectacular botanical diversity, measured in terms of species, habitats and unique life forms, such as the Afro-montane pachycauls (Figure 7.1). The impact of human population growth is resulting in both increased habitat loss and fragmentation. Current estimates suggest that Africa suffers an annual forest clearance of between 1.3 and 3.7 million (ha), largely driven by subsistence farming (Boahene, 1998). This trend will continue as human populations and their associated demands on ecosystems grow; the population of Africa may double from 728 million in 1995 to 1.49 billion by 2025 (WRI, 1996). Since about 75% of Africans live in rural areas there will be a continued need for new crop and grazing lands (Pagiola *et al.*, 1998).

Figure 7.1   Planting of the Succulent Garden at the Nairobi Botanic Garden, a garden designed to highlight the diversity of succulent plants indigenous to East Africa.

Botanic gardens, and other *ex situ* facilities for biodiversity, can play important roles in the management of biodiversity, both through propagation, as a support to threatened species recovery planning, and as public venues for education (Maunder, 1994). Such facilities are often accessible to an urban public and may contain important residual fragments of natural habitats e.g. the rainforest at Entebbe Botanic Garden and upland *Croton* L. (Euphorbiaceae) forest at the Nairobi Arboretum. Botanic gardens contain unique resources for horticulture, botany and public education that can play an important role in national biodiversity initiatives. However, the identification of a specified local and/or national role for these facilities is fundamental to the security of a botanic garden. There is still a tendency for botanic gardens to assume that a collection of species is the first priority and, accordingly, risk a future as isolated enclaves without a political constituency. Through promoting botanic gardens as supports to landscape or ecosystem-scale conservation, development and education, new collaborative relationships can be forged.

Although Africa contains biodiversity-rich nations (*sensu* Mittermeier *et al.*, 1998), the continent has disproportionately few facilities for plants. Currently there are only 59 botanic gardens in 19 African countries, with 17 (29%) being located in the Republic of South Africa alone (UNEP, 1995). African botanic gardens are often perceived as colonial artefacts and currently persist with low levels of political and financial investment. This can be attributed to a number of factors including levels of historical investment, a strong focus on the primary need for protected areas and poorly developed national infrastructures for biological sciences.

Opportunities from the CBD

The Convention on Biological Diversity (CBD) is the major legislative influence on the conservation of biodiversity, directly influencing national activities through the requirement to produce and implement Biodiversity Action Plans (Glowka *et al.*, 1994) (see Appendix 1). Outlined under Article 6 (General Measures for Conservation and Sustainable Use) is the requirement for each Contracting Party to develop a National Biodiversity Strategy Action Plan or programme. Statements on the role and value of *ex situ* conservation should feature in these documents. The CBD recognises the value of *ex situ* conservation (Article 9) with an emphasis on undertaking these activities 'preferably in the country of origin' and as a support to the 'recovery and rehabilitation of threatened species and for their reintroduction into their natural habitats'. In addition, the Convention regulates and controls the collection of material for *ex situ* conservation 'so as not to threaten ecosystems and *in situ* populations of species, except where special temporary *ex situ* measures are required'. Clear recommendations are made on the need for funding 'the establishment and maintenance of *ex situ* conservation facilities in developing countries'.

1. Key trends in East African plant conservation

The retention of indigenous plant diversity will be increasingly dependent upon approaches that sustain local economic development, and maintain biodiversity in settled areas and the viability of isolated habitat fragments. The retention of wild habitats represents an 'opportunity cost' in terms of land allocation and competition with agriculture (Norton-Griffiths and Southey, 1995); on the other hand, the degradation of ecosystem resources also represents a long-term, and often under-valued, cost (Bojo, 1996).

We see the most significant points for botanic gardens and similar agencies in relation to the requirements of the CBD and for managing East African botanical diversity as:

- Biodiversity-rich natural habitats continue to be lost and fragmented, through conversion to land uses which support relatively low levels of indigenous biodiversity.

- Data on declines in plant diversity are not widely available for Sub-Saharan Africa, accordingly candidates and management targets for conservation investment will be difficult to accurately identify.

- The dramatic and continued reduction in populations of high value species for reasons other than habitat loss, particularly for high value (both local and global) medicinals/timber species where overharvesting threatens wild populations e.g. *Prunus africana* (Hook. f.) Kalkman (Rosaceae) (see Jaenicke *et al.*, Chapter 20).

- Protected areas and forest reserves are subject to increasing levels of encroachment by settlement and cultivation (Sanchez and Leakey, 1997). Hence, protected areas are increasingly becoming ecological islands, and many of their enclosed species' populations will increasingly face issues of viability. However, it is evident that modified habitats can contribute to maintaining connectivity between protected areas (Siegfried *et al.*, 1998).

- The main economically valued use of biodiversity has been through tourist viewing. Accordingly, biodiversity in accessible parks earns more than biodiversity in the general landscape. There is an urgent need to diversify economic opportunity based on biodiversity, including plant resources, in areas outside protected areas (Lusigi, 1994; Sibanda and Omwega, 1996).

- Despite differential pricing systems, East Africa's protected areas and wild landscapes are still visited by relatively few of the countries' own citizens. At the same time, it is striking that Africa has a very low number of biodiversity display facilities as alternatives, and accessible venues for the public to investigate their natural heritage.

How can botanic gardens, existing and proposed, respond to the above challenges? Can new roles be established as support to national strategies?

2. East African responses to *ex situ* conservation

The value of *ex situ* conservation and biodiversity restoration along with the need to invest in developing *ex situ* facilities has been recognised in national policy and strategy documents for Uganda (Bukenya-Ziraba, 1998; Government of Uganda, 1992, 1998; Opio-Odongo *et al.*, 1998); Kenya (Government of Kenya and UNEP, 1992; Government of Kenya, 1998; Mugabe *et al.*, 1998) and Tanzania (Kaiza-Boshe *et al.*, 1998). For instance, the National Environmental Policy for Tanzania (Government of Tanzania, 1994, cited in Kaiza-Boshe *et al.*, 1998) recognises the need for an integrated approach to biodiversity conservation:

- Conservation of biological diversity is to be attained in the wild (*in situ* conservation) in protected areas, outside protected areas and on the farm, in gene banks and botanical gardens (*ex situ* conservation), and laboratories relevant to conservation of biological diversity. *Ex situ* conservation and biotechnology are least developed in Tanzania;

- Conservation of biological diversity outside of protected areas and *ex situ* establishments is not well addressed;

- Facilities for *ex situ* conservation and biotechnology are under-developed;

- Lack of human capacity and technology in collecting and storing biological specimens has forced the country to store much of its genetic materials and biological specimens in gene banks and collection centres outside the country.

3. Botanic gardens and biodiversity restoration

The retention, and indeed restoration, of plant diversity in East Africa will take place primarily at three levels:

1. The management of individual species;
2. A focus on diversity hotspots of restricted areas;
3. The management of biodiversity across extensive landscapes.

Figure 7.2   Whilst botanic gardens have traditionally focused on the display and management of species they have an important role to play in promoting and supporting the conservation of ecosystems.

All three activities will cover a range of land uses, including intensively utilised and populated areas.

Whilst botanic gardens have traditionally focused on species as the unit of conservation management, we propose a modified approach. Despite there being a current lack of accurate threatened species lists for the majority of African nations, there is an increasing need to focus on over-used utility species of great social value (e.g. medicinals) and retaining important plant habitats (Figure 7.2). The restoration of habitats, particularly areas adjacent to protected areas, provides a means of enabling a significant number of species to recover and provide a cost-effective response to the biodiversity crisis. Restoration of fragmented tropical habitats will become a major tool for biodiversity retention in many areas. Whilst the practical application of restoration ecology in East Africa is in its early stages (Chapman and Chapman, 1999; Lyaru, 1999; see Robertson *et al.*, Chapter 14), its application means that biodiversity facilities, such as botanic gardens, no longer respond only to the inadequately documented symptoms of habitat loss (e.g. threatened species), but can also contribute directly to slowing the causative processes.

Habitat conversion, as the major threat to biodiversity, not only represents the direct loss of a habitat, but also, through the extent and rate of degradation, retards the rate of natural colonisation and succession (Dobson *et al.*, 1997). Accordingly, it could be argued that protected areas, isolated in areas of increasingly degraded habitat, only slow the rate of habitat erosion and species loss. Models of habitat loss suggest that the process produces an 'extinction debt', a pool of species destined for extinction unless the habitat is restored (Sinclair *et al.*, 1995; Tilman *et al.*, 1994). Falk (1996) has proposed the idea of 'Restoration Priority Areas' – areas of degraded land requiring restoration on a large scale. In an East African context, these restoration areas could be established as directed by biodiversity needs (such areas would surround or lie within important protected areas or linkages), and development needs (cultural restoration and restoration of degraded agricultural systems and watersheds).

Figure 7.3   Botanic gardens can play a valuable role in displaying otherwise inaccessible biodiversity. Here a specimen of *Gigasiphon macrosiphon* (Harms) Brenan (Leguminosae), endemic to the coastal forests of Kenya and Tanzania, makes a magnificent ornamental tree in the Nairobi Botanic Garden.

The need for habitat restoration is demonstrated through the increasingly fragmented forest areas in the Eastern Arc Mountains, such as the Taita Hills, Kenya, re-gazetted agricultural areas, such as Mgahinga Gorilla Sanctuary, Uganda, and increasingly degraded community wildlife areas, such as the Mwaluganje Elephant Sanctuary, Kenya. The highly fragmented and endemic-rich coastal forests of Kenya and Tanzania (Burgess *et al.*, 1998; Burgess and Clarke, 2000) represent an urgent opportunity for restoration of degraded forest patches and establishment of linking corridors (Figure 7.3). One such restoration project, on the Kenyan North Coast, is described by Robertson *et al.*, (Chapter 14).

The CBD identifies responsibility for the conservation and utilisation of biodiversity to the range country and to local communities. We propose that *ex situ* facilities, including botanic gardens, should demonstrate the values of large protected areas, while offering the attractions of an accessible display, with awareness and conservation messages co-ordinated between the two.

Strategic development of botanic gardens

Botanic gardens need a broad and stable client base for long-term financial and political viability. To achieve viability, a botanic garden should develop a long-term management strategy, achieved through three basic steps:

1. Determination of the basic long-term goals (strategic and operational);

2. Adoption of courses of action; and

3. Allocation of resources (financial, human, physical and intellectual).

The first step is to identify stakeholders for the facility. Stakeholders are any person or party that can affect or be affected by the activities and policies of the botanic garden. These can be identified through the two processes of internal and external analysis. The role of a botanic garden should also directly relate to stated national and regional biodiversity issues. The challenge is to establish valued roles that match national needs and dispel the anachronistic 'palm tree and bandstand' image of old botanic gardens (see Table 7.1). Botanic gardens need to establish secure roles that are complementary to the existing biodiversity agencies (e.g. protected areas, forest services, museums, etc.) and that utilise the core competencies of display and horticulture. The roles need to be scientifically valid and respond to the essential needs of resource use and habitat conservation.

Case Study: The evolution of the Amani Botanical Garden, Tanzania

Botanic gardens will modify their roles as cultural, economic and scientific contexts change and such changes will influence the available strategic opportunities for the facility. The Amani Botanic Garden (ABG) was established by the Biologisch-Landwirtschaftliche Institut of the then German colonial authorities in 1902 (Figure 7.4). A century later, where does this garden fit into a modern and independent Tanzania? What does it want to achieve and with whom does it work? The original collections were developed to serve the colonial economy through trials for potential crops, with an emphasis on medicinals, fruits, dyes, cosmetics, timber, etc. The value of the collections, and the role of the facility, has been reduced by a number of factors: fluctuations in boundary demarcations, loss of horticultural support and associated documentation. Perhaps, most importantly, the original reason for the establishment of the collections is no longer valid i.e. does Tanzania still need this facility for testing exotic plantation crops?

Table 7.1 Challenges for tropical botanic gardens

Issue/challenge	Strategic responses
1. Botanic gardens have limited internal (national) constituency amongst influential decision-makers: the political and professional support is often strongest from overseas institutions	• Recognise the need to be 'supply driven' and responsive to stated national priorities while building and strengthening a national constituency for long-term support • Recognise that the global community should pay some of the cost for achieving global objectives • Integrate plant conservation/biodiversity with other 'supply driven' agendas so that decision-makers recognise congruency of agendas • Develop effective local outreach and public educational display
2. Plant resources, and botanic garden core competencies, have been treated as open access resources with little monetary value	• Work to put value on plant resources • Work to put value on core competencies of botany, horticulture, public display and education • Strengthen both incentives and capacity for responsible stewardship/management of biological/plant resources in the public and private sectors
3. Benefits of conservation are long-term, often judged against significant short-term costs	• Work to improve short-term benefits and reduce short-term costs for stakeholders • Encourage stakeholder interest in long-term benefits
4. Land use and resource use practices are the main driving forces of biodiversity loss	• Focus on making conservation a competitive land/resource use option for both natural and human dominated landscapes • Develop and apply skills in restoration ecology and habitat management
5. Challenges will continue to increase, not decrease e.g. continued land hunger in the tropical regions	• Promote instruments influencing long-term behaviour rather than short-term responses • Promote the establishment of processes for integration and conflict management e.g. develop skills in restoration ecology • Structure long-term financial and technical assistance

Continued

Table 7.1 Continued	
Issue/challenge	**Strategic/responses**
6. Botanic gardens have relatively poor results from conservation intervention to date	• Difficult to establish specific targets in terms of desired endpoints for the traditional management focus on threatened species due to poor base-line data on species' conservation status
	• Set targets in terms of reversing known negative trends and enhancing positive ones, particularly with respect to causative land use and biological resource use practices, rather than focusing only on the symptoms e.g. threatened species
	• When work does focus on a threatened species give priority to taxa of quantifiable social, economic or ecological utility
7. Two main constraints to success are (a) economic, political and institutional, and (b) technical	• Focus on finding ways to influence economic, political and institutional climate for biodiversity conservation and the security of botanic gardens
	• Focus on retaining and developing human capital and skills base
	• Focus on botanic garden core competencies

The role and future of the ABG was assessed at three levels:

i) Institutional strategy

ABG is a facility of the Forest Department and part of the East Usambara Conservation Area Management Programme (EUCAMP). The strategy must support the fundamental objectives of the EUCAMP strategy, namely:

- protection and conservation of biological diversity and catchment forests
- collaboration and participation of local people and relevant organisations
- innovation – developing new pilot approaches to forest management and research
- developing systems to ensure long-term sustainability, particularly for the financial and institutional elements of forest conservation.

Figure 7.4   The Amani Botanic Garden was established as a trial ground for tropical crops. It could be argued that this role should be revised to take into account the challenges facing a facility in the midst of a globally important biodiversity hotspot.

ii) What services can the botanic garden provide to local, national and regional stakeholders?

A botanic garden can identify stakeholders through assessing the working environment by identifying the prevailing:

- social, cultural and demographic forces in the area

- technological influences, including products, processes, communications and transport

- economic influences, including government fiscal and monetary policies, local incomes and living standards, etc.

- governmental, legal and regulatory influences

- biodiversity issues and priorities as outlined in country studies and biodiversity action plans.

Box 7.1   **The long-term Mission of the Amani Research Facility and Botanic Garden (ARFBG)**

Mission:

To support the conservation of biological diversity and catchment forests, through habitat management, applied conservation biology research and collaboration with local communities. ARFBG to be recognised as a leading facility for the study and conservation of Afro-montane forest biodiversity; ARFBG to be recognised as a leading training centre in East Africa for biodiversity management skills.

Objectives:

- Protection and conservation of biological diversity and catchment forests, with a focus on the biodiversity management needs of the Usambara and the Eastern Arc Mountains specifically, and threatened tropical biodiversity hotspots in general, through applied ecological and conservation biology studies. The Amani facility should provide a complementary resource for existing national and regional activities, e.g. training in biodiversity management and the provision of field gene bank-botanic garden facilities for the National Gene Bank.

- Collaboration and participation of local people and relevant organisations to develop strong local patterns of partnership to improve the local quality of life for rural communities.

- Innovation and development of new pilot approaches to forest management and research, to work with the Tanzania Forest Research Institute (TAFORI) and other relevant agencies to develop practical responses to the management of fragmented biodiversity hotspots – representing tools of international relevance.

- Developing systems to ensure long-term sustainability, particularly for the financial and institutional elements of forest conservation, whilst a number of local initiatives are successfully developing links with local communities. The long-term security of the facility will depend upon some degree of external financial and professional support.

iii) What are the internal organisational competencies and resources?

The following questions must be addressed:

- What are the resources and competencies that need to be developed and utilised?
- What areas of weakness need to be addressed in a future strategy and its implementation?
- How to evaluate the delivery and performance of botanic garden products?
- How to evaluate financial performance?
- How to assess investment requirements?

The future of ABG as a contemporary facility should not be based on the historical rehabilitation of collections and facilities. The focus should be on the regional and national role of a scientific facility in an area of declining biodiversity and increasing pressure on natural resources (Box 7.1). In addition, some of the exotic collections now pose a local and potentially regional threat to indigenous biodiversity. A number of plant species from ABG have become naturalised with some of these having become invasive and potentially damaging to the local ecology (see Appendix 5).

The rehabilitation of ABG could follow several paths. The historical rehabilitation based on the old plantations will require a considerable recurrent budget to manage and reinstate existing plots (a team of at least ten people would be required) and retain the ongoing risk of invasive species; this represents an open-ended management cost with limited opportunities for external funding and research. The development of ABG as a tourism venue could potentially serve the national tourism industry; however, it is likely that such a facility would develop collections with an exotic and ornamental bias and would not directly support national imperatives for conservation and community-based initiatives. It is strongly recommended that ABG develop as a facility serving a broad range of national and international clients.

It is difficult to justify the rehabilitation of a classical colonial collection. Indeed, a traditional botanic garden with a focus on exotic collections has little relevance to the biodiversity and people of the Usambaras. The opportunity exists to develop a facility utilising the existing buildings, and importantly the international image of this 'lost garden in the mountains', that can support the conservation of the Eastern Arc environment.

Conclusions

A vision for botanic gardens in East African conservation

The CBD provides a policy framework that explicitly links the role of national *ex situ* facilities with the conservation of wild populations and ecosystems. However, if the existing East African protected area networks are to continue to support biodiversity at current or historical levels new approaches are required to connect these increasingly isolated areas. Our vision is that a new kind of botanic garden facility can meet the obligations of the CBD, support landscape-level conservation, and establish local communities as significant partners in protected area and natural resource management (Box 7.2).

This vision contains scope for a large number of functions that will be directly driven by local need and potential. Importantly, these new facilities would squash the artificial *in situ* and *ex situ* divide, open up funding opportunities (e.g. the Global Environment Facility), and provide quantifiable evidence of the value of facilities in terms of local employment, areas and resources restored, and taxa managed for recovery (Cunningham, 1994; Maunder, 1998). These proposed roles for botanic gardens need to be undertaken in the context of economic analysis and proper business planning (Lutz and Caldecott, 1996).

To develop a relevant role for botanic gardens and attain political and financial security, a number of processes are needed:

i) The starting point is consultation with the identified clients and stakeholders; such surveys identify client needs. The botanic garden is not an autonomous refuge, but needs a secure client base for viability. There is an urgent need to incorporate botanic gardens into national conservation infrastructures and, in particular, foster effective links with protected area agencies.

ii) The capacity building problems of tropical botanic gardens and associated poor development history are directly related to institutional environment and governance-related impediments. A botanic garden, and the wider plant conservation infrastructure, needs to identify specific national and local incentives and sanctions that can be used as connectors between the central government institutions (the legal and policy infrastructure) and the public sector (the financial infrastructure). Whilst government money is unlikely to support many facilities, a botanic garden needs a secure legal infrastructure that encourages links to the private sector, an increasingly important source of funds.

iii) The botanic garden – as a facility straddling the issues of land degradation, species conservation, and sustainable development – can only operate efficiently through reconciliation between government legislation and non-governmental organisations in terms of institutional legitimacy and enforceability (see Berg, 1993). This would suggest that botanic gardens should place further emphasis on developing skills in political advocacy, situation analysis, and stakeholder mobilisation. The traditional conservation focus on threatened species should be supplemented with a focus on tackling land use and social issues, the driving forces behind biodiversity loss.

iv) Botanic gardens, as a public service, will need to explore tripartite arrangements that link state, civil society/indigenous institutions and the private sector, and, to ensure financial security and flexibility, botanic gardens need to recognise the success of microenterprises (Dia, 1996). Microenterprises absorb large numbers of workers and encourage the wider participation of local investors.

Box 7.2 **The role of botanic garden facilities in supporting landscape level conservation** (derived from Maunder *et al.*, 1999)

Goal: The diversification of natural product based enterprises on lands outside protected areas, for the particular benefits of local people, in support of landscape level conservation.

Objectives:

- to promote awareness and educate local people and visitors about local species and ecology, protected areas, ecological dynamics and the dependence of human communities on natural resources
- to meet conservation management objectives for key local species
- to offer employment opportunities and create a skilled and effective workforce
- to forge productive relationships between conservation managers and neighbouring communities

Requirements:

- facility structure and activities based on a rigorous and viable business plan derived from participatory processes
- role directly linked to nationally and locally recognised priorities for conservation, capacity building and income generation
- formal agreements between partners, e.g. communities, financiers, managers, local authorities, technical advisory groups
- managed on sound scientific principles in pursuit of species' conservation objectives
- access to best technical expertise, whether in-country or from outside;
- culturally appropriate (for implementing its objectives) for both international visitors and national citizens, with priority to the latter
- supported through strategic alliances at the local, national and international levels

Benefits:

1. Conservation impact

- national and local requirements of the CBD are delivered
- diversification of enterprises based on wild landscapes
- functional links between a biodiversity display facility, local protected areas and surrounding lands would promote landscape-level natural resource planning and conservation

2. To the local community

- employment of local people
- development of a skilled workforce
- creation of new revenue sources, e.g. microenterprises
- provision of managed plant and animal resources
- centre for broad environmental debate and action, e.g. Agenda 21 activities
- demonstration of best practice in land and natural resource management

Continued

Box 7.2 Continued

3. For conservation knowledge and awareness

- showing key species in semi-natural conditions to many citizens, and overseas tourist visitors
- opportunities to see local species of particular significance due to cultural value, rarity, etc.
- plants and animals seen at closer range, with sightings guaranteed on short visits
- displays used to explain conservation of larger natural areas and wild populations

4. To species conservation

- establishing best technical expertise used in planning and management
- species propagated for local reintroduction or as part of larger propagation programme
- species being kept for habitat maintenance
- habitat continuity beyond the park boundaries for some species (not necessarily those on display) achieved
- potential for acting as *ex situ* and in-country propagation facilities for more cost-effective propagation and reintroduction

5. To national park and landscape level conservation

- greater awareness of the values of national parks and their roles in landscape scale resource conservation
- strong functional links between the community and park managers
- an area-based approach to interpretation and awareness, rather than on a park-by-park basis with no interpretation of intervening areas through which visitors pass
- greater awareness among visitors and stakeholders of the benefits of a landscape level approach
- scope for testing, demonstrating and practising methods of habitat restoration;

6. For commercial opportunities

- commercial investment based on relatively small areas
- protected area authority could become an investor in the facility
- potential for attracting new local investors into biodiversity conservation because of site specificity and smaller scale
- scope for investment in related recreational facilities, accommodation, services
- creation of new revenue sources e.g. microenterprises, or sales of locally produced goods to overseas outlets

Tropical botanic gardens, as one important resource in the portfolio of plant conservation tools, can play a vital role in supporting plant conservation. However, this will be dependent upon identifying contemporary roles that link national biodiversity and social priorities to core competencies of horticulture and display.

Note: EUCAMP was formerly known as the East Usambaras Catchment Forest Project. It was renamed in 2000 to mark the next phase of programme development in 1999.

Acknowledgements

The authors would like to thank their colleagues in East Africa for the discussions and debates that have helped direct this paper, in particular, Quentin Luke, Perpetua Ipulet, Stella Simiyu and William Wambugu (National Museums of Kenya), Ann Robertson, Malindi, Kenya, staff of the Entebbe Botanic Garden, Uganda and EUCAMP, Tanzania. The study on Amani Botanic Garden was undertaken on behalf of the East Usambara Catchment Forest Project with financial support from Finnish Development Corporation (FINNIDA).

References

Berg, E.J. (1993). *Rethinking Technical Co-operation: reforms for capacity building in Africa.* United Nations Development Programme, New York, USA.

Boahene, K. (1998). The challenge of deforestation in tropical Africa: reflections on its principal causes, consequences and solutions. *Land Degradation and Development* **9**: 247–258.

Bojo, J. (1996). The costs of land degradation in sub-Saharan Africa. *Ecological Economics* **16**: 161–173.

Bukenya-Ziraba, R. (1998). Germplasm Collection and Conservation in Uganda: Prospects and Constraints, pp. 91–106. In: P. Adams and J.E. Adams (eds). *Conservation and Utilisation of African Plants. Conservation of Plant Genes III.* Vol. 71. Missouri Botanical Garden Press, Missouri, USA.

Burgess, N.D. and Clark, G.P. (eds). (2001). Coastal Forests of Eastern Africa. IUCN Forest Conservation Programme. IUCN Gland, Switzerland and Cambridge, UK.

Burgess, N.D., Clarke, G.P. and Rodgers, W.A. (1998). Coastal forests of Eastern Africa: status, endemism patterns and their potential causes. *Biological Journal of the Linnean Society* **64**: 337–367.

Chapman, C.A. and Chapman, L.J. (1999). Forest restoration in abandoned agricultural land: a case study from East Africa. *Conservation Biology* **13**: 1301–1311.

Cunningham, A.U. (1994). Integrating local plant resources and habitat management. *Biodiversity and Conservation* **3**: 104–115.

Dia, M. (1996). *Africa's Management in the 1990s and Beyond: reconciling indigenous and transplanted institutions.* The World Bank, Washington DC, USA.

Dobson, A.P., Bradshaw, A.D. and Baker, A.J.M. (1997). Hopes for the future: restoration ecology and conservation biology. *Science* **277**: 515–522.

Falk, D.A. (1996). Choosing a future for ecological restoration, pp. 211–215. In: D.L. Peterson and C.V. Klimas (eds). *The Role of Restoration in Ecosystem Management.* Society for Ecological Restoration, Madison, Wisconsin, USA.

Glowka, L., Burhenne-Guilman, F., Synge, H., McNeely, J.A. and Gündling, L. (1994). *A Guide to the Convention on Biological Diversity.* Environmental Policy and Law paper no. 30. IUCN, Gland, Switzerland.

Government of Kenya and UNEP (1992). *The Costs and Benefits and Unmet Needs of Biological Diversity in Kenya.* Government of Kenya (National Biodiversity Unit and National Museums of Kenya), UNEP and ODA, Nairobi, Kenya.

Government of Kenya (1998). *First National Report to the Conference of the Parties (COP): National Biodiversity and Action Plan.* Ministry of Environmental Conservation, Nairobi, Kenya.

Government of Uganda (1992). *Uganda Country Study on the Costs, Benefits and Unmet Needs of Biological Diversity Conservation.* Government of Uganda (National Biodiversity Unit), Uganda

Government of Uganda. (1998). *First National Report to the Conference of the Parties (COP): first conservation report on the conservation of biodiversity in uganda.* Government of Uganda, Uganda.

Kaiza-Boshe, T., Kamara, B. and Mugabe, J. (1998). Biodiversity management in Tanzania, pp 121–152. In: J. Mugabe and N. Clark (eds). *Managing Biodiversity: National Systems of Conservation and Innovation in Africa.* African Centre for Technology Studies (ACTS), Nairobi, Kenya.

Lusigi, W.J. (1994). Socioeconomic and ecological prospects for multiple-use of protected areas in Africa. *Biodiversity and Conservation* **3**: 449–458.

Lutz, E. and Caldecott, J. (1996). *Decentralization and Biodiversity Conservation.* The World Bank, Washington DC, USA.

Lyaru, H.V.M. (1999). Seed rain and its role in the recolonization of degraded hill slopes in semi-arid central Tanzania. *African Journal of Ecology* **37**: 137–148.

Maunder, M. (1994). Botanic gardens: future challenges and responsibilities. *Biodiversity and Conservation* **3**: 97–103.

Maunder, M. (1998). *Development and Conservation Strategy for the Amani Botanic Garden, Tanzania: potential for establishing a regional biodiversity facility for the Eastern Arc Mountains.* Report undertaken for the Ministry of Forestry and Bee Keeping, Tanzania under assignment from FINNIDA. The Royal Botanic Gardens, Kew, UK.

Maunder, M., Stanley-Price, M. and Soorae, P. (1999). The role of in-country *ex situ* facilities in supporting species and habitat recovery: some perspectives from East Africa, pp. 31–47. In: T.L. Roth, W.F. Swanson and L.K. Blattman (eds). *Linking Zoo and Field Research to Advance Conservation. Proceedings of the Seventh World Conference on Breeding Endangered Species.* The Cincinnati Zoo and Botanical Garden, Cincinnati, USA.

Mittermeier, R.A., Myers, N., Thomsen, J., de Fonseca, G.A.B. and Olivieri, S. (1998). Biodiversity hotspots and major tropical wilderness areas: approaches to setting conservation priorities. *Conservation Biology* **12**: 516–520.

Mugabe, J., Marekia, N. and Mukii, D. (1998). Biodiversity management in Kenya. In: J. Mugabe and N. Clark (eds). *Managing Biodiversity: National Systems of Conservation and Innovation in Africa*, pp. 91–120. African Centre for Technology Studies (ACTS), Nairobi, Kenya.

Norton-Griffiths, M., and Southey, C. (1995). The opportunity costs of biodiversity conservation in Kenya. *Ecological Economics* **12**: 125–139.

Opio-Odongo, J.M.A., Tumushabe, G.W. and Kakuru, W. (1998). Biodiversity management in Uganda. In: J. Mugabe and N. Clark (eds). *Managing Biodiversity: National Systems of Conservation and Innovation in Africa*, pp. 153–181. African Centre for Technology Studies (ACTS), Nairobi, Kenya.

Pagiola, S.J., Kellenberg, K., Vidaeus, L. and Srivastava, J. (1998). Maintaining biodiversity in agricultural development. *Finance and Development* **35**: 38–41.

Sanchez, P.A. and Leakey, R.R.B. (1997). Land use transformation in Africa: three determinants for balancing food security with natural resource utilization. *European Journal of Agronomy* **7**: 15–23.

Sibanda, B.M.C. and Omwega, A.K. (1996). Some reflections on conservation, sustainable development and equitable sharing of benefits from wildlife in Africa: the case of Kenya and Zimbabwe. *South African Journal of Wildlife Research* **26**: 175–181.

Siegfried, W.R., Benn, G.A. and Gelderblom, C.M. (1998). Regional assessment and conservation implications of landscape characteristics of African national parks. *Biological Conservation* **84**: 131–140.

Sinclair, A.R.E., Hik, D.S., Schmitz, O.J., Scudder, G.G.E., Turpin, D.H. and Larter, N.C. (1995). Biodiversity and the need for habitat renewal. *Ecological Applications* **5**: 579–587.

Tilman, D., May, R.M., Lehman, C.L. and Nowak, M.A. (1994). Habitat destruction and the extinction debt. *Nature* **371**: 65–66.

UNEP (1995). *Global Biodiversity Assessment.* Cambridge University Press, Cambridge, UK.

WRI (1996). *World Resources: A Guide to the Global Environment. The Urban Environment.* World Resources Institute/Oxford University Press, Oxford, UK.

Chapter 8

Coastal Plant Conservation in East Africa:

the work of the National Museums of Kenya in the sacred Kaya forests of coastal Kenya

KENYA

Anthony N. Githitho
Coastal Forest Conservation Unit, National
Museums of Kenya, Kilifi, Kenya

Introduction

There is increasing international interest in the link between traditional culture and natural resource management. The protection of sacred sites due to religious and cultural traditions has often resulted intentionally or otherwise, in the preservation of relatively undisturbed areas rich in biological diversity (Okafor and Ladipo, 1995). In a number of countries, governments have extended formal legal protection to such areas as part of their national portfolios of protected areas. The Convention on Biological Diversity (CBD) recognises the need for 'protection and encouragement of customary use of biological resources in accordance with traditional cultural practices that are compatible with conservation and sustainable use requirements' (Article 10 (see Appendix 1); Glowka *et al.*, 1995). Increasingly, however, these traditional uses and values have been forgotten and their value for conservation must be actively promoted within a contemporary social and economic context.

The Mijikenda Kaya Forests

The sacred Kaya Forests of Kenya are situated in the coastal plain and hills of Kenya, occurring within a zone roughly 250 km long and up to 50 km inland from the sea. There are over 60 known Kaya or sacred grove sites in the contiguous coastal districts of Kwale, Mombassa, Kilifi and Malindi. Kaya forest patches are relatively small, varying in size between 50 and 400 hectares (ha), and represent largely residual patches of the once extensive and diverse lowland forest system of the Zanzibar-Inhambane Regional Mosaic (White, 1983). Kaya forest also occurs in Tanzania north of the Pangani River.

The forests owe their existence directly to the beliefs and culture of the coastal Mijikenda ethnic groups. According to local traditions, the forests historically sheltered small fortified villages when the Mijikenda first appeared in the region three centuries or more ago. These Kayas were a defence against northern enemies who pursued them to this region ('Kaya' means homestead in many Bantu languages). It is presumed that as conditions became more secure, particularly since the late nineteenth century, the groups left the forest refuges and began to clear and cultivate away from the Kayas (Figure 8.1). However, the sites of the original settlements or Kayas are often still marked by forest clearings maintained by the communities, led by their Elders, as sacred places and burial grounds. Cutting of trees and destruction of vegetation around these sites was prohibited, the main aim being to preserve the

Figure 8.1　A huge specimen of *Cycas thouarsii* Gaudich. (Cycadaceae) within a Kaya forest near Diani, Kenya. This cycad is not native to East Africa and possibly represents a relict from an old coastal trading site.

surrounding Kaya forest as a screen around the clearings to maintain secrecy and sacredness. Consequently, while the surrounding areas were gradually converted to farmland, the Kaya sites remained as forest patches of varying size. Studies have shown that these patches are of high biodiversity conservation importance, both regionally and internationally (Burgess *et al.*, 1998; Robertson, 1986; Robertson and Luke, 1993).

Destruction and loss of Kaya forests

Over the past three or four decades there has been an erosion in both the local communities, knowledge and respect for traditional values. This is due to economic, social, cultural and other related changes in the society that have reduced social cohesion and eroded the shared values of the local communities. This has been coupled with rising demands for both forest products and land for agriculture, mining, tourism and other activities resulting from a regional increase in population and international tourism. One result has been the destruction of the small Kaya forests and groves. The Kaya forest patches may occur in fertile areas with relatively dense populations. The forests were also logged for valuable hardwood timber species. Along the south coast, several Digo Kayas were located next to the beaches. These have now been converted into intensive hotel and housing developments (Figure 8.2 and Figure 8.3) and some were included in planned settlement schemes. Damaged Kayas include Kaya Chale, Diani, Kinondo, Tiwi and Waa (Table 8.1). For instance, Kaya Chonyi, the sacred forest of the Chonyi Mijikenda group, has been reduced in forest cover to less than a fifth of its reported original extent through local agricultural encroachment over the last 150 years.

Table 8.1   Habitat loss for selected Kaya sites

Kaya	Estimated original area (ha)	Estimated forest loss (%)	Main cause
Jibana	140	30	Agricultural expansion
Chonyi	200	80	Agricultural expansion
Rabai	>1,000	30	Agricultural expansion
Kambe	80	25	Agricultural expansion
Kauma	100	30	Agricultural expansion
Diani	30	50	Property development
Chale	50	50	Property development
Kinondo	45	30	Property development

Conservation value of the Kayas

The Kayas are botanically diverse and have a high national, regional and international value for biodiversity (Hawthorne, 1993), and are important components of the East African coastal biodiversity hotspot (Myers *et al.*, 2000). Two surveys undertaken by the National Museums of Kenya (NMK) and funded by The Worldwide Fund for Nature (WWF) are of particular importance as they focused on coastal forests in particular (Robertson, 1986; Robertson and Luke, 1993).

Figure 8.2   The coastal forests of East Africa are an internationally recognised centre of plant diversity and endemism. However, the coast of Kenya is under immense commercial pressure as it develops as an international tourism destination.

Figure 8.3   Surviving Kaya forests are under intensive pressure from commercial development. This Kaya has been damaged during construction for a private residence.

The first study was undertaken in two sections:

1. Compilation of a species inventory for the Kaya forests to determine plant diversity derived from existing data and specimens at the East African Herbarium and new field collections

2. Compilation of a list of Kaya forests, identifying locations conservation status and recording local attitudes towards conservation.

Despite the limited time and finances, the study highlighted which forests appeared most interesting and recommended that there was a need for urgent and long-term actions to be taken to conserve the sites. Discussions were held with both traditional Kaya elders, as well as government officials. The report included recommendations that the main Kaya forests should be gazetted as National Monuments to provide them with legal status and protection.

The Kaya floristic survey generated increased interest in coastal forests and led to the recommendation by WWF that the survey be continued and extended to all the Kenyan coastal forests. This was eventually realised in the development and implementation of the WWF 'Kenya Coastal Forest Status, Conservation and Management Project' undertaken by the National Museums of Kenya (Robertson and Luke, 1993). Most of the larger forest blocks along with smaller forests and woodland patches were surveyed with selection assisted by maps, aerial surveys, previous field experience and local informants. Relatively well-known areas where some plant collections had been made, such as Shimba Hills, were lowest in priority with potentially interesting unexplored areas given priority. Many Kayas surveyed during the 1986 study were also revisited and others were visited for the first time. Considerable knowledge of species distribution and status had been acquired during the first Kaya surveys. Subsequently, only unfamiliar or newly recorded species were collected, with revised site checklists prepared for each site (Robertson and Luke, 1993).

As a result of the Coastal Forest Survey (CFS) undertaken between 1988 and 1991, a checklist of all known vascular plants of the coastal districts was produced. An analysis of the data collected confirmed the conservation importance of the Kayas. The Kayas cover about 2,000 ha, or about 2.5% of the remaining Kenyan coastal forest, yet are of great conservation importance. The CFS developed a measure of relative conservation value, taking into account the number of species endemic to the coastal forest zone of Kenya with five or fewer known locations, and species with fewer than five localities in Kenya but extant elsewhere in Africa. This was then used as a measure of percentage rarity per plot or Kaya grove (Table 8.2). In total, seven out of the 20 sites with the highest conservation status in coastal Kenya were Kayas forests (Table 8.2). Studies by Beentje (1988) have demonstrated that more than half of Kenya's rare plants are found in the Coastal Region. The disproportionately large number of rare trees recorded for the Kayas may indicate that the surviving Kaya forests cover a broad range of habitat and micro-climatic conditions that influence the taxonomic diversity represented within these fragments.

Table 8.2 **Proportion of rare plant species in selected Kayas of high conservation value** (from Robertson and Luke, 1993)

Kaya	Forested area (ha)	Recorded number of plant species	% Rare*
Jibana/ Pangani	250	354	20
Kinondo	30	112	14
Dzombo	295	361	10
Kivara	130	170	4
Muhaka	130	278	9
Mrima	290	271	9
Rabai	850	425	5

* Includes both 'Globally Rare' (< 5 localities and known only from the Coastal Forest Survey area) and 'Rare in Kenya' (< 5 localities in Kenya but occurring elsewhere in Africa) (Robertson and Luke, 1993).

Legal protection for Kaya forests

The CFS report proposed that the Kaya forests should be gazetted as National Monuments. It was proposed that they be cared for by the NMK through the establishment of a dedicated unit. During this period and independently of the WWF survey (1988–91), a local Member of Parliament from Kwale District proposed a motion in Parliament that all Kayas should be gazetted by the NMK as prohibited areas and as special places of prayer.

The Kenya Government, at the instigation of the NMK, began formally placing Kayas under the care of the NMK in January 1992. This was finally achieved through the gazetting of 22 Kayas and sacred groves in Kwale District as National Monuments. This gave the NMK the authority to protect the groves under the Antiquities and Monuments Act (Figure 8.4). Four sites in Kilifi District on the north coast had earlier been declared Forest Areas under the Forests Act. This legal mechanism places less stress on the traditional cultural aspect, but is equally powerful as a conservation measure. The gazettements as National Monuments were based on approximate areas and map locations for each site as there was inadequate time to demarcate them on the ground. The declaration covered a variety of land tenures, including central government, local authority, and private land.

Figure 8.4 Kaya forests, that have been gazetted under the Kenyan government's Antiquities and Monuments Act, are protected from development through co-operative management with local communities and the use of conspicuous signage.

The Coastal Forest Conservation Unit

A proposal was prepared in 1990 by WWF International in consultation with the NMK and the CFS Project Executants. Its purpose was to attract funding to set up the Coastal Forest Conservation Unit (CFCU), within the NMK to care for Kayas gazetted as National Monuments. WWF International approved funding and the Project began in 1992 with a Project Executant and a small staff of five, including two field assistants, led by a Project Officer. Importantly, the Unit's office was located at Ukunda on the south coast near to Kaya sites. In 1994/95 the Project received support from the British government's Department for International Development and WWF-UK under the Joint Funding Scheme. The additional support enabled the Unit to broaden its mandate from plant surveys and protection to include awareness and community conservation activities, and to increase its coverage to include the North coast. Project funding amounted to US$1 million over five years up to the year 2000. As a result of this increased support, the CFCU is working with at least 45 of the over 60 known Kaya sites. The project now has two offices, one each on the North and South coasts, with a total field staff of 17.

The goal of the project is to conserve the coastal Kaya forests and sacred groves for their biodiversity and cultural significance and to conserve the cultural heritage of the local communities associated with the Kaya forests. The objectives of the CFCU under its broadened mandate include:

- the revision of the Antiquities and Monuments Act so as to strengthen its use for conservation of coastal forests and associated Kaya culture
- to increase the number of Kaya forests and sacred groves legally gazetted as National Monuments
- to increase the body of scientific and social data relevant to management for conservation of coastal forests and associated culture
- to achieve an increased awareness of the need for and participation in Kaya conservation by communities adjacent to Kayas
- to decrease pressure from local communities and external agents on Kaya land and resources
- To create an enabling institutional framework and capacity for Kaya conservation

The long-term objective is to gazette all Kayas and sacred groves and to introduce or enhance sustainable management processes. The project was charged with the task of continuing investigations into the coastal flora and fauna, and to assist the Kaya Elders and local communities to protect the Kayas for their biological and cultural values.

The laws pertaining to Kaya conservation were reviewed to strengthen the existing Antiquities and Monuments Act, with respect to Kayas and similar cultural sites of biodiversity value. For example, The Antiquities and Monuments Act needs to explicitly include areas of cultural significance, or cultural landscapes, as one of its protected categories. Legal measures are also required to prevent transfers and other transactions in respect of gazetted monuments. The Act should provide restrictions of sale or transfer of sites by enabling the placement of enforceable charges or declarations of interest in respect to those sites in land registries. A lawyer was appointed to co-ordinate the legal review process in order to strengthen the existing Antiquities Act. Progress has been slow due to a number of factors beyond the Project's control including an institutional review process within the NMK that will influence the institution's legal mandate.

Gazettement of sites

Gazettement involves the local community, through consultation, in demarcating a permanent boundary for each Kaya. Initially, a declaration is published in the Government gazette (an official government newspaper) providing 30 days for any objections to be received. If there are no objections within that time the site is confirmed as a National Monument and conspicuous signs are placed around the boundary to secure the site. The majority of the more important sites (in terms of biodiversity) are now gazetted as National Monuments; from a total of 45 CFCU priority sites, 36 sites are now National Monuments, with an additional four already gazetted as Forest Reserves. The CFCU has proposed the Kayas as United Nations Educational, Scientific and Cultural Organisation (UNESCO) World Heritage Sites and they have been placed on the candidate list as part of the registration process. World Heritage Sites are internationally recognised sites of outstanding universal value as natural and cultural heritage. Listing as a World Heritage Site provides opportunities to attract funding for conservation activities from the international community.

Research support to Kaya management

The CFCU research objectives focus on building an information database to aid prioritisation and planning for forest conservation. Botanical surveys of Kayas by CFCU continue to update and improve the records established by earlier CFS studies. New and important records continue to be made. For example, in one year, 1997/98, a total of ten new records were made for the coastal area, including three new records for Kenya and two possible new species. Kaya managers now have access to extensive and verified information on the plant diversity resident in Kaya forests. However, research activities in other areas such as ecology, cultural values or sociological issues, are poorly developed mainly due to resource constraints. For instance, an ecological monitoring system for coastal forests is not yet in place. It is planned to develop and implement site-specific conservation guidelines.

The CFCU has been able to facilitate research institutions and individual researchers in projects related to coastal forests although it had very limited funds for this. Examples include studies of Important Bird Areas with Birdlife International and studies on the conservation attitudes of women at selected Kaya sites. However, collaboration with universities has not developed as well as would have been expected. The Unit has had no research funds to support fieldwork by both faculty members and students from national universities that are also hard pressed for research funds.

Figure 8.5   Mbinda J. Munge checking a propagation trial for *Gigasiphon macrosiphon* (Harms) Brenan (Leguminosae) at the Coastal Forest Conservation Unit's nursery in Ukunda, on the Kenyan south coast. This species is endemic to the East African coastal forests and highly threatened in the wild.

An indigenous tree nursery was set up at the CFCU's south coast office, in order to establish effective propagation methods for rare and threatened Kaya species. The nursery also produces plants for sale, particularly those with some ornamental or medicinal value. Popularising these plants for horticultural use was perceived as a useful strategy to promote their *ex situ* conservation and promote the planting of indigenous tree species in the community and for hotel landscaping (Figure 8.5).

Increased awareness of the need for and participation in Kaya conservation

The Environmental Education and Awareness Programme is a crucial component of the CFCU's activities. Its purpose is to stimulate interest in the conservation of the Kaya forests including the associated culture and traditions. These activities are co-ordinated by the Environmental Education

Figure 8.6   A training session within a Kaya with local forest guardians examining the conservation of these valuable fragments of forest, as part of the NMK-Darwin course in Plant Conservation Techniques for East Africa.

Officer, but involve all field staff. They are targeted at local communities, schools, colleges and the general public; they involve meetings and workshops (Figure 8.6), school visits and programmes and collaboration with national electronic and print media. The most recent review of the project indicated that more work needed to be done in the area of local community awareness activities in Kaya areas. Most of the community contacts had tended to be with Kaya Elders and it was noted that basic knowledge about the Kayas and CFCU among members of the local community was not as high as expected. It was important to raise community awareness in order to increase villager participation in conservation activities. Approaches used include community meetings involving songs and drama.

The provision and development of alternative wood and other resources for communities neighbouring the Kayas was proposed as a strategy to help decrease utilisation pressure on the forests. CFCU will contribute to this strategy by facilitating the Forest Department's work in helping local communities at selected locations to develop alternative wood resources (woodlots). A bee-keeping project is also being undertaken around a Kaya in a semi-arid area. The direct impact of these activities in reducing pressure may be difficult to prove, but they are valuable as contributions to community development through the generation of income for local communities. The project has also provided support for a system of local community guards who are given a token payment for keeping watch on the forest sites. Being locals, they have proved to be a highly effective network for monitoring the way the forests are being used by the local community and reporting cases of destruction.

Developing an institutional framework and local capacity for Kaya conservation

Institutional development cannot be separated from the issue of sustainable conservation management and the conservation of Kayas must extend beyond the life of any external and short-lived support. To ensure long-term and effective conservation, appropriate laws and management procedures need to be underpinned by both institutional structures and effective organisational procedures. These can include local and regional community organisations, co-ordinating agencies, research organisations and independent funding mechanisms to provide long-term financial support. The establishment or strengthening of these structures was seen as crucial for the survival of the Kaya forests.

It is likely that the NMK will continue to play the key role as a co-ordinating organisation. CFCU is also facilitating the strengthening of local Kaya Elders Committees as focal organisations by assisting in their registration for formal recognition by the Kenyan government where possible. The question of future funding for CFCU beyond the WWF project is currently receiving serious attention as the parent organisation, NMK, does not have the resources to maintain current CFCU activities. The need for a Trust Fund is being investigated.

Lessons Learnt

Significant progress has been made in a number of areas, particularly in the gazettement of sites and botanical surveys. However, continuing and increased effort is needed for a number of components, such as education and awareness raising, research and monitoring, institutional development, capacity building and legal review. An additional period of funding may be necessary if the CFCU is to complete its work, particularly with respect to capacity building. It is now widely accepted that conservation and development programmes need a considerable time to mature properly. Nonetheless, some lessons have been learnt during the Project's existence, which are useful guides for future efforts. Local communities, in order to protect traditional sites, may often need to enter into various types of partnerships with external bodies including government and non-government institutions. These partnerships, with the additional resources that they often bring, may help to compensate for the erosion of traditional protection systems. For example, it may be possible to provide for the deployment of local guards where the traditional religious and cultural taboo systems are no longer effective protection measures. The sustainability of these measures should, however, be questioned. Gazettement by itself is not adequate as the relevant Acts tend to be weak and the enforcement capacity of state organisations generally poor. The gazettement of sites with the placement of conspicuous signs can be a significant deterrent against the destruction of forests and transactions relating to Kaya land. Gazettement is accompanied by conspicuous placement of Monument notices at gazetted sites. Combined with public vigilance and activism this has often prevented Kaya land from being sold or interfered with.

As noted above, the protection of sacred sites is heavily dependent on social and cultural values and the cohesion of local communities. It is, however, neither practical, nor indeed, desirable to turn the clock back and reconstitute local community composition and population patterns as they were many years ago. The approach adopted in various conservation programmes, including the Kayas, is to conduct an education and awareness programme both among the local communities, and further afield, using various media. While this cannot completely restore cultural traditions associated with the Kayas, it serves to revive interest in the Kayas within various groups of people for various reasons. Local communities that are near or adjacent to Kayas may still benefit from awareness raising programmes.

The community guards system has been highly effective for monitoring local developments, where the community guards are chosen by the Elders and provided with a token payment as recognition of their effort. The system has not been without problems. One issue is that the guards wanted to be identified with NMK or CFCU rather than the local community, thereby alienating themselves from local communities.

Developing local and regional institutions for Kaya conservation may take considerable time; in addition, addressing pressing local economic and developmental needs constitutes a major challenge. The problem is enormous, but the CFCU is making a small contribution to local economies by supporting community projects including farm forestry and bee-keeping in some Kaya areas. Local economic development is, however, linked closely with national macroeconomic conditions and, in this sense, conservation of these sites is affected by the general state of the economy. A key issue in the next few years is the need for sustainable funding sources. This needs to be coupled with the development of strong institutions to meet the challenges of Kaya conservation and management.

References

Beentje, H.J. (1988). Atlas of the rare trees of Kenya. *Utafiti* **1**: 71–123.

Burgess, N.D., Clarke, G.P. and Rodgers, W.A. (1998). Coastal forests of Eastern Africa: status, endemism patterns and their potential causes. *Biological Journal of the Linnean Society* **64**: 337–367.

Glowka, K., Burhenne-Guilmin, F., Synge, H., McNeely, J.A. and Gündling, L. (1994). *A Guide to the Convention on Biological Diversity.* Environmental Policy and Law paper No. 30. IUCN, Gland, Switzerland.

Hawthorne, W.D. (1993). East African coastal forest botany, pp. 57–102. In: J.C. Lovett and S.K. Wasser (eds). *Biogeography and Ecology of the Rainforests of Eastern Africa.* Cambridge University Press, Cambridge, UK.

Myers, N., Mittermeier, R.A., Mittermeier, C.G., Da Fonseca, G.A.B. and Kent, J. (2000). Biodiversity hotspots for conservation priorities. *Nature* **403**: 853–858.

Okafor, J.C. and Ladipo, D.O. (1995). Fetish groves in the conservation of threatened flora in southern Nigeria, pp. 167–180. In: L.A. Bennun, R.A. Aman and S.A. Crafter (eds). *Conservation of Biodiversity in Africa. Local Initiatives and Institutional Roles.* National Museums of Kenya, Nairobi, Kenya.

Robertson, S.A. (1986). *Preliminary Floristic Survey of the Kaya Forests of Coastal Kenya.* Unpublished report to the Director, National Museums of Kenya and WWF International. National Museums of Kenya, Nairobi, Kenya.

Robertson, S. and Luke, W.R.Q. (1993). *Kenya Coastal Forests: The Report of the NMK/WWF Coast Forest Survey WWF Project 3256: Kenya, Coast Forest Status, Conservation and Management.* Unpublished report to the Director, National Museums of Kenya and WWF. National Museums of Kenya, Nairobi, Kenya.

White, F. (1983). *The Vegetation of Africa: a descriptive memoir to accompany the UNESCO/ AETFAT/UNSO Vegetation Map of Africa.* Natural Resources Research. UNESCO, France, Paris.

Chapter **9**

Atlantic Rainforest Restoration by the Rio de Janeiro Botanic Garden Research Institute

BRAZIL

Luiz Fernando D. de Moraes
Rio de Janeiro Botanic Garden Research Institute,
Brazil

Cintia Luchiari
Rio de Janeiro Botanic Garden Research Institute,
Brazil

José Maria Assumpçã
Rio de Janeiro Botanic Garden Research Institute,
Brazil

Rafael Puglia-Neto
Brazilian Institute of Environment and Renewable
Resources, Brazil

Tânia Sampaio Pereira
Rio de Janeiro Botanic Garden Research Institute,
Atlantic Rain Forest Programme, Brazil

Introduction

Botanic gardens and biodiversity conservation

Reintroduction programmes and skills in cultivating plants are examples of the linkage botanic gardens can provide between *in situ* and *ex situ* conservation. The provisions on conservation and sustainable use, as set out by the Convention on Biological Diversity (CBD) (Glowka *et al.*, 1994; see Appendix 1), have directed the objectives of the Rio de Janeiro Botanic Garden Research Institute. Our concern about threatened plant resources is the guide for all our activities.

The Atlantic rainforest

One of the world's most threatened natural environments is the Brazilian Atlantic Rainforest. This forest covers the highlands and lowlands along the Brazilian coast, the 'Mata Atlântica', and comprises about 8,000 endemic plant and 567 vertebrate species, making it the fourth most important biodiversity hotspot in the world (Myers *et al.*, 2000). These hotspots contain 'exceptional concentrations of endemic species undergoing exceptional loss of habitat' (Myers *et al.*, 2000).

The remnant vegetation now represents only about 7.5% of the original forest, which was estimated to cover 1,227,600 km^2 (Myers *et al.*, 2000). These remnant fragments have suffered frequent threats since Brazil was discovered (Guedes-Bruni, 1998). The Atlantic Coastal Rainforest was cleared firstly for timber extraction, then for agriculture (especially sugar cane and coffee crops) and more recently these areas were converted to cattle pastures.

Today, besides the extensive floristic surveys that indicate plant community composition and stand structure for the remnant Atlantic Forest (Guedes-Bruni, 1998), there is an urgent need to restore environmental sustainability using the best methods to reconcile economic development with natural resources conservation. If plant community development (the establishment of natural forest regeneration through secondary succession) is predictable, it may be feasible to manage natural succession processes to promote restoration (Palmer, 1997). Rio de Janeiro Botanic Garden's researchers started a series of surveys on the remnant forests of Rio de Janeiro State. Management techniques for the majority of our natural systems are not well known. As Palmer (1997) states, the science of ecological restoration – that is, the development and testing of a body of theory for repairing damaged ecosystems – is in its infancy. Any conservation action is welcome, though, especially for this threatened ecosystem. The problem then, is how to catalyse natural successional processes.

Succession models are used to predict how more extensive restoration projects can achieve their goals. The principal objective seems to be the creation of a species mix and environmental conditions that permit the site to become self-sustainable, if that is possible (Parker, 1997). The first step was to recognise which species pioneer natural regeneration sites; secondly, to develop techniques to propagate them. Both of these would give a greater understanding of how successional processes take place in these remnant forests.

The Atlantic Rainforest Research Program

The Atlantic Rain Forest Research Program was created in 1988 by researchers from the Rio de Janeiro Botanic Garden, to undertake floristic surveys for the most important forest remnants. The Program has undertaken surveys in Conservation Units in Rio de Janeiro State: Macaé de Cima Ecological Reserve (municipality of Nova Friburgo), Paraíso Ecological Station (in Guapimirim), Poço das Antas Biological Reserve (Silva Jardim) and Itatiaia National Park (Itatiaia). In 1993, as a result of concern about the conservation status of the Atlantic Rainforest, the Program established a research project in the Poço das Antas Reserve to study the restoration of the reserve's degraded areas. This Revegetation Project aims to establish techniques that could accelerate natural forest regeneration. According to Parrotta *et al.* (1997), plantations can facilitate forest succession at the understorey level through modification of both physical and biological site conditions. The first goal was to create and test a model of mixed plantations using indigenous tree species.

Plantations are supposed to provide good conditions to allow the tree species to compete with grasses. However, the aim of the project was initially to develop the best techniques to restore degraded areas rather than the most cost effective. The planting model used by the Revegetation Project is similar to the one used for Kageyama *et al.* (1992), and is based on the successional processes of tropical forests and their relationship to natural tree-fall gaps in the forest (Denslow, 1980; Martinez-Ramos, 1985). Four successional stages were used to select species: pioneer, early secondary, old secondary and climax (Budowski, 1965). Several authors have worked on classifying tree species into successional groups for the tropical forests. Ferretti *et al.* (1995) summarised some important features for selecting species for experimental plantations, based on Budowski (1966), Denslow (1980) and Martinez-Ramos (1985). This information is presented in Table 9.1.

Table 9.1 **Separating tree species into successional groups according to their growth characteristics and reproductive habit.** Source: Ferretti *et al.* (1995)

Characteristics	Successional Group			
	Pioneer (P)	Early Secondary (ES)	Late Secondary (LS)	Climax (C)
Growth	very fast	fast	medium slow	slow or very slow
Wood	very light	light	medium hard	hard
Tolerance to shade	very intolerant	intolerant	tolerant in juvenile stage	tolerant
Regeneration strategy	soil seed bank	seedlings bank	seedlings bank	seedlings bank
Dissemination of seeds	birds, bats, wind (long distances)	birds, bats, wind (long distances)	wind principally	gravity, mammals, birds
Size of seeds or fruits dispersed	small	medium	small to medium	large
Seed dormancy	photo- or thermo-induced	absent	absent	innate (embryo immaturity)
Age at first reproduction	early (1–5 yrs.)	intermediate (5–10 yrs.)	relatively late (10–20 yrs.)	late (> 20 yrs.)
Dependence on specific pollinators	low	high	high	high
Lifespan	very short (< 10 yrs.)	short (10–25 yrs.)	long (25–100 yrs.)	very long (>100 yrs.)

Besides the floristic survey, other information was considered to be essential for species selection. The results of studies on plant-animal interaction, secondary succession, population dynamics and tree phenology were used to indicate the most suitable species to include in the restoration plots (Table 9.1).

The Revegetation Project is based on the concept of ecological restoration, which can be defined as the recreation of ecologically viable communities by protecting and fostering ecosystems with the potential for natural regeneration. Therefore, ecological restoration is concerned fundamentally with ecosystem function, not only its structure. This necessarily entails long-term monitoring to assess if this has been achieved. To achieve this goal, the planting design must mimic the remnant forest structure and composition.

Poço das Antas Biological Reserve

Poço das Antas Biological Reserve (5,160 ha) is located in the central coastal area of Rio de Janeiro State, in south-eastern Brazil. The topography of 65% of the area of the reserve has a slope of between 0–20%, 25% is covered by hills up to 100 metres (m), and another 10% comprises hills ranging from 100 m to 200 m high. Soil types are: ultisols (41% of the area); an association of typic humaquet, haplaquet and fluvent entisols (38% of the area); organic soils (histosols; 19%) and inceptisols (2%). Mean annual rainfall is around 2,200 millimetres (mm), well distributed throughout the year. There is a light dry season from May to August.

Whilst the area is legally protected, degraded areas and abandoned cattle pastures (mostly covered by *Panicum maximum* Jacq., *Melinis minutiflora* P. Beauv. (an invasive exotic, remaining from abandoned pastures) and the indigenous *Imperata brasiliensis* Trin. (all Gramineae), comprise 40% of the reserve area. Exotic grasses are effective and aggressive competitors against indigenous species and, by encouraging fire, they retard colonisation by tree species (D'Antonio and Vitousek, 1992) (see Appendix 5). Over the last 17 years, frequent fires have seriously impacted forest regeneration. Coarse aerial photograph analysis indicates a net loss of forested area of 3% for the burnt areas (close to the reserve boundaries); however, the core zone of the reserve, which has rarely burned, shows a net gain of forested area of about 15% (Dietz, pers. com.). Accordingly, the experimental plots with indigenous tree species were set up to form a roadside 'buffer' strip of vegetation, in order to prevent fire from burning degraded areas and to allow indigenous forest regeneration to occur.

The reserve's forest fragments represent the most important habitat remnant for the 'Endangered' (*sensu* IUCN, 1994) golden lion tamarin, *Leontopithecus rosalia* L., a primate species endemic to the lowland Atlantic Forest of Rio de Janeiro. A population viability study recommended a minimum viable population size of 2,000 animals living in the wild for the successful conservation of the animal (Seal *et al.*, 1990). According to the authors, the total habitat needed for 2,000 animals is about 25,000 ha within 100 years. Current habitat area is around 10,000 ha, spread over forest fragments isolated from each other, comprising a population of about 800 individuals living in the wild. The golden lion tamarin habitat needs to be expanded as a matter of urgency.

The Revegetation Project

The implementation of the Revegetation Project followed the steps below:

i) Planting design
A study of the vegetation cover of the reserve showed areas in a gradient of successional stages, ranging from grasslands (resulting from human

P	ES	P	ES	P	ES	P	ES	P	ES	P	ES	P
P	C	P	C	P	C	P	C	P	C	P	C	P
P	ES	P	ES	P	ES	P	ES	P	ES	P	ES	P
P	LS	P	LS	P	LS	P	LS	P	LS	P	LS	P
P	ES	P	ES	P	ES	P	ES	P	ES	P	ES	P

↑

25

rows

80 columns →

Trees are planted in a grid design according to their Successional group

Legend for Successional group: C   Climax   LS Late secondary
                                     P   Pioneer   ES Early Secondary

Spacing: 2.0 × 2.5 m
Planting density: 2,000 saplings/ha

Figure 9.1   Model planting design with mixed native species at Poço das Antas Biological Reserve, Rio de Janeiro, Brazil.

disturbance) to mature forests. The restoration action selected to recover those degraded areas was planting native tree species that could successfully compete with the invasive grass species.

Since one of the first goals was to shade out invasive grasses, the planting design dictated that at least 50% of the saplings should be composed of fast growing species (pioneers (P) and early secondary (ES) successional group species), to provide shade in the short term and to prepare for later successional stages.

The saplings are distributed in such a way that eight fast growing species are planted around each shade tolerant species. Figure 9.1 shows a model for a planting design of 2,000 saplings/ha (spacing of 2.0 m × 2.5 m). Another density used was 2,500 saplings/ha (2.0 m × 2.0 m). The most important aspect of this model is the relationship between tree species of different successional stages within the suggested planting layout.

ii) Selection of species
Studies of secondary succession at the reserve indicated those species that seemed to act as pioneers at the beginning of the forest regeneration process. A relatively high growth rate was considered a high priority feature, since it may indicate good competitiveness with invasive species. Other features, such as importance for local wildlife and high dispersion potential, were considered in the selection process. This information was recorded

from field observations. Observations of seed germination and seedling development at the nursery also supported species selection. Some species with a high potential as colonisers could not be used because of failure of seed germination.

iii) Seed collection

The introduction of a two-year observation period at the reserve gave data on the phenology of candidate tree species and generated a calendar for the fruiting periods. This information guided subsequent seed collection. Seeds were collected both for seedling production and for laboratory research, to determine the best temperature and humidity conditions for seed storage and germination. In order to encourage local community involvement, a plant nursery was built in the reserve. Local people were hired to undertake the tasks of seed collection and seedling cultivation.

To conserve sufficient genetic diversity in the populations of species to be restored to the plots, Kageyama and Gandara (1993) suggest the use of an Effective Population Size (Ne) to represent the population from where the seeds must be collected. Since this project has long-term conservation goals, and inbreeding coefficient increases rapidly in populations of small effective size, a Ne=50 was used (Kageyama and Gandara, 1993). Assuming that most tropical tree species are allogamic (cross-pollinated), we have each individual representing Ne=4, which means collecting seeds from 12 individuals within the reserve.

iv) Site selection and preparation

There are two representative geomorphologic units at the reserve: wetlands (lowland areas, ranging from periodically to permanently flooded) and drylands (represented by hills with slopes ranging mostly from 20–100%, reaching 150–200 m in height). Table 9.2 shows the characteristics of the trial plots.

Because the experimental site is located in a Conservation Unit, no chemical products could be used for weed control. The prohibition of chemical weed control is a legislative issue. However, the use of certain herbicides could help reduce future maintenance costs of the experimental site. Soil analysis indicated low fertility soils for the reserve as a whole. Since one of the criteria adopted for species selection was their fitness to edaphic conditions, no fertilisers were used for the plantings. This would also have raised the costs of the trials prohibitively.

v) Planting

Since 1994, 12 experimental plots (varying from 0.8 ha to 1.5 ha each) have been set up both in the wetlands and drylands. This paper reports the results for three plots: Trials I and II in the wetlands, and Trial III in the drylands.

Table 9.2 **Characteristics of the trial plots**			
Characteristic	**Trial I**	**Trial II**	**Trial III**
Planting date	March 1994	May 1996	January 1995
Site location	lowland	lowland	upland
Slope class (%)	0–5	0–5	30–45
Site preparation	manual	mechanisation	manual
Soil type	Typic Humaquet	Haplaquet and Fluvent (Entisol)	Ultisol
Size of plot (ha)	1.0	1.0	0.9
Spacing (m)	2.0 × 2.5	2.0 × 2.0	2.0 × 2.7
No. of saplings planted	2,000	2,500	1,621
Dominant invasive species*	Pm, Ib	Pm	Mm
Maintenance period (years after planting)	2	1	1

*Dominant invasive species (all Gramineae): Pm = *Panicum maximum* Jacq.; Ib = *Imperata brasiliensis* Trin.; Mm = *Melinis minutiflora* P. Beauv.

According to the planting design, the late successional tree species (LS and C species) were planted at the same time as the pioneer species (ES and P species). This was also intended to reduce costs. Two spacings were chosen to favour early canopy closure: 2.0 m × 2.5 m and 2.0 m × 2.0 m. The first one represented a density of 2,000 plants/ha, and the second, 2,500 plants/ha. These values were based on densities of 1,000 to 1,200 trees/ha, characteristic of mature tropical forests at the beginning of the successional process (Hubbel and Foster, 1983), and taking into account the fact that 50% of early pioneer species die within 5–10 years after planting. Saplings were considered to be ready for planting when they reached 40 cm in height. This was attained within about 4–6 months after seed germination for early secondary and pioneer species and 6–8 months for late secondary and climax species. The area to be planted was not fenced nor physically isolated in any way, but a 5m wide strip was kept weeded to protect the plantation from fire. One month later, planting was monitored to evaluate survival rates. Replacement of dead trees was carried out only where more than 10% of the total amount of plants had died. Weeding of the plots was maintained until the invasive grass species were considered to be under control by shading. Vine control was not a major problem during site preparation before planting.

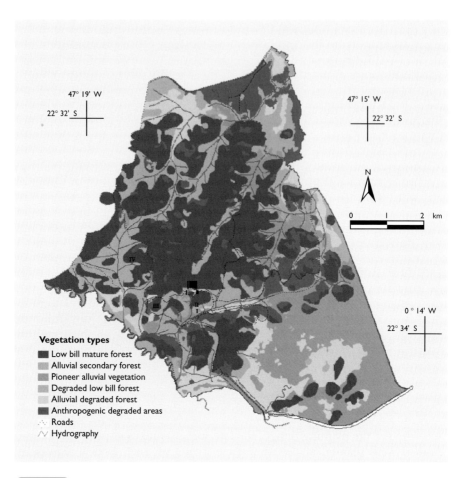

Figure 9.2 Location of the study area, Poço das Antas Biological Reserve, Rio de Janeiro, Brazil, 1995.

Monitoring of the plantings was carried out annually for the first five years to evaluate plant development (the first measurements were taken one year after planting). Measurements of survival rate, height and crown diameters were taken. Digital maps of vegetation (Figure 9.2), topography, slopes and soils were generated from a Geographical Information System (GIS) and were used to support site selection and the choice of management techniques. Further, the maps allowed for the definition of management units and long-term planning.

Results

This paper discusses the results of three experimental planting (Trials I, II and III). However, it should be noted that they do not represent treatments of the same experiment. Data for survival rates, average height and relative growth rates are shown in Tables 9.3, 9.4, 9.5 and Figures 9.3, 9.4 and 9.5 for Trials I, II and III, respectively. Relative growth rate was calculated to give some indication of how fast the plants of each species established.

Table 9.3 Trial I – survival and growth rates for planted indigenous tree species in temporarily flooded grasslands at the Poço das Antas Biological Reserve

Species	Survival rate[1] (%) (N)	Average height (m)			
		1996[2]	1997 (GR'96–'97)	1998 (GR '97–'98)	1999 (GR '98–'99)
Pioneer:					
Acnistus arborescens (L.) Schltdl. (Solanaceae)	82 (100)	2.14	3.14 (46.73)	2.43 (–22.61)	2.94 (20.99)
Mimosa bimucronata (DC) Kuntze (Leguminosae, Mimosoidae)	97 (400)	4.10	5.39 (31.46)	5.98 (10.95)	6.34 (6.02)
Early Secondary:					
Cytharexylum myrianthum Cham. (Verbenaceae)	90 (100)	2.74	3.68 (34.31)	3.58 (–2.72)	5.80 (62.01)
Inga edulis Mart. (Leguminosae, Mimosoidae)	93 (400)	1.80	3.38 (87.78)	5.57 (64.79)	6.43 (15.44)
Lonchocarpus virgilioides (Vogel) Benth. (Leguminosae, Papilionoidiae)	85 (100)	1.34	1.84 (37.31)	2.50 (35.87)	2.66 (6.40)
Pseudobombax grandiflorum (Cav.) A. Robyns (Bombacaceae)	92 (100)	2.74	3.45 (25.91)	4.43 (28.41)	4.95 (11.74)
Tibouchina granulosa (Desr.) Cogn. (Melastomataceae)	70 (200)	2.12	3.68 (73.58)	5.18 (40.76)	6.04 (16.60)
Late Secondary:					
Guarea guidonea (L.) Sleumer (Meliaceae)	93 (200)	1.41	2.42 (71.63)	3.30 (36.36)	3.88 (17.58)

[1] Measured one year after planting, N – total number of individuals;
[2] First measurements were taken two years after planting.
GR – Growth Rate (%); GR = $(H_n – H_{n-1}/H_{n-1}) \times 100$; where H – Height; n – plant age when measurement was taken (in years).

Table 9.4 **Trial II – Survival and growth rates for planted indigenous tree species in temporarily flooded grasslands at the Poço das Antas Biological Reserve**

Species	Survival rate[1] (%) (N)	Average height (m)		
		1997	1998 (GR[2] '97–'98)	1999 (GR '98–'99)
Pioneer:				
Margaritaria nobilis L. f. (Euphorbiaceae)	98.2 (168)	1.13	1.95 (70.44)	2.94 (52.77)
Mimosa bimucronata (DC.) Kuntze (Leguminosae, Mimosoidae)	100.0 (168)	3.37	5.46 (62.02)	6.44 (17.95)
Early Secondary:				
Cytharexylum myrianthum Cham. (Verbenaceae)	100.0 (279)	2.88	4.58 (59.03)	5.72 (24.89)
Inga affinis DC. (Leguminosae, Mimosoidae)	96.5 (286)	1.55	3.23 (108.39)	4.31 (33.44)
Inga laurina (Sw.) Willd. (Leguminosae, Mimosoidae)	100.0 (317)	1.57	2.70 (71.97)	3.38 (25.19)
Pseudobombax grandiflorum (Cav.) A. Robyns (Bombacaceae)	98.9 (265)	1.78	3.01 (69.10)	4.30 (42.86)
Late Secondary:				
Guarea guidonea (L.) Sleumer (Meliaceae)	100.0 (120)	1.07	2.14 (100.00)	3.31 (54.67)
Jaracatia spinosa (Aubl.) A.DC. (Caricaceae)	95.8 (72)	1.59	2.71 (22.75)	4.14 (38.11)
Climax:				
Calophyllum brasiliense Cambess. (Clusiaceae)	100.0 (58)	2.33	2.86 (72.57)	3.95 (50.77)
Copaifera langsdorfii Desf. (Leguminosae, Caesalpiniodae)	99.0 (192)	0.76	1.42 (86.84)	2.40 (69.01)

[1] Measured one year after planting, N – total number of individuals;
[2] GR – Growth Rate (%); GR = $(H_n - H_{n-1}/H_{n-1}) \times 100$; where H = Height; n – plant age when measurement was taken (in years).

Trial I was planted in March 1994, Trial II in May 1996 and Trial III in March 1995. March is at the end of the rainy season and May in the beginning of the dry season. Since rainfall data do not indicate a deficit of water for the reserve soils, those periods seemed to be the best for planting, and the lower temperatures inhibited grass development.

Table 9.5 **Trial III – survival and growth rates for planted indigenous tree species in sloping well-drained areas, at the Poço das Antas Biological Reserve**

Species	Survival rate[1] (%) (N)	Average height (m)			
		1996	1997 (GR[2] '96–'97)	1998 (GR '97–'98)	1999 (GR '98–'99)
Pioneer:					
Aegiphila sellowiana Cham. (Verbenaceae)	90.9 (130)	3.40	5.46 (60.59)	5.67 (3.85)	5.03 (–11.29)
Gochnatia polymorpha (Less.) Cabrera (Compositae)	93.4 (450)	2.12	3.57 (68.40)	4.40 (23.25)	4.63 (5.23)
Early Secondary:					
Piptadenia gonoacantha (Mart.) J.F. Macbr. (Leguminosae, Mimosoidae)	95.0 (611)	2.40	3.89 (62.08)	4.85 (24.68)	5.24 (8.04)
Sparattosperma leucanthum (Vell.) K. Schum. (Bignoniaceae)	99.3 (140)	1.79	2.94 (64.25)	3.14 (6.80)	2.81 (–10.51)
Late Secondary:					
Piptadenia paniculata Benth. (Leguminosae, Mimosoidae)	97.2 (105)	2.09	3.47 (66.03)	2.64 (–23.92)	1.10 (–58.33)
Tabebuia chrysotricha (Mart. ex DC.) Standl. (Bignoneaceae)	99.2 (119)	1.68	2.20 (30.95)	2.89 (31.36)	2.94 (1.73)

[1] Measured one year after planting, N – total number of individuals;
[2] GR – Growth Rate (%); GR = $(H_n - H_{n-1}/H_{n-1}) \times 100$; where H – Height; n – plant age when measurement was taken (in years).

The results indicate that the planting design used can encourage fast-growing canopy cover to suppress invasive grasses. The high growth rates of *Mimosa bimucronata* (DC.) Kuntze (Leguminosae) for wetlands (Tables 9.3 and 9.4) and *Aegiphila sellowiana* Cham. (Verbenaceae), and *Gochnatia polymorpha* (Less.) Cabrera (Compositae) for the upland, dry areas (Table 9.5) show that the use of native pioneers in the restoration of degraded areas is quite feasible. In Trial I, *Inga edulis* Mart. (Leguminosae) reached the highest average height, but *M. bimucronata* showed remarkable growth rates, as well as *Guarea guidonea* (L.) Sleumer (Meliaceae) and *Tibouchina granulosa* (Desr.) Cogn. (Melastomataceae) (Table 9.3 and Figure 9.3) *M. bimucronata, I. edulis* and *T. granulosa* attained the greatest average height in five years and therefore would seem to be the species most likely to contribute to a fast canopy closure.

Trial II, like Trial I was set up in a lowland area, and showed very similar development for *M. bimucronata, Pseudobombax grandiflorum* (Cav.) A. Robyns (Bombacaceae) and *Cytharexylum myrianthum* Cham. (Verbenaceae) (Table 9.4

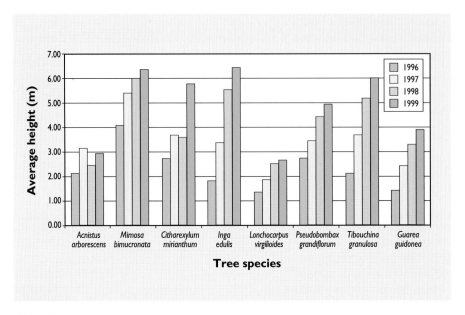

Figure 9.3   Average heights for species in Trial I. – Trees planted in 1994.

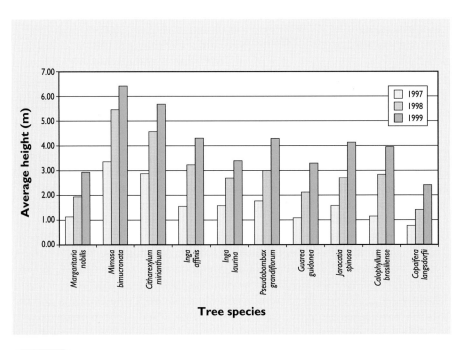

Figure 9.4   Average heights for species in Trial II. – Trees planted in 1996.

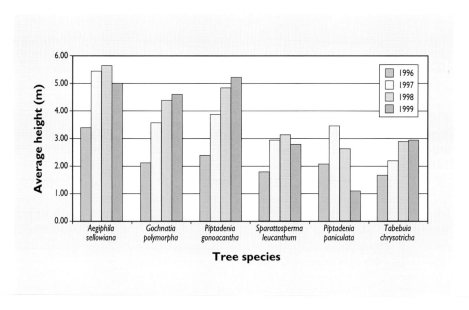

Average heights for species in Trial III. – Trees planted in 1995.

and Figure 9.4). *Inga affinis* DC. (Leguminosae), (Caricaceae) *Jaracatia spinosa* (Aubl.) A.DC. and *Calophyllum brasiliense* Cambess. (Clusiaceae) reached high values of average height and growth rates in Trial II, contributing both to a more diverse canopy and a fast canopy closure. It should be stressed that the good growth rates shown for late secondary and climax species in the last measurement, in 1999, three years after planting, indicated that fast growing species such as *M. bimucronata* can very rapidly provide suitable conditions for the more shade tolerant species such as *Copaifera langsdorfii* Desf. (Leguminosae) to grow. Mechanisation during site preparation seems to be an important factor contributing to the performance of this trial as a whole.

Trial II, planted later than Trials I and III, was the first real chance to use the planting model properly, since there were enough saplings for the different successional groups and a larger list of species to select from, using information gathered from field research. The most remarkable results for Trial II were the survival rates, all above 95% (Table 9.4). Knowles and Parrotta (1995), who carried out research on the restoration of areas degraded by mining in the Amazon, selected species for restoration with a survival rate above 75%.

For Trial III, *Piptadenia gonoacantha* Mart. J.F. Macbr. (Leguminosae) and *G. polymorpha* showed the highest values for growth rate (Table 9.5), with *P. gonoacantha* reaching the highest average height (Figure 9.5). The initial results of this trial mistakenly led to the decision to stop the maintenance of the trial plot after one year from planting, since the species planted reached good average heights and it was observed that significant seedling recruitment from the local soil seed bank was taking place. However, despite growing fast, the species selected for trial did not contribute to rapid canopy closure, as a dense crown was not produced. This allowed gaps where the grass, *M. minutiflora*, could recolonise, preventing natural succession taking place. For the last year (four years after planting), *A. sellowiana, Sparattosperma leucanthum* (Vell.) K. Schum. (Bignoniaceae) and *Piptadenia paniculata* Benth. showed negative relative growth rates (Table. 5). This can be explained by the mortality of a few individuals for the first two species (15% of planted individuals for each species died), and the mortality of most individuals (90%) of *P. paniculata*. Reasons for these mortalities were not investigated. At the time of planting for Trial III, there were no saplings of climax or other late secondary species available in a significant amount for planting. Other species were planted in this same area, but as they were not in significant numbers, they are not represented in Trial III. The use of more species may have made this trial more successful.

The high survival rates for the late secondary and climax species in all three trials, and in some cases relatively high growth rates would seem to indicate that some species tolerate full sunlight, despite not being colonisers.

During the period covered in this paper, temperature and rainfall mostly followed the reserve pattern, except for February, 1997, which was a hot and unusually dry month. However, since this period was only a month, it is assumed that it did not have a significant impact on plant development.

Project Costs

The total cost for setting up Trial I plantation, including seedling cultivation, site preparation, planting and maintenance, was approximately US$2,600/ha. This amount was reduced to US$2,200 when mechanisation was adopted (Trial II). Mechanisation in site preparation retards the resprouting of grass species, limiting the maintenance period to the first year. Without this, Trial I, sited also in the wetland needed to be weeded by hand for the first two years. The trial planted on the sloped area (Trial III) cost less (US$2,100/ha) because, as mentioned above a decision was made to maintain the plot for the first year only as a result of the rapid establishment of the saplings, which it was assumed would suppress the growth of the grass species *M. minutiflora* in subsequent years.

Conclusions and Lessons Learnt

First of all, our results show that planting with native species has an important effect on controlling invasive grasses, indicating that they can be used successfully to initiate natural forest regeneration. Species selection is of great importance for successful establishment of the planting, as well as to reduce the costs. Besides characteristics such as rapid canopy establishment, species selection based on their attractiveness to wildlife (ability of plantations to provide habitat and food) must be considered (Parrotta *et al.*, 1997). Species selected included those dispersed by animals, especially by bats, birds and primates, namely *A. sellowiana*, *I. edulis*, *I. affinis*, *I. laurina* (Sw.) Willd., *G. guidonia, Lonchocarpus virgilioides* (Vogel) Benth. (Leguminosae), *Acnistus arborescens* (L.) Schltdl. (Solanaceae), *Copaifera langsdorffii* and *Cytharexylum myrianthum.*

Mechanisation for site preparation should be used whenever local conditions (flat areas, soils suitable for mechanisation) allow it, because turning the upper soil layers is likely to improve natural seedling regeneration by stimulation of the soil seed bank, thus catalysing forest succession processes. Soil seed bank contents are of great interest in tropical forest succession studies (Chandrashekara and Ramakrishnan, 1993; Hopkins *et al.*, 1990; Quintana-Ascencio *et al.*, 1996; Rico-Gray and García-Franco, 1992; Saulei and Swaine, 1988; Swaine *et al.*, 1997; Young *et al.*, 1987). Seed bank information supplements restoration management practice. With the first weeding out of the grass species, seeds of pioneer species present in the soil seed bank, especially *Trema micrantha* (L.) Blume (Ulmaceae) and *Cecropia* Loefl. spp. (Cecropiaceae), were stimulated to germinate by the increased light levels. As a matter of fact, it indicates that those species need not be planted, as their seeds seem to show a long viability in the soil seed bank. Maintenance would help these recruited tree species to establish and contribute to shading out invasive species. Mechanisation also had a major impact on suppressing invasive grasses probably by breaking up rhizomes. It is assumed that planted saplings could then compete successfully with grass species.

Other trials set up in the wetland areas were not monitored for different reasons. A single fire, caused by a car pulling off the road into one of the plantings, seriously burned two of the one-hectare experimental plots ten months after planting. A survey was made four months later to evaluate damage to surviving plants, and most of the species were resprouting (80–90% of affected individuals). A comparison between a burnt plot and an unaffected one of the same age (4.5 years) showed no significant difference in the development of the same species.

When the first trials were started, there was a lack of knowledge about species classification into successional groups. Today, field experience supports reclassification and the Revegetation Project is working with a list of about 60 tree species which have been tested in other trials, in both lowland and upland sites. Particular attention must be paid to secondary species, which contribute to the high plant diversity in tropical forests. Further research is required to develop techniques that will reduce the costs of establishing plantations. It is also necessary to gain a better understanding of how these plantations of indigenous species favour understorey tree development.

It is over seven years since the first trial was established and we have seen some very positive developments. Plantations with native species demonstrate the potential to restore degraded areas at the Reserve, which means providing a larger habitat for the golden lion tamarin. The Rio de Janeiro Botanic Garden Research Institute has acquired the knowledge and expertise in restoration ecology, and more broadly in 'integrated conservation strategies' (BGCI, 1995). This is a rapidly expanding field of research in Brazil. Our partners intend to fund the long-term monitoring of the plantings for conservation purposes. In addition to this, the nursery built at the Reserve can provide seedlings not only for the Revegetation Project plantations in the Reserve, but for other small restoration projects. It will therefore be partially sustainable through self-financing. The presence of a local plant nursery, producing seedlings of native tree species, could also stimulate other institutions or landholders with an interest in restoring deforested areas, by the sale of surplus saplings. Our experience could be used as a model for *in situ* conservation in other areas, for example the Tijuca National Park, which has been threatened with frequent forest landslides and human encroachment for settlement.

Acknowledgements

We would like to thank all the Staff of the Poço das Antas Biological Reserve/IBAMA for full collaboration on office and field issues; Shell do Brasil S.A., Petrobras, MacArthur Foundation and CNPq (Brazilian Research Council) for financial support and Margaret Mee Botanic Foundation for administrative support. We would also like to thank Mike Maunder, Clare Hankamer, and an anonymous reviewer for the suggestions to this paper.

References

BGCI. (1995). *A Handbook for Botanic Gardens on the Reintroduction of Plants to the Wild.* Botanic Gardens Conservation International, London, UK.

Budowski, G. (1965). Distribution of tropical American rain forest species in the light of successional process. *Turrialba* **15**: 40–42.

Chandrashekara, U.M. and Ramakrishnan, P.S. (1993). Germinable soil seed bank dynamics during the gap phase of a humid tropical forest in the Western Ghats of Kerala, India. *Journal of Tropical Ecology* **9**: 455–467.

D'Antonio, C.M. and Vitousek, P.M. (1992). Biological invasions by exotic grasses, the grass/fire/cycle, and global change. *Annual Review in Ecological Systematics* **23**: 63–87.

Denslow, J. (1980). Gap partitioning among tropical rain forest trees. *Biotropica* **12**: 47–55.

Ferretti, A.R. *et al.* (1995). Classificação das Espécies Arbóreas em Grupos Ecológicos para Revegetação com Nativas no Estado de São Paulo. *Florestar Estatístico* **3** (7).

Glowka, L., Burhenne-Guilmin F., Synge, H., McNeely, J.A. and Gündling, L. (1994). *A Guide to the Convention on Biological Diversity.* Environmental Policy and Law paper No. 30. IUCN, Gland, Switzerland.

Guedes-Bruni, R.R. (1998). *Composição, estrutura e similaridade florística de dossel em seis unidades fisionômicas de Mata Atlântica no Rio de Janeiro.* Ph.D. thesis, USP, São Paulo, Brazil.

Hopkins, M.S., Tracey, J.G. and Graham, A.W. (1990). The size and composition of soil seed-banks in remnant patches of three structural rainforest types in North Queensland. *Australian Journal of Ecology* **15**: 43–50.

Kageyama, P.Y., Freixêdas, V.M., Geres, W.L.A., Dias, J.H.P and Borges, A.S. (1992). Consórcio de espécies nativas de diferentes grupos sucessionais em Teodoro Sampaio-SP, pp. 527–533. *II Congresso Nacional sobre Essências Nativas.* SP, Inst. Flor. São Paulo, Brazil.

Kageyama, P.Y. and Gandara, F.B. (1993). Dinâmica de Populações de Espécies Arbóreas: implicações para o manejo e a conservação, pp. 1–7. *III Simpósio de, Ecossistemas da Costa Brasileira,* ACIESP.

Knowles, O.H. and Parrotta, J.A. (1995). Amazon forest restoration: an innovative system for native species selection based on phenological data and field performance indices. *Commonwealth Forestry Review* **74**(3): 16–24.

Martinez-Ramos, M.C. (1985). Claros, ciclos vitales de los arboles tropicales y regeneratión natural de las selvas altas perenifolias. In: A. Gomez-Pompa and R.S. Del Amo (eds). *Investigaciones sobre la regeneratión de selvas en Vera Cruz.* INIRB, Ed. Alhambra Mexicana, Mexico.

Myers, N., Mittermeier, R.A., Mittermeier, C.G., Fonseca, G.A.B. and Kent, J. (2000). Biodiversity hotspots for conservation priorities. *Nature* **403**: 853–858.

Palmer, M.A. (1997). Ecological theory and community restoration ecology. *Restoration Ecology* **5**(4): 291–300.

Parker, V.T. (1997). The scale of succession models and restoration objectives. *Restoration Ecology* **5**(4): 301–306.

Parrotta, J.A., Turnbull, J.W. and Jones, N. (1997). Catalyzing native forest regeneration on degraded tropical lands. *Forest Ecology and Management* **99** (1, 2) 1–7.

Quintana-Ascencio, P.F., González-Espinosa, M., Ramirez-Marcial, N., Dominguez-Vázquez, G. and Martínez-Icó, M. (1996). Soil seed banks and regeneration of tropical rain forest from milpa fields at the Selva Lacandona, Chiapas, Mexico. *Biotropica* **28**(2): 192–209.

Rico-Gray, V. and García-Franco, J.G. (1992). Vegetation and soil seed bank of successional stages in tropical lowland decidous forest. *Journal of Vegetation Science* **3**: 617–624.

Saulei, S.M. and Swaine, M.D. (1988). Rain forest seed dynamics during succession at Gogol, Papua New Guinea. *Journal of Ecology* **76**: 1133–1152.

Seal, U.S., Ballou, J.D. and Padua, C.V. (eds) (1990). *Leontopithecus*: Population Viability Workshop. Belo Horizonte, Brazil. Conservation Breeding Specialist Group, Apple Valley, Minnesota, USA.

Swaine, M.D. *et al.* (1997). Ecology of forest trees in Ghana. London University of Aberdeen – ODA Forestry Series no. 7, UK.

Young, K.R., Ewel, J.J. and Brown, B.J. (1987). Seed dynamics during forest succession in Costa Rica. *Vegetatio* **71**: 157–173.

Chapter 10

Plant Conservation Initiatives in Singapore

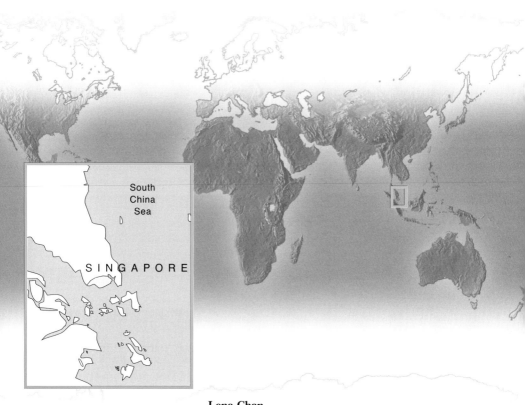

Lena Chan
National Parks Board, Singapore

Introduction

The year 1819 marked the beginning of modern Singapore with its founding by entrepreneur and naturalist Sir Stamford Raffles. One hundred and eighty years later Singapore has been transformed from a largely forested island to a highly urbanised city republic, principally dependent on its human rather than its natural resources. With a total population of 3,612,000 in 1998, Singapore's population density stands at 5,578 people per km².

The Republic of Singapore comprises the main island of Singapore and some 63 offshore islands, amounting to a total land area of 647.5 km². It is situated between latitudes 1°09'N and 1°29'N and longitudes 103°36'E and 104°25'E, and experiences an equatorial climate. Its topography is monotonous, generally flat with the highest peak, Bukit Timah, reaching a height of 162.5 m.

Singapore contains a variety of vegetation types including lowland tropical rainforest, freshwater swamps, secondary forest, mangrove swamps and coastal vegetation. Singapore has been through a dramatic change in land use and landscape character. In 1819, the island was almost 100% forest; today only about 2% of Singapore harbours good dryland tropical lowland rainforest and about 6% comprises a diverse range of other ecosystems. It has changed from a forest-covered island, through a period where agriculture was the dominant land use, to one of the world's most densely populated urban regions. This transition is documented through a wide range of official and unofficial publications, books, reports, and papers. For instance, Corlett (1992) demonstrated that the area under cultivation reached a maximum in 1935 of just under 60% of the total land area. Thereafter, urbanisation took over as the major land use.

Studies by Turner (1994) found that from a vascular flora of 2,277 species, a total of 594 species have become locally extinct, with a further 508 species threatened with local extinction; 957 are classed as rare and 218 as common. The original botanical inventories were undertaken in the late nineteenth century after a significant period of forest clearance; accordingly, it is possible that additional unrecorded species have been lost from Singapore. On the other hand, species that were recorded as extinct in the Singapore Red Data Book (Ng and Wee, 1994) because there were no herbarium specimens or records for the past 50 years, have been recorded in recent surveys.

Conservation in Singapore

The challenge lies in maintaining and enhancing this rich plant diversity in a largely closed system where there is a multitude of other competing land uses. To meet the environmental challenges of the 1990s, the government of Singapore established the Singapore Green Plan (SGP) (Ministry of Environment, 1993). This document contains the vision for a green city:

- with high standards of public health, with clean air, land, water and a quiet living environment;
- conducive to gracious living, with people concerned about and taking a personal interest in the care of both the local and global environment;
- that will be a regional centre for environmental technology.

The National Parks Board (NParks), inaugurated in 1990, is the statutory body in Singapore responsible for urban landscape and nature conservation. In July 1996, NParks and the Parks and Recreation Department merged with a shared mission: 'We Make Singapore Our Garden'. NParks was given the mandate to manage national parks and nature reserves that are dedicated, set aside and reserved for all or a number of the following purposes:

- the propagation, protection and preservation of the flora and fauna of Singapore;
- the study, research and preservation of objects and places of aesthetic, historical or scientific interest;
- the study, research and dissemination of knowledge in botany, horticulture, biotechnology and natural and local history.

The Singapore Botanic Gardens (SBG) (Figure 10.1) and Fort Canning Park are also administered by NParks. NParks presently manages more than 1,625 ha of parks, open spaces, and park connectors; and more than 4,000 ha of roadside vegetation. NParks' remit includes the management of 2,839 ha of nature reserves as Singapore's most significant natural heritage and as the Republic's largest refuge for indigenous flora and fauna. As Singapore's designated Scientific Authority on Nature Conservation, NParks advises the Ministries of the Singapore Government on policies pertaining to nature conservation. In addition, NParks is also the lead agency responsible for the implementation of nature conservation programmes, including implementation of Singapore's responsibilities to the Convention on Biological Diversity (CBD) (see Appendix 1).

Within NParks, the Nature Conservation Branch (NCB) is responsible for the formulation of policies pertaining to nature conservation and the co-ordination of national conservation efforts. The NCB collaborates closely with the SBG Herbarium and is administratively under SBG. The role of SBG in the history of nature conservation dates back to 1849 when it was first established in its present grounds (Burkill, 1918).

Figure 10.1   The main entrance to Singapore Botanic Gardens.

Conservation Action in Singapore

Conservation planning and management in Singapore has to take into account three major factors:

1. Singapore is effectively a city island, with its terrestrial land area posing a physical constraint and, accordingly, conservation planning in the Singapore context, is the management of biodiversity within a finite and restricted land area.

2. The area of natural vegetation is less than 10% of Singapore's land area.

3. Surviving habitats throughout Singapore are found in fragmented pockets under pressure from changing land use and exotic species (see Appendix 5).

Under the SGP, the government had set aside an area not less than 3,130 ha for nature conservation. This area comprises representative areas of the ecosystems extant in Singapore, designated as 'Nature Areas' (Ministry of the Environment, 1993).

A multidisciplinary approach is needed for an effective plant conservation strategy in Singapore as the pace of urban development is rapid. Working within the constraints of a small, highly urbanised area with only fragments of natural habitat, Singapore has to plan its conservation efforts incorporating the following actions:

a) *in situ* conservation, including the identification of habitat areas and the establishment of park connectors;

b) survey and monitoring of Nature Areas;

c) the design and production of databases to manage data on Singapore's biodiversity;

d) species level management including reintroduction and translocation

e) habitat restoration.

Description of actions

a) *In situ* conservation

The retention of areas of extant habitat is given priority through the establishment of legally protected Nature Reserves. Some 3,000 ha had been identified as reserves by 1993, selected according to the following criteria:

• The site should have a natural environment in terms of its landscape and wildlife.

• The site should be ecologically stable and have the ability to support a large variety of wildlife. The presence of endangered or rare species, or taxa unique to Singapore, would be an added merit.

• The site should have the potential for recreation, education and scientific research.

• The site should be able to coexist with adjoining, or nearby developments.

These reserves contain approximately 85% of the extant indigenous species in less than 3% of Singapore's land area. The bulk of Singapore's biodiversity is concentrated in a small area of tropical lowland dipterocarp forest and freshwater swamp forest. Aerial photographs and Development Guide Plans are used to locate areas of natural vegetation. Preliminary field surveys are then carried out to record indigenous species present, and complete, as far as is possible, inventories of these species in Singapore.

A number of naturally vegetated areas, often small plots as tiny as 4 ha, represent the diverse range of ecosystems found in Singapore, e.g. mangroves, secondary forests, marshland, sandy beaches, and rocky coastline. Green spaces called park connectors, have been established to join naturally vegetated areas, that would otherwise be isolated, to counter the effects of increased and prolonged habitat fragmentation.

b) Survey and monitoring of Nature Areas

Singapore has a heritage of natural history records, therefore, the opportunity exists to chart changes in the status of species and habitats over time. As the areas of natural vegetation are small and fragmented, it is expected that significant changes in species abundance and distribution will occur over time, and the value of surveying and monitoring these areas cannot be over-emphasised. A co-ordinated and planned survey provides many advantages. Surveys are undertaken for the different taxonomic groups and habitats over similar time periods; methodologies are comparable, and the accuracy of the data collected can be checked. Surveys have included a wide range of taxonomic groups, including: vascular plants, mammals, birds, reptiles, amphibia, fish, freshwater prawns and crabs, butterflies, stick and leaf insects, dragonflies, and water beetles.

Since the Nature Reserves occupy the largest portion of the Nature Areas, and are expected to harbour the richest biodiversity, they were selected for comprehensive survey first. Between 1992 to 1997, a biological survey was carried out in the Nature Reserves where a total of 1,634 species of vascular plants were recorded in the nineteenth century and, 443 of which have not been seen in the last ten years, constituting a loss of 27% (Chan and Corlett, 1997). Short-term recruitment of trees has been studied in a 2 ha permanent plot in the Bukit Timah Nature Reserve. In contrast to the recorded decline of bird and mammal populations, the results for the short-term population dynamics of most tree species in the 2 ha plot indicate that they are not in rapid decline (Ercelawn *et al.*, 1998).

c) The design and production of databases to manage data on Singapore's biodiversity

To manage biological data it is essential to have a cost-effective and practical database system for long-term handling and analysis of biological records (Pearce and Bytebier, see Chapter 4). These include herbarium records, inventories of flora and fauna, contact lists of scientific experts, published species literature on biodiversity, locations of ecosystems and distribution on a Geographical Information System (GIS), photographic records of flora, fauna, ecosystems, and land use data. One of the major problems of databases is the reliability of taxonomic information (see Robertson *et al.*, Chapter 14). NParks will be launching its National Biodiversity Reference Centre; some of the databases, namely flora, fauna, and list of scientific experts, have been set up while the others are in preparation.

Networking and the establishment of regional databases between countries with similar species will strengthen biodiversity conservation efforts. Two such initiatives include: the Association of South East Asian Nations (ASEAN) Regional Centre for Biodiversity Conservation and the South East Asian loop of BioNET-International, a technical co-operation network for sustainable development through capacity building in taxonomy (ASEANET).

The plant taxonomic inventory for Singapore is well known, in addition to a number of recent studies including The Singapore Red Data Book that have assessed the conservation status of the flora (Ng and Wee, 1994; Tan, 1995). The Herbarium and the Nature Conservation Branch of NParks have initiated a joint project to reassess the conservation status of threatened plants in Singapore and to formulate a strategy for their management. To achieve these objectives, the data on herbarium specimens in the SBG Herbarium and the herbarium of the National University of Singapore for all listed threatened species are being entered onto the Botanical Research and Herbarium Management System (BRAHMS) database in the SBG Herbarium. With these data, we are able to reassess the conservation status of the species on a quantitative basis (i.e. by utilising data such as number of localities where the plant species were found, dates of last collection, number of specimens, names of collectors, etc.) and are able to sort by locality to identify areas of high diversity (Pearce and Bytebier, see Chapter 4). This work is followed up by species surveys to assess the viability of the populations of these species (by recording the population size and structure, whether there are mature individuals and regeneration with seedling establishment taking place, girth sizes of large individuals, etc.). A sample of the data sheet used for the species survey is included (Figure 10.2). With these quantitative data, management strategies can be drawn up for the management of the particular species (by implementing active habitat management to ensure reproducing populations) and for the habitat (including the maintenance of accurate up to date information on which areas contain threatened species).

d) Species management

When threatened plants are located and only a few specimens are found in the wild, cuttings, seeds or seedlings are collected for *ex situ* conservation. The material is propagated in nurseries and then reintroduced back to appropriate and secure habitats or populations established in the SBG or other NParks administered areas. In some cases, species may be restricted to a few isolated populations in sites that are not threatened in the foreseeable future. In these cases, efforts are made to ensure that the optimum growth conditions for these species are maintained. For instance, there are only a few known locations for the fern *Dipteris conjugata* Reinw. (Dipteridaceae) and NParks staff have removed the overgrown secondary vegetation that was impeding its growth (Figure 10.3).

The tissue culture laboratory of SBG is used to propagate plants that are difficult to propagate by conventional horticultural techniques, such as cuttings, or for plants with reproductive problems. Tissue culture can be a time-consuming and relatively expensive process, so it is used when no other propagation method works. For instance, the laboratory is applying tissue culture techniques to the propagation of the mangrove *Kandelia candel* (L.) Druce (Rhizophoraceae); there is only one known tree in Singapore and the flowers abort before the fruit is formed. Tissue culture is also being attempted for *D. conjugata* as it is found only in very few sites and tissue culture has proved an effective method for propagating ferns.

Species Population & Propagation Form

Species:

Family:

Locality:

GPS Location:

Area size:

Plant present?   NO/YES

Population Size:

No. of seedlings:

10	20	30	40	50	60	70	80	90	100
10	20	30	40	50	60	70	80	90	100
10	20	30	40	50	60	70	80	90	100

No. of saplings/juveniles:

10	20	30	40	50	60	70	80	90	100
10	20	30	40	50	60	70	80	90	100
10	20	30	40	50	60	70	80	90	100

No. of reproductive individuals:

10	20	30	40	50	60	70	80	90	100
10	20	30	40	50	60	70	80	90	100
10	20	30	40	50	60	70	80	90	100

dbh of largest individual (cm)

Qty tagged

Taken:

Herbarium specimen? YES/NO		no.
Photo? YES/NO		no.
Materials for propagation? YES/NO	Fruit/Seed YES/NO	no.
	Cutting YES/NO	no.
	Whole Plant YES/NO	no.
	Tissue for micropropagation YES/NO	no.

State of habitat:

pristine	
disturbed past, no threat at present	
area likely to be disturbed	
currently being disturbed	

Reason:

Data recorded by: Date:

Figure 10.2   A sample survey form for recording quantitative data on threatened species.

Figure 10.3 A staff member of the National Parks Board clearing secondary growth that shaded the vulnerable fern, *Dipteris conjugata* Reinw. (Dipteridaceae). Photo: L. Chan.

As a result of Singapore's small size and the extent of habitat destruction, it is often considered appropriate to salvage specimens of trees and shrubs that would not be considered valuable elsewhere. Specimens of trees are only translocated when they represent a species that is threatened with immediate habitat destruction. It is a last resort measure since it is usually a very costly exercise. Two case studies of translocation are reported:

1. **The salvage of *Leea angulata* Korth. ex Miq. (Leeaceae).** This species is described as extraordinary by Corner (1988), as it 'develops first as a thorny scrambler, like a wild rose or bramble, and subsequently becomes an ungainly tree'. NParks carried out this salvage operation because, at the time, this species was only known from one site (all known sites of the three indigenous species of *Leea* were being monitored). Although a bulldozer uprooted the tree, it is now growing healthily in SBG. Subsequent to this, a number of potential new sites for this species have been located;

2. **The translocation of** *Dracaena maingayi* **Hook.f. (Agavaceae).**
   A specimen of *D. maingayi*, which is considered as 'vulnerable' in Singapore (Tan, 1995), was discovered in an area earmarked for development. The rarity of such a large specimen warranted that the 17 m tall tree, estimated to be about 100 years old, should be translocated to the SBG where it now grows.

Another example of intervention (but not leading to translocation) is that of the mangrove, *Dolichandrone spathacea* (L.f.) Schumann (Bignoniaceae). Cuttings, seedlings and seeds were collected since the sites where the trees are found are not threatened. The material was obtained from a few sites and planted in the SBG so that a wider gene pool of this species could be conserved *ex situ* where they are now thriving.

The salvage of plants destined for destruction by development, whilst a very drastic conservation measure, allows the retention of specimens of aesthetic value and the potential utilisation of indigenous genotypes for reintroduction and restoration projects.

e) Habitat restoration

Turner *et al.* (1996) recorded 448 historical plant records from old accounts and herbarium collections, and of those 220 recorded taxa are still present in forest fragments, a persistence of about 50% over a century. However, a large proportion of the surviving taxa were represented by very small populations. For instance, half of the trees recorded were represented by only one or two individuals. The survey also recorded the highest levels of loss amongst understorey plants, climbers and epiphytes. These fragments have suffered from the impact of competitive exotic weeds and unusually large populations of seed predators, such as squirrels. More recent surveys carried out by the staff of the SBG Herbarium and NCB have led to the re-discovery of species that were recorded as extinct by Turner (1996).

Based on the above findings, it was decided that these important fragments of forest required intensive management. A restoration ecology project is currently being carried out on the 4 ha SBG Rainforest, a small patch of residual lowland dipterocarp forest within the management of SBG. The project has taken the following steps:

1. A checklist of species recorded in the plot was established, derived from published and unpublished data and herbarium records that date back to 1870.

2. An on-going inventory of the species found in the plot was carried out to identify species loss and changes in forest structure.

3. Exotic species are manually weeded out periodically to encourage recovery of natural vegetation. Volunteers are involved in this process.

4. Indigenous species that had been recorded previously in the plot are reintroduced from individuals of similar provenance.

5. Since the herbaceous plants are presently the least represented and as their presence would increase the humidity of the undergrowth, they are planted first.

Lessons Learnt

Cropper (1993) wrote in the preface of his book, 'No matter where you are in Australia, you are probably close to a population of some rare plant'. This is also applicable to Singapore, as it is possible that threatened, rare, previously unrecorded and interesting species will be found in small, urbanised sites, even in secondary vegetation in the centre of the city. For example, a total of 22 plant species not seen for over 50 years and considered extinct in Singapore were rediscovered between 1996 and 1997.

There is much that can be learnt from past publications and Singapore has an invaluable heritage of plant collections and field botanists. Initial research, prior to embarking on a project mapping threatened species, should make full use of such publications and field notes. For example, several specimens of *L. angulata* were found near a harbour site after reading *Wayside Trees of Malaya* (Corner, 1988). Under a section on *L. angulata*, Corner describes 'several compact coppiced trees near Keppel Harbour, Singapore'. Surveys have also proved invaluable in the rediscovery of species listed on the 'Species Lost to Singapore' list. For example, a half-day survey of a 4 ha plot carried by four experienced field staff revealed that eight out of the ten species documented were species thought to be extinct in Singapore.

Future Steps

With over twenty years of championing the Garden City Campaign, Singapore has accumulated important data and developed expertise in the field of urban conservation. Much of the above phenological and demographic data and surveys will be continued. Further permanent plots will be established so that long-term data collection can support the long-term management of species and fragmented habitats. The challenge for nature conservation in Singapore lies in drawing on a wide range of expertise in the sustainable management of the fragmented areas of natural vegetation. These fragments will act as refuges for the rich biodiversity of this highly urbanised small island city state.

Acknowledgements

The author would like to thank Dr Tan Wee Kiat, Dr Leong Chee Chiew, Dr Chin See Chung and Dr Ruth Kiew for commenting on the text.

References

Burkill, I.H. (1918). The establishment of the botanic gardens, Singapore. *Gardens' Bulletin, Straits Settlement* **2**: 55–72.

Chan, L. and Corlett, R.T. (eds) (1997). Biodiversity in the nature reserves of Singapore. *Gardens' Bulletin* **49**: 147–425.

Corlett, R.T. (1992). The ecological transformation of Singapore, 1819–1900. *Journal of Biogeography* **19**: 411–420.

Corner, E.J.H. (1988). *Wayside Trees of Malaya.* (3rd edition) Volumes 1 and 2. Malayan Nature Society, Kuala Lumpur, Malaysia.

Cropper, S.C. (1993). *Management of Endangered Plants.* CSIRO, Melbourne, Australia.

Ercelawn, A.C., LaFrankie, J.V., Lum, S.K.Y. and Lee, S.K. (1998). Short-term recruitment of trees in a forest fragment in Singapore. *Tropics* **8**: 105–115.

Ministry of the Environment (1993). *The Singapore Green Plan – Action Programmes.* Ministry of the Environment, Singapore.

Ng, P.K.L. and Wee, Y.C. (eds) (1994). *The Singapore Red Data Book. Threatened Plants and Animals of Singapore.* Nature Society, Singapore.

Tan, H.T.W. (ed.) (1995). *A Guide to the Threatened Plants of Singapore.* Singapore Science Centre, Singapore.

Turner, I.M. (1994). The taxonomy and ecology of vascular plant flora of Singapore: a statistical analysis. *Botanical Journal of the Linnean Society* **114**: 215–227.

Turner, I.M., Chua, K.S., Ong, J.S.Y., Soong, B.C. and Tan, H.T.W. (1996). A century of plant species loss from an isolated fragment of lowland tropical rain forest. *Conservation Biology* **10**: 1229–1244.

Integrating Forestry and Biodiversity Conservation in Tropical Forests in Trinidad

Caribbean Sea

TOBAGO

TRINIDAD

VENEZUELA

Colin Clubbe
Royal Botanic Gardens, Kew, UK

Seuram Jhilmit
Forest Resource Inventory and Management Section (FRIM), Ministry of Agriculture, Land and Marine Resources, Trinidad and Tobago

Introduction

The twin-island state of Trinidad and Tobago forms the most southerly pair of islands of the Caribbean chain. Trinidad is only 13 km from the coast of Venezuela (Figure 11.1). Tobago lies approximately 32 km north-east of Trinidad. Located approximately 10° north of the equator, Trinidad has a land area of 4,828 square km² whilst Tobago is 300 km². The population is now approximately 1.3 million compared to a registered population of 477,800 in 1942. Current annual population increase is estimated at 1.7%. Trinidad and Tobago gained independence from the UK in 1962, and became a republic within the Commonwealth in 1976.

Trinidad and Tobago lie on the South American Continental Shelf. Geologically, Trinidad is an extension of South America and separation from the continental mainland occurred in recent geological times, about 5–6,000 years ago. It is likely that the northern part of Trinidad was linked to Tobago during the Pleistocene, but separation was earlier, possibly 11,000 years ago. They experience a typical tropical climate with distinct wet and dry seasons. Average annual rainfall is strongly influenced by topography and in Trinidad ranges from 1,200 mm in the west and south-western sections to 3,500 mm in the north and north-eastern parts of the island. The mean annual temperatures at low elevations are 21°C at night and 29°C during the day, with seasonal variations of 2–3°C.

Plant communities range from deciduous and evergreen seasonal forests (Figure 11.2) at lower altitudes to montane (cloud) forest and elfin woodland at higher elevations (maximum elevation of 915 m). Edaphic communities comprise a rich mosaic of habitats from swamp forest to mangrove woodland, and marsh forest to savannah (Beard, 1946). High species richness has been recorded across a range of taxa: 2,160 species of plants with 110 thought to be endemic; 420 species of birds; 100 species of mammals (mostly bats); 85 reptiles; 25 amphibians and 630 species of butterflies.

Trinidad and Tobago has a rich forestry history dating back to 1765 when a 6,000 acre (2,428 ha) block of forest on Tobago's main ridge was set aside for watershed protection and thus became the western hemisphere's first Forest Reserve. The responsibility for managing and protecting Trinidad and Tobago's forests lies with the Forestry Division of the Ministry of Agriculture, Land and Marine Resources (MALMR), which in common with many developing countries, is under-resourced. Although 44.8% of the land area of Trinidad and Tobago is classified as forest, much of this has been degraded in recent years and increasing concern has been expressed about the quality and species composition of much of the natural forest because of over-exploitation, shifting cultivation and fire (Chalmers, 1992).

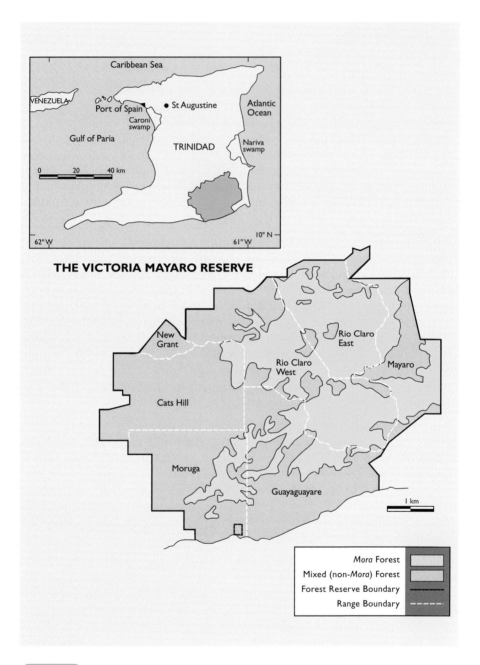

Figure 11.1 Map of Trinidad in relation to the coast of South America (Venezuela) showing the location of the Victoria Mayaro Reserve and details of the range boundaries and forest types found within the Reserve.

Figure 11.2   Interior of moist, lowland evergreen seasonal forest in the Victoria Mayaro Reserve: an area of rich biodiversity.

Trinidad and Tobago signed the Convention on Biological Diversity (CBD) at the Earth Summit in Rio in 1992 and ratified it on 1 August 1996. The Environmental Management Act, passed by Parliament in 1995, established the Environmental Management Authority (EMA) as the body with the responsibility for co-ordinating and co-operating with other agencies and governmental ministries to meet the objectives of the Act. The EMA is the national focal point for the CBD. An interim first national report to the CBD has been lodged with the CBD Secretariat, as required under Article 26 (Glowka *et al.*, 1994; Appendix 1). The CBD makes all national reports available via its website at www.biodiv.org/world/reports.asp/. The EMA in collaboration with the MALMR are preparing a proposal for funding from the Global Environment Facility (GEF) to formulate its National Biodiversity Strategy and Action Plan, a priority activity under Article 6 (General Measures for Conservation and Sustainable Use) of the CBD .

Loss of Biodiversity

The world's tropical forests are being subjected to increasing pressure from expanding populations leading to over-exploitation and often to complete destruction. Rates of deforestation in the tropics remain high despite global concern (FAO, 1999; Laurance, 1999; Myers, 1994). Estimates of the annual rate of forest loss in developing countries are 13.7 million ha between 1990–1995, compared with 15.5 million ha annually between 1980–1990. Whether this represents a significant downward trend in deforestation will only become apparent once the comparable data sets from the Global Forest Resources Assessment 2000 are fully analysed (FAO, 1999). Tropical humid forests originally covered 14–18 million km^2, and about half of this original area remains. Much of this loss is recent, and about 1 million km^2 is cleared every 5–10 years. Burning and selective logging severely damages several times the area cleared (Pimm and Raven, 2000). Laurance (1999) identified four factors that are emerging as key drivers of destruction in the tropics: human population pressure, weak government institutions and poor policies, increasing trade liberalisation and industrial logging.

Tropical deforestation has severe biological and economic consequences. The biological richness of tropical forests is well known. More that half of the world's species are estimated to occur in tropical forests, which occupy less that 7% of the earth's land surface. Much conservation effort has been focused on trying to determine centres of diversity. On a global scale, the World Conservation Union (IUCN)/Worldwide Fund for Nature (WWF) Centres of Plant Diversity project identified 234 areas of particularly high diversity, concentrated in north-western South America, Central America, tropical Africa, the eastern Mediterranean and South East Asia/Malesia (WWF and IUCN, 1994–1997). Myers (1988) identified ten tropical forest hotspots, which harboured 34,400 endemic species of plants, some 13% of the world's flora, in an area of 292,000 km^2 or 0.2% of the earth's land surface. On a regional scale, the Caribbean Islands as a biogeographic unit have been ranked third in a listing of the world's 25 most important biodiversity hotspots, based primarily on plant endemism and degree of threat through habitat loss (Myers *et al.*, 2000). Containing 2.3% of the world's endemic plant species the Caribbean Islands biogeographic unit, which includes Trinidad and Tobago, has lost nearly 90% of its primary vegetation (Mittermeier *et al.*, 1998; Myers *et al.*, 2000). The economic consequences of tropical forest destruction are equally severe. Forests provide important goods, both timber and non-timber forest products and an important range of ecological services. The economic value of services such as climate regulation, carbon sequestration, retention of water and flood control, and the conservation of soil, whilst often undervalued, is considerable. Costanza *et al.* (1997) estimated the total world value of ecosystem services to be US$33 trillion (see Maunder and Clubbe, Chapter 3; Box 3.1).

Economic value of Tropical Forests

The challenge posed by Article 6 of the CBD (General Measures for Conservation and Sustainable Use) is to integrate the conservation of biological diversity and sustainable use of its components into relevant sectoral and cross-sectoral plans, programmes and policies (Article 6b) and to develop national biodiversity strategies, plans or programmes which reflect this (Article 6a). The sustainable management of tropical forests must be a cornerstone of such policies. Sustainable forest management entails balancing the economic, environmental and social functions and values of forests for the benefit of present and future generations – a complex and challenging task in the face of the earth's rapidly expanding human population and the increasing demands for forest products and services (FAO, 1999).

Sustainability is now viewed as one of the guiding principles for development. It is one of the core objectives of the CBD and Article 10 specifically addresses the sustainable use of components of biological diversity. The sustainability of any system of natural forest management must ensure that forest integrity is maintained into the future. Sustainable use is an ecosystem-, rather than a species-oriented, definition focusing as it does on ensuring that the use of components of biological diversity does not lead to the long-term decline of that biological diversity (Glowka *et al.*, 1994). Consequently, we need to separate considerations of sustainable yield of a timber species, which does not necessarily consider its interrelationship with other species, from ideas of sustainable use of timber which is compatible with the maintenance of the long-term viability of supporting and dependent species and ecosystems. When assessing the sustainability of the natural forest management systems in Trinidad we posed ourselves two research questions:

1. Are these management practices sustainable for timber production (sustainable yield)?
2. Are these management practices sustainable for biodiversity (sustainable use)?

Sustainable yield has traditionally been a forestry question whilst sustainable use lies at the core of forest conservation. Sustainable forestry must achieve both these goals and recent efforts have been directed towards the development of practical management methodologies and guides to achieve this (Higman *et al.*, 1999). This approach has been highlighted by the World Commission on Forests and Sustainable Development (WCFSD) who conclude that 'we must urgently choose a path that respects the ecological values of forest while recognising their role in social and economic development' (WCFSD, 1999).

Natural Forest Management in Trinidad and Tobago

Natural forest management has been defined as 'controlled and regular harvesting, combined with silvicultural and protective measures to sustain or increase the commercial value of subsequent stands, all relying on natural regeneration of native species' (Schmidt, 1987). Three types of natural forest management have been practised in Trinidad and Tobago for varying numbers of years. These are the Shelterwood System, a complicated improvement felling system practised at Arena Forest and Mount Harris, and not included in this analysis, and two types of Selection Systems: Open Range Management and the Periodic Block System, which are the subject of this analysis (Clubbe and Jhilmit, 1992).

Open Range Management (ORM) has been practised since the beginning of the twentieth century and has been applied widely to most of Trinidad's forests. Timber species are allocated to one of four classes based on their commercial value: Class 1, the most valuable, to Class 4 being of little commercial value. The only real control exercised in this system is a girth limit on certain species in the more valuable classes. Trees cannot be felled if they are below the girth limit for the species. Girth limit considerations fall into three categories:

- a girth limit and the presence of a suitable replacement tree (of a given size) of the same species (within a specified distance from the one being felled) for many of the more valuable Class 1 and Class 2 species e.g. *Cedrela odorata* L. (Meliaceae) (Class 1) (girth limit 244 cm) and *Mora excelsa* Benth. (Leguminosae) (Class 2) (girth limit 183 cm);

- a girth limit, but without the requirement for a replacement tree e.g. *Carapa guianensis* Aubl. (Meliaceae) (Class 2) and *Sterculia caribaea* R. Br. (Sterculiaceae) (Class 3) (girth limit 152.5 cm for both species);

- no girth limit for some Class 3 and most Class 4 species e.g. *Trichilia smithii* C. DC. (Meliaceae) (Class 3) and *Spondias mombin* L. (Anacardiaceae) (Class 4).

The forest is effectively open for felling on a permanent basis and the system allows woodworkers to remove trees as they reach their girth limit.

The Periodic Block System (PBS), introduced in 1948, is a felling system based on area control. Blocks of forest 150–300 ha are delineated and harvested over a two-year period (Figure 11.3). After harvesting, the blocks are closed for a period of 25–30 years. Consequently there are blocks of different ages and at different stages of recovery throughout the range. Individual trees for felling within the PBS are selected by a team of skilled silvicultural markers who

Figure 11.3   Felling of *Mora excelsa* Benth. (Leguminosae) within a Periodic Block System compartment in the Victoria Mayaro Reserve.

systematically walk through a block and physically mark each tree to be felled with a painted number and a forestry stamp. Strict girth limits are not used, although size is taken into account as one of the criteria applied. A series of ecological criteria are used. These take factors into account such as crown closure (degree of canopy openness), tree maturity, frequency of occurrence and presence of a replacement individual, its value as a seed tree or its value for wildlife (e.g. keystone species, such as *S. mombin*, a Class 4 species) and stream bank protection (Figure 11.4). On average 7–10 trees are removed per ha (Figure 11.5).

The Victoria Mayaro Forest Reserve (VM Reserve) is located in the south-east corner of Trinidad and comprises 51,915 ha of mainly evergreen seasonal forest (Figure 11.1). Beard (1946) classified the forest as a *Carapa-Eschweilera* association. Two forms (faciations) of this association are found within the VM Reserve, one where *M. excelsa* (mora) is absent, the *Pentaclethra macroloba-Sabal* faciation, and one where mora is both dominant and gregarious, the *Mora* faciation. *Mora* forests are characterised by the extreme dominance of *M. excelsa* which can account for 80–90% of the canopy trees and form an almost continuous canopy at 36–43 m, although individual trees may reach 58 m. The topography of the VM Reserve is undulating, 15–60 m above sea level.

Figure 11.4   Forestry Division officers in the Periodic Block System forest of the Victoria Mayaro Reserve discussing the impact of timber extraction during the wet season.

Figure 11.5   *Mora* logs awaiting transportation to a local sawmill.

The VM Reserve has had an interesting history of land use that lends itself well to study as a model system for tropical forest dynamics and sustainable utilisation. Large areas of primary forest remain unexploited whilst various areas of the reserve are harvested for timber. Most of the production forest has been managed under ORM whilst approximately 10,000 ha have been managed under the PBS since 1960. Some areas have been clear-felled and replaced by teak (*Tectona grandis* L.f. (Verbenaceae)). The reserve, therefore, represents a mosaic of natural forest, production forest in various stages of recovery and teak plantations. The following analyses are based on work conducted within the *Mora* forest only.

Methodology

An inventory of the indigenous forests of Trinidad and Tobago (approximately 220,900 ha) was undertaken between 1978–1980. Forest typing was done by photo-interpretation of aerial photographs using Beard's (1946) forest classification, and a synthesis map was produced showing their distribution. Field data were collected to estimate the total stock and quality, species distribution and volume for future forest management and optimal utilisation of forest resources. Tree data (species, girth, height and form) were collected from 20 m by 50 m contiguous plots arranged along belt transects, approximately 1 km long and 0.5 km apart (FRIM, 1980).

After the inventory, the Forest Resource Inventory and Management (FRIM) section of the Forestry Division started a long-term monitoring programme of Trinidad's forests. Sets of 1 ha permanent sample plots (PSP) were set up throughout a range of forest types across the island in both primary (unlogged) forest and in areas logged under the PBS. Within these plots all trees with a diameter at breast height (dbh at 1.3 m) of 20 cm or greater were identified, marked, their dbh and height measured and any malformations noted. For the 25 most important tree species, including *M. excelsa* and *C. guianensis*, the lower dbh limit was 10 cm. Temporary subplots (5 m x 5 m) were set up within the 1 ha PSP to monitor sapling and seedling regeneration of the 25 most important timber tree species (FRIM, 1978). This on-going activity by FRIM is a key contribution to the needs of Article 7 of the CBD (Identification and Monitoring). No monitoring was undertaken within ORM forests by FRIM.

For this study, we used a combination of existing FRIM data and data we specifically collected in the field in 1993, in an attempt to answer our two research questions: are the management practices sustainable for both timber production (sustainable yield) and for biodiversity (sustainable use)? To investigate sustainable yield we undertook an analysis of those data collected as part of the 1978–80 inventory, as well as examining the timber extraction records held by the Forestry Division. The more important conservation question is whether the extraction of *Mora* and other timber species represents the sustainable use of forest resources and is not leading to a decline in overall biological diversity. In order to investigate this, we undertook an inventory of all plant species within these forest types to assess the status of the broader plant community. The methodology adopted used standard plot techniques (see Alder and Synnott, 1992; Bullock, 1996; Kent and Coker, 1994).

The established forestry PSPs were used where possible for our investigations of the total plant community. Existing 1 ha PSPs were used in the primary forest of the Guayaguayare area of the VM Reserve (PSP 79, 80, 87, 88), in PBS Block 1 within the Rio Claro West range, and PBS block 7 located in the Rio Claro East range of the VM Reserve. Since no PSPs had been established in ORM forests, a temporary sample plot (TSP) was established in the Moruga range using the same methodology to ensure comparability (Figure 11.1).

Each plot was divided into 10 m × 10 m temporary subplots, using small wooden stakes and tape. Each 100 square metre (m^2) subplot was allocated a number between 1 and 100, and those subplots for sampling determined from random number tables. At the time of sampling, subplots were further subdivided using small canes to facilitate systematic field measurements and prevent double counting. Within each subplot a total plant inventory was undertaken. Plants were identified and counted in the field. Abundance was measured as the number of individual stems within each 100 m^2 subplot. Epiphytes were counted from the ground with the aid of binoculars where necessary. Individual bromeliad plants were treated as one individual for abundance, although they may be clonal. The same rule was applied to counting orchids, and to the few grasses and sedges which were found on the forest floor. Hemi-epiphytes and lianas were treated as separate individuals if they retained no connection with each other. Again these could be branches of an original single plant.

The determination of an acceptable number of samples was done on the basis of species accumulation curves (Kent and Coker, 1994). Species accumulation curves levelled off to an acceptable extent after 8 subplots, so that eight subplots were assessed for each forest type (Figure 11.6). Each subplot took one field day to complete by a team of three people. Much of the field identification was based on non-flowering characters, since little is in flower in a tropical forest at any one time. Sampling was undertaken during the main rainy season (August–January) to try to maximise the likelihood of reproductive material being present as an aid to identification. Key plants were tagged and re-visited over the following two years to try to follow development and so aid identification. Specimens were collected and taken to the National Herbarium of Trinidad and Tobago for further identification and verification. Some specimens remained unknown at the species level with identification possible only to family or genus level (Flora of Trinidad and Tobago, 1928–1992).

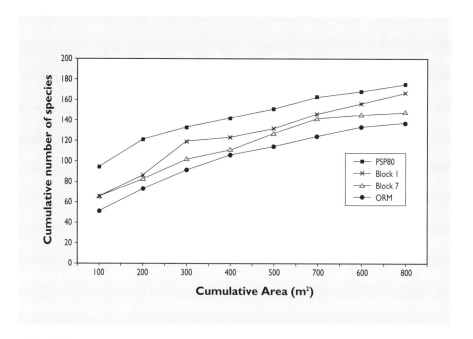

Figure 11.6   Species accumulation curves for three forest types where plot PSP 80 was located in primary forest, Blocks 1 and 7 in a Periodic Block System (PBS) and a third plot was located in an Open Range Management (ORM) system.

Results and Analysis

Table 11.1 summarises the known logging history and other key forestry data for the four forest management conditions that form the core of this analysis carried out in 1993. These data were extracted from FRIM's inventory records. Block 7 was logged in 1967–68 and Block 1 in 1986–87. Both of these blocks were closed after logging. Logging was continuous in the ORM forest in Moruga until a temporary ban was introduced in 1989. The results shown thus represent potential recovery times of 25 and six years respectively for the two blocks in the PBS forest and four years since the intensive logging was stopped in the ORM forest.

Basal area estimates are not strictly comparable since different sampling methods were used to obtain these estimates. However, they do provide information that allows us to make observations of recovery trends. Estimates of the total tree basal area for the primary forest range from 46–76 m² ha⁻¹, with PSP 80 at 46.4 m² ha⁻¹ being at the lower end of this range. Both PBS blocks show an increase in basal area after closure and have almost recovered

Table 11.1 **Comparisons of logging history, estimates of total tree basal area (trees with a diameter at breast height >10 cm), and of proportions of *Mora excelsa* Benth. (Leguminosae) in a range of forest compartments in the Victoria Mayaro Forest Reserve**

Parameter	Forest Type Sampled			
	Primary Forest (Guayaguayare >891 ha)	**PBS – Block 7 (Rio Claro East, 162 ha)**	**PBS – Block 1 (Rio Claro West, 152 ha)**	**ORM forest (Moruga >836 ha)**
Logging history	Never logged	Logged in 1967–1968	Logged in 1986–1987	Logged continuously until 1989
Estimated post-harvest basal area (m² ha⁻¹)	Not applicable	25.2[b]	24.1[b]	Not applicable
% *Mora* remaining after logging	Not applicable	62%[b]	75%[b]	Not applicable
Time since logging stopped	Not applicable	25 years	6 years	4 years
Estimated total tree basal area (m² ha⁻¹) at time of 1993 survey	46–76[a]	40.8[b]	42.7[b]	28[c]
% *Mora* at time of 1993 survey	77–90%[a]	80%[b]	82%[b]	60%[c]

[a] Estimates in primary forest based on 4 PSPs in Guayaguayare (PSP 79, 80, 87, 88).
[b] Estimates in Periodic Block Management (PBS) forest based on 5% strip enumeration.
[c] Estimates in Open Range Management (ORM) forest based on 2% strip enumeration.

to a point comparable with PSP 80 in terms of total basal area and proportion of *M. excelsa*. The recovery rates in Block 1 look remarkable and warrant further investigation. These results point towards the PBS being sustainable for both general timber production and for the major species, *M. excelsa*. The estimated total basal area of trees in the ORM forest at Moruga and the proportion of *M. excelsa* stems are comparable with the post-harvest values in the blocks. Extraction has taken place in the ORM forest for more than 30 years, as opposed to two years for the PBS blocks. This has left the forest looking in a much poorer state than a typical block just after logging has been completed. It remains to be seen if the forest can recover from this point. PBS blocks seem to be recovering well from a forestry standpoint (basal area increasing and a re-establishment of *Mora* as the forest dominant species) whilst providing a sustainable yield of timber. The ORM forest at Moruga shows signs of serious deterioration.

Table 11.2 shows some of the summary data for each of the four forest types within the VM Reserve based on the plant inventory work undertaken. Means and standard errors were calculated from the eight sub-plots measured. The significance of any differences between forest management types was analysed using one-way analysis of variance (ANOVA) using Microsoft Excel. The results from the total plant inventory indicate significantly higher species richness in the more recently logged plots. The mean number of species per plot increases significantly with greater disturbance ($F_{3,28}$ = 5.70, p=0.0045). The mean density of all species follows this trend and shows a highly significant increase in the disturbed forests ($F_{3,28}$ = 13.70, p=0.0001). ORM and PBS Block 1 both show higher species richness and plant density than either PBS Block 7 or the primary forest. However, many of the 'extra' species that were found within the more recently logged forest types were species associated with disturbance, rather than true forest understorey species. The increase in density in the more recently logged plots was accounted for by 'gap species', such as *Heliconia spatho-circinada* Aristeg. (Heliconiaceae) as well as species which seem to have responded to the greater light availability. In particular, some of the aroid climbers including *Monstera obliqua* Miq. (Araceae) and *Anthurium pentaphyllum* (Aubl.) G. Don (Araceae) responded by producing large numbers of seedlings.

Since we were concerned with maintaining ecosystem integrity, it was important to look at the composition and distribution of plants in relation to the 'starting point' as illustrated by the unlogged, primary forest. Ideally, we needed to be able to compare species composition between sites at a community level, rather than merely list numbers of species. Similarity relations at a community level can be expressed mathematically as an index of similarity or as a community coefficient. The literature on community similarity is wide and a large number of indices have been developed (see Kent and Coker, 1994; Mueller-Dombois and Ellenberg, 1974). Most indices are based on the presence-absence relationship between the number of species common to two areas, or communities, and the total number of species recorded (e.g. Sørensen's Index). However, the similarity of two communities is not only a function of the number of common or unique species, but also of the relative number of individuals of these species. Indices of similarity can be modified to include these abundance figures. In this way, indices of community similarity can be applied to both qualitative and quantitative data. A widely used index of similarity, which takes abundance into account, is a modification of Sørensen's Index applied by Motyka (IS_{mo}):

Table 11.2 **Summary data derived from the total plant inventory of the four forest types**

Parameter	Forest Type Sampled			
	Primary Forest – PSP80 (Guayaguayare)	**PBS Forest: Block 7 – (Rio Claro East)**	**PBS Forest: Block 1 – (Rio Claro West)**	**ORM Forest: (Moruga East Plot)**
Logging history	No logging	Logged in 1967–1968	Logged in 1986–1987	Logged continuously until 1989
Recovery time (to 1993 survey)	Not applicable	25 years	6 years	4 years
Number of 100m² sub-plots sampled	8	8	8	8
Total Species richness (eight sub-plots)	137	148	175	166
Mean no. species per 100m² sub-plot ±SE	61.0 ± 3.5	69.4 ± 3.0	82.0 ± 3.1	75.3 ± 5.0
Mean density per 100m² sub-plot ±SE	595.3 ± 76.5	939.8 ± 53.3	1,267.6 ± 91.5	1,054.8 ± 77.5
Total number of gap species	14	27	44	55
Mean no. gap species per 100m² sub-plot ±SE	6.6 ± 0.7	11.3 ± 1.2	16.5 ± 2.6	16.6 ± 3.1
Mean density of gap individuals per 100m² sub-plot ±SE	63.4 ± 16.9	122.4 ± 19.8	187.6 ± 53.5	156.8 ± 32.3
Total number of epiphyte species	16	15	16	9
Mean no. epiphyte species per 100m² sub-plot ±SE	6.3 ± 0.7	5.8 ± 0.9	7.3 ± 0.7	2.9 ± 0.4
Mean density of epiphytes per 100m² sub-plot ±SE	69.6 ±19.9	99.0 ±18.1	134.5 ±18.2	7.3 ±1.8

Where SE = Standard Error.

$$IS_{mo} = (2Mw/(Ma+Mb)) \times 100$$

where:

Mw = the sum of the smaller abundances (expressed as number of individuals in this case) of the species common to the two communities (not the sum of both values since the smaller value is contained in both communities)

Ma = the sum of the abundances of all species in community A.

Mb = the sum of the abundances of all species in community B.

Motyka's Index of community similarity was calculated for a variety of species combinations derived from the total plant inventory (Tables 11.3–11.5). Initially (IS_{mo}) was calculated for all species recorded from the four forest types. As can be seen from Table 11.3 there is not much difference in the values across the forest types and including all species together in the analysis is not a very good discriminator, although both PSP 80 and Block 7 compared to Moruga are lower (>10%).

Table 11.3 **Motyka's Index of Community Similarity for all species (N=275 species)**

Forest Type	Primary: PSP 80	PBS: Block 7	PBS: Block 1	ORM: Moruga
Primary: PSP 80	–			
PBS: Block 7	49%	–		
PBS: Block 1	51%	56%	–	
ORM: Moruga	39%	36%	56%	–

Table 11.4 **Motyka's Index of Community Similarity for gap species (N=80 species)**

Forest Type	Primary: PSP 80	PBS: Block 7	PBS: Block 1	ORM: Moruga
Primary: PSP 80	–			
PBS: Block 7	37%	–		
PBS: Block 1	41%	47%	–	
ORM: Moruga	26%	46%	51%	–

Table 11.5 **Motyka's Index of Community Similarity for epiphytes (N=25 species)**

Forest Type	Primary: PSP 80	PBS: Block 7	PBS: Block 1	ORM: Moruga
Primary: PSP 80	–			
PBS: Block 7	79%	–		
PBS: Block 1	69%	84%	–	
ORM: Moruga	11%	8%	8%	–

However, when particular ecological guilds were examined, some important differences became evident. The first group looked at were 'gap species' – those species that invade gaps in the canopy and can be indicators of disturbance. In the Trinidad forests, gap species include many species of the families Heliconiaceae, Marataceae, Costaceae, Melastomataceae and Piperaceae that occupy riparian areas and tree-fall gaps in the undisturbed forest. These species rapidly colonise disturbance areas and are particularly common along extraction trails and forest edges in the logged forest. The mean number of gap species per plot increased significantly from 6.6 in the primary forest to 16.6 in the ORM forest ($F_{3,28}$ = 4.97, p=0.007) (Table 11.2). A large number of non-forest plants, including species of Gramineae, Cyperaceae, Leguminosae and Euphorbiaceae normally associated with disturbed habitats were found only in the Moruga ORM plots. However, the mean density of gap species was not significantly different across the forest types ($F_{3,28}$ = 2.47, p=0.08), indicating the patchy nature of their distribution.

The trend in similarity indices for these gap species suggests a higher similarity between the ORM and more recently logged PBS forest ie PBS 1, that may be indicative of a greater disturbance due to logging (Table 11.4). The lowest similarity is between the Moruga open range forest (the most disturbed forest) and the primary, unlogged forest (PSP 80), as might be expected. The most striking differences, however, were encountered when the epiphyte community was examined (Table 11.5). Epiphytes are an important ecological group within the forest ecosystem (Figure 11.7). In this forest they comprise 25 species from four families of vascular plants; Orchidaceae (13 species), Bromeliaceae (7), Cactaceae (1), Araceae (1) and three species of ferns. Their populations appear to have crashed in terms of numbers of individuals and species representation in the open range forests and this ecological group could provide a good indicator of forest disturbance and openness (Table 11.2). The similarity indices suggest that the epiphyte flora in PBS Block 7 had recovered to 79%, after 25 years, PBS Block 1 to 69%, after six years, whereas the ORM forest at Moruga had only a 11% similarity of epiphyte flora, compared with the primary forest (Table 11.5). This important difference in the nature of the epiphyte community was also reflected in the results of the ANOVA which showed significant differences in both the overall number ($F_{3,28}$ = 7.35, p=0.0009) and density ($F_{3,28}$ = 511.96, p=0.0006) of epiphyte species (Table 11.2).

Although further investigations need to be undertaken to identify the causes of these differences, the crash in epiphyte populations may reflect a degree of damage to the canopy that has destroyed the humid microclimate in which these epiphytes thrive. It may also reflect a breakdown in pollination and/or regeneration processes. The constant disturbance in the ORM forest may interfere with epiphyte seed dispersal (largely wind dispersed) in the canopy and germination and seedling establishment processes. On the strength of these data, epiphytes may prove to be a good indicator species of forest health within the VM Reserve and elsewhere, and should be monitored closely. Other ecological groups may provide useful indicators in other forest types.

Figure 11.7 *Guzmania lingulata* Mez (Bromeliaceae), one of the most widespread epiphytes within the Victoria Mayaro Reserve.

Future Steps

Inventory and monitoring are resource-intensive activities, both in terms of time and the expertise needed to undertake them effectively. The number and range of plant species recorded for the areas sampled within the Victoria-Mayaro Reserve may seem daunting for the non-specialist, although the size of the general flora of this area is smaller than many equivalent lowland forests of continental South America. Accurate field identification is vital and therefore the need for field identification skills is particularly important. Currently these skills lie within a small group of dedicated professionals who are severely under-resourced. There is a desperate need to further build capacity in the National Herbarium of Trinidad and Tobago as well as the Forestry Division. There is need for good documentation of biodiversity and the wider development of databases linked to analytical tools such as Geographic Information Systems (GIS). Only in this way can these data be made available in a form that can aid the planning process. There is a need to revitalise Trinidad's botanic garden and to enable it to take on a broader conservation role, particularly in respect of rare and threatened plant species (CBD, Article 9) and in public education and awareness (CBD, Article 13). All these activities require skilled personnel and sustained financing. Current levels of funding are very low. In the past, inventory work programmes have been halted because of lack of funds for such basic items as petrol for field transportation. It is important that issues like these are considered during the drafting of the National Biodiversity Strategy and Action Plan currently being prepared by the Environmental Management Authority.

Data presented here and elsewhere clearly show that the ORM system is not sustainable for biodiversity. There is also some doubt whether it is sustainable for timber production. It is clear that this form of management should not be continued. The Forestry Division has already closed the Moruga East Range to logging because of concerns over the quality of the forest. The data presented here support this decision. The challenge now is the rehabilitation and restoration of these forests. Restoration ecology is one of the newest and most challenging disciplines in ecology, particularly in tropical ecosystems (Brown and Lugo, 1994). This also begs the broader question of strategic land use planning, an issue of particular concern to Small Island Developing States where land area is small and ecological processes much more interdependent (Bass and Dala-Clayton, 1995).

Indications from these data are that the PBS of management is sustainable for some components of the plant community. The recovery indicated for these plant species, and the general appearance of the forests are very encouraging. Complementary studies in these forests have indicated that the PBS management system is sustainable for fruit-feeding butterflies and for a bromeliad-dwelling tree frog (Wood *et al.*, 1999). Whereas, different taxa may respond differently to particular management regimes, we have good reason to consider that the relatively low impact approach to forest utilisation of the Periodic Block System is sustainable for biodiversity.

Continued monitoring of the wider plant community is important to assess whether the management of Trinidad's forest is sustainable over the long term. It is also important to introduce monitoring of other taxa in order to understand more fully the effects of this management at the ecosystem level. These taxa could be chosen for their ecological role or for their conservation value. In this way the commitments to CBD Articles 7 (Identification and Monitoring), 8 (*In situ* Conservation) and 10 (Sustainable Use of the Components of Biological Diversity) can be achieved.

Acknowledgements

A field project of this nature involves many people and we would like to thank all those who gave logistical and practical support to us during this project. Our particular thanks go to Winston Johnston of the National Herbarium of Trinidad and Tobago for his enthusiasm for fieldwork, sharing his considerable knowledge of the flora and for verification of all the field identifications. Thanks to all of the FRIM team who worked tirelessly in the field with us and ensured that there was always petrol in the landrover to get us back, and for organising those wonderful fish broths in the forest. The first author is grateful for the many profitable and interesting discussions with C. Dennis Adams of the Natural History Museum in London, with Yasmin Comeau, curator of the National Herbarium of Trinidad and Tobago, and other colleagues at the University of the West Indies about the ecology and flora of Trinidad and Tobago; to Michael Gillman from the Open University for ecological discussions in the field and at home; and to Deborah Seddon for help in the field and being a wonderful companion.

References

Alder, D. and Synnott, T.J. (1992). *Permanent Sample Plot Techniques for Mixed Tropical Forest.* Tropical Forestry Papers No. 25. Oxford Forestry Institute, University of Oxford, UK.

Bass, S. and Dalal-Clayton, B. (1995). *Small Island States and Sustainable Development: Strategic Issues and Experiences*, 3, *Environmental planning Issues, No. 8.* Environmental Planning Group, International Institute for Environment and Development, London, UK.

Beard, J.S. (1946). *The Natural Vegetation of Trinidad.* Clarendon Press, Oxford, UK.

Brown, S. and Lugo, A.E. (1994). Rehabilitation of tropical lands: a key to sustaining development. *Restoration Ecology* **2**: 97–111.

Bullock, J. (1996). Plants, pp. 111–138. In: W. Sutherland (ed.). *Ecological Census Techniques.* Cambridge University Press, Cambridge, UK.

Chalmers, W.S. (1992). *Trinidad and Tobago National Forest Action Programme.* Report of the Country Mission Team, FAO/CARICOM, Port-of-Spain, Trinidad.

Clubbe, C.P. and Jhilmit, S. (1992). A case study of natural forest management in Trinidad, pp. 201–209. In: F.R. Millar and K.L. Adam. (eds). *Wise Management of Tropical Forests.* Oxford Forestry Institute, University of Oxford, Oxford, UK.

Costanza, R. *et al.* (1997). The value of the world's ecosystem services and natural capital. *Nature* **387**: 254–257.

FAO (1999) *State of the World's Forests 1999.* Food and Agriculture Organisation of the United Nations, Rome, Italy. Also available on the Internet at www.fao.org/forestry/ FO/SOFO/SOFO99/sofo99-e.stm.

Flora of Trinidad and Tobago (1928–1992). Government Printery, Port-of-Spain, Republic of Trinidad and Tobago.

FRIM (1978). *Sampling Manual.* Prepared for the Government of Trinidad and Tobago by Institutional Consultants (International Ltd.) in co-operation with the Forestry Division, Ministry of Agriculture, Land and Fisheries, Trinidad and Tobago.

FRIM (1980). *Inventory of the Indigenous Forest of Trinidad and Tobago.* 11 volumes. Prepared for the Government of Trinidad and Tobago by Institutional Consultants (International Ltd.) in co-operation with the Forestry Division, Ministry of Agriculture, Land and Fisheries, Trinidad and Tobago.

Glowka, L., Burhenn-Guilmin, F., Synge, H., McNeely, J. and Gündling, L. (1994). *A Guide to the Convention on Biological Diversity.* Environmental Policy and Law paper no. 30. IUCN, Gland, Switzerland.

Higman, S., Bass, S., Judd, N., Mayers, J. and Nussbaum, R. (1999). *The Sustainable Forestry Handbook.* Earthscan, London, UK.

Kent, M. and Coker, P. (1994). *Vegetation Description and Analysis.* John Wiley, London, UK.

Laurance, W. F. (1999). Reflections on the tropical deforestation crisis. *Biological Conservation* **91**: 109–117.

Mittermeier, R.A., Myers, N., Thomsen, J.B., da Fonseca, A.B. and Olivieri, S. (1998). Biodiversity hotspots and major tropical wilderness areas: approaches to setting conservation priorities. *Conservation Biology* **12**(3): 516–520.

Mueller-Dombois, D. and Ellenberg, H. (1974). *Aims and Methods in Vegetation Ecology.* John Wiley, London, UK.

Myers, N. (1988). Threatened biotas: "hotspots" in tropical forests. *Environmentalist* **8**: 1–20.

Myers, N. (1994). Tropical deforestation: rates and patterns, pp. 27–40. In: K. Brown, and D.W. Pearce (eds). *The Causes of Tropical Deforestation: the economic and satistical analysis of factors giving rise to the loss of the tropical forests.* University College Press, London, UK.

Myers, N., Mittermeier, R.A., Mittermeier, C.G., da Fonseca, G.A.B. and Kent, J. (2000). Biodiversity hotspots and conservation priorities. *Nature* **403**: 853–858.

Pimm, S.L. and Raven, P. (2000). Extinction by numbers. *Nature* **403**: 843–844.

Schmidt, R. (1987). Tropical rain forest management: a status report. *Unasylva* **156** (39): 2–17.

Wood, B.C. *et al.* (1999). A sustainable tropical forest management system in Trinidad. *Conference Proceedings of the International Sustainable Development Conference* at Leeds, March 25–16, 1999: 428–434.

World Commission on Forest and Sustainable Development (1999). *Our Forests ... Our Future: Report of the World Commission on Forest and Sustainable Development.* Cambridge University Press, Cambridge, UK.

WWF and IUCN (1994–1997). *Centres of Plant Diversity: A guide and strategy for their conservation.* 3 Volumes. IUCN Publications Unit, Cambridge, UK.

Chapter 12

The Role of Tropical Botanical Diversity in Supporting Development

Ghillean T. Prance
The Eden Project, Cornwall, UK

Introduction

The fact that the tropics are so botanically diverse is both an advantage and a disadvantage for the support of development. Diversity of species and vegetation types means that there are a great variety of plants from which to choose for utilisation by development projects. On the other hand, many tropical ecosystems are so complex that their management is too difficult for developers to cope with and the trend has been to replace the natural vegetation with simpler communities, such as cattle pasture, or a monoculture of a timber or fruit crop. This latter approach can often lead to failure of the crop because it is so different from the natural ecosystem. For example, the failure of many cattle pastures along the Trans-Amazon Highway or the failure of the rice-growing in the Jarí Project in Pará Brazil. The greatest challenge is to develop productive systems of management that are sustainable in the long-term and it is here that the use of botanical diversity can be of great help. Using various examples from Brazil, I hope to show that the maintenance and use of diversity is the best option for much of the tropics.

How to Cope with Diversity

The tropics are well known for their remarkable diversity of species and ecosystems. Table 12.1 gives a few statistics of tree species diversity in various areas of tropical rainforest where there can be over 300 species of ≥10 cm diameter at breast height (dbh) within a single hectare. With this number of species there cannot be many individuals of any one species and many are represented by a single individual in a hectare of forest (see Table 12.2 where 73 species have just one individual within the hectare studied). This scattered distribution makes it difficult to exploit any particular tree species whether for timber or for a non-timber forest product (NTFP). For example, an Amazonian rubber cutter has to cut several kilometres of trail to visit enough trees in a day to make it worth gathering the latex. The ecosystems that have a lower species diversity, such as the oligarchic forests described below, have a much greater potential for exploitation. Some tropical savannahs are equally diverse and their destruction for agriculture is causing as much species and habitat loss as deforestation of rainforest. This is particularly true in the cerrado region of the Planalto of Central Brazil. Increasing population size and migration is putting much greater pressure on the species diverse ecosystems of Brazil such as the Amazon rainforest and the cerrado or savannah region of the central high plains. The challenge is how to use these areas sustainably, which is certainly not happening at present.

Table 12.1 **Diversity of trees ≥10 cm diameter at breast height (dbh) in some 1 ha rainforest plots**

Location	No. of species	No. of stems ≥10 cm dbh	Reference
Amazonian Ecuador Serra Grande, Bahia, Brazil	307	693	Valencia *et al.* (1994)
Mishana, Peru	289	858	Gentry (1988)
Chocó, Colombia	258	675	Faber-Langendoen and Gentry (1991)
Papua New Guinea	228	693	Wright *et al.* (1997)
Gunung Mulu, Sarawak	214	778	Proctor *et al.* (1983)
Cocha Cashu, Peru	201	673	Gentry and Terborgh (1990)
Oveng, Gabon	131		Reitsma (1988)
Queensland, Australia	108	957	Phillips *et al.* (1994)
Alto Ivon, Bolivia	94	649	Boom (1987)

In recent years, there have been several studies of the extent to which indigenous peoples use the species diverse rainforest (Prance *et al.*, 1987). For example, the Chácabo Indians of Bolivia have uses for 82% of the species and 92% of the individual trees of ≥10 cm dbh on a hectare of forest in Bolivia studied by Boom (1989). The Ka'apor of Maranhão state in Brazil use all 136 species of trees and lianas ≥10 cm dbh in the hectare plot studied by Balée (1986) and Bennett (1992a) found that the Quijos Quichua people of Ecuador use over 90% of the species and 96% of the individual trees on the plots that he sampled. These studies were made only of the tree species employed by these tribes. Indigenous people also use many other plants in the herb and shrub layer of the forest as well as many lianas and some epiphytes. Bennett (1992b) demonstrated this when he recorded the uses of epiphytes, lianas and parasites used by the Shuar of Ecuador. Toledo *et al.* (1995) listed 1,330 species used by the indigenous people of Mexico. These data are enough to show that indigenous peoples have put diversity to use and in most cases have retained much of the original ecosystem. This suggests that development systems that use diversity rather than destroy it are more likely to succeed in the long-term. The indigenous agroforestry systems mentioned below are based on diversity rather than monoculture. The ecology of many tropical ecosystems indicates that diversity, not monoculture, is the norm. This diversity can be used in agroforestry systems and can also provide specific new uses of plants that can be further developed.

Sustainable use of tropical areas will depend heavily on the study of local ecological conditions and the interpretation and application of that information. For example, the few sustainable use systems encountered in

Table 12.2 **Structure of the species composition of a hectare of Central Amazonian rainforest containing 164 species of trees of ≥10 cm diameter at breast height, showing the most frequent 24 species** (data from L.V. Ferreira, pers. com.)

Species	Family	No. of trees
Oenocarpus bacaba Mart.	Arecaceae	53
Couepia obovata Ducke	Chrysobalanaceae	33
Pourouma paraensis Huber	Cecropiaceae	24
Protium grandifolium Engl.	Burseraceae	24
Guarea carinata Ducke	Meliaceae	23
Goupia glabra Aubl.	Celastraceae	21
Eschweilera odora (Poepp. ex O. Berg) Miers	Lecythidaceae	19
Ferdinandusa Pohl	Rubiaceae	19
Maximiliana maripa (Aubl.) Drude	Arecaceae	17
Nectandra rubra (Mez) C.K. Allen	Lauraceae	15
Licania heteromorpha Benth.	Chrysobalanaceae	13
Humiria balsamifera Aubl.	Humiriaceae	10
Euterpe precatoria Mart.	Arecaceae	10
Duguetia A. St.-Hil. sp. 1	Annonaceae	10
Iryanthera tricornis Ducke	Myristicaceae	9
Bertholletia excelsa Bonpl.	Lecythidaceae	9
Qualea paraensis Ducke	Vochysiaceae	9
Trattinickia rhoifolia Willd.	Burseraceae	9
Bocageopsis R.E. Fr. sp. 2	Annonaceae	9
Helicostylis podogyne Ducke	Moraceae	8
Swartzia corrugata Benth.	Papilionaceae	8
Calophyllum brasiliense Cambess.	Guttiferae	7
Micropholis (Griseb.) Pierre sp. 2	Sapotaceae	7
Virola Aubl. sp. 1	Myristicaceae	7

Species with 6 trees	4
Species with 5 trees	12
Species with 4 trees	10
Species with 3 trees	17
Species with 2 trees	24
Species with 1 tree	73

Amazonia in the upland, non-flooded forest, apart from well-planned extractivist activities, are mostly the agroforestry systems of the indigenous people. (See Denevan *et al.* (1984), Denevan and Padoch (1988) and Prance (1998) for the Bora Indians in Peru, and Hecht and Posey (1979) and Posey (1985) for the Kayapó Indians of Brazil). In these and many other examples their agricultural systems are based on diversity and on a good mixture of woody and herbaceous crops. Their fields often closely resemble a natural succession in a light gap. For example, the Bora Indians of Peru grow, in addition to more generally known crops such as banana *Musa* L. spp. (Musaceae) and manioc

(*Manihot esculenta* Crantz (Euphorbiaceae)), such native fruit trees as *Pourouma cecropiifolia* Mart. (Cecropiaceae) (uvilla) and *Theobroma bicolor* Bonpl. (Sterculiaceae) (macambo). These species diverse systems are sustainable in marked contrast to many monoculture systems that have been introduced into the Amazon region such as those based on pasture grass, timber species, such as *Gmelina* L. (Verbenaceae), or bananas. The principal reasons for this are the greater protection from pests and pathogens and the greater protection of the soil structure and nutrients offered by a species diverse agroforestry system.

In the upper Amazon in Peru, agroforestry systems very similar to indigenous ones, have been adopted by local people. Padoch (1990), Padoch *et al.* (1985) and Padoch and de Jong (1989) have described one of these systems at the village of Santa Rosa on the Ucayali River above Iquitos. They listed 36 plant species that were available for marketing during the study period (Table 12.3). Among them are internationally used crops, such as coffee, oranges and sugar cane, but the majority are local species such as cocona (*Solanum sessiliflorum* Dunal (Solanaceae)), sachamangua (*Grias peruviana* Miers (Lecythidaceae)) and umari (*Poraqueiba sericea* Tul. (Icacinaceae)), species that are really in the process of domestication. Many of these local species, originally cultivated by indigenous tribes, have now entered in a small way into the market economy. One of the best ways in which diversity can support development is through such mixed crop agroforestry systems.

A concept for using species diverse forests that arose in Brazil is that of extractive reserves (see Allegretti, 1990; Prance, 1989; Schwartzman, 1989; Schwartzman and Allegretti, 1987). In these reserves the local people are allowed and encouraged to extract products, but not to clear-cut except for small areas for their agricultural crops. Such reserves have been established in the Brazilian states of Acre, Amapá, Maranhão and Rondônia, largely due to the action of rubber tappers (see Table 12.4). The extractive reserves have been useful to slow down deforestation, but they have certain limits as a permanent solution for sustainable use of the Amazon rainforest (see, for example: Browder, 1990, 1992; Homma, 1994; Prance, 1994). Most of the Amazonian extractive reserves have relied on two products only: rubber latex (*Hevea brasiliensis* Müll. Arg. (Euphorbiaceae)) and Brazil nuts (*Bertholletia excelsa* Bonpl. (Lecythidaceae)). The price of Brazil nuts has fluctuated greatly on the world market, and Brazil no longer subsidises Amazon rubber at prices higher than the world market. The result is that the extractors receive very poor returns from this practice. One of the challenges for the future is to develop a market for a much wider range of possible extractive products (see Table 12.5). In addition, most extracted products have traditionally been marketed through a long chain of intermediaries so that the extractor gets paid very little. There have been a few efforts to shorten the chain of intermediaries through more direct buying. For example, the Body Shop Company, UK, buys Brazil nut oil directly from the Kayapó Indians for use in cosmetic products. The 'Fairtrade scheme' (See the UK Fairtrade website: www.fairtrade.org.uk.) which ensures a just deal for producers is an important development to assist local producers.

Table 12.3 **Resources available for marketing in 1986 from the agroforestry of the Peruvian village of Santa Rosa** (adapted from Padoch and de Jong, 1989)

Product	Unit used
Anona (*Annona* L. sp.)	fruits
Barbasco (*Lonchocarpus* Kunth sp.)	trunks
Bijao (*Heliconia* L. sp.)	plants
Breadfruit (*Artocarpus altilis* (Parkinson) Fosberg)	fruits
Caimito (*Pouteria caimito* (Ruiz & Pav.) Radlk.)	fruits
Coffee (*Coffea arabica* L.)	Kgs
Cocona (*Solanum sessiliflorum* Dunal)	plants
Coconilla (*Solanum* L. sp.)	plants
Daledale (*Calathea allouia* (Aubl.) Lindl.)	plants
Guaba (*Inga edulis* Mart.)	fruits
Guaba pelusa (*Inga* Mill. sp.)	fruits
Guava (*Psidium guajava* L.)	fruits
Huasai (*Euterpe precatoria* Mart.)	trunks
Huicungo (*Astrocaryum huicungo* Dammer ex Burret)	trunks
Jute seeds (*Urena lobata* L.)	Kgs
Leche caspi (*Couma macrocarpa* Barb. Rodr.)	trunks
Lemon (*Citrus aurantifolia* (Christm.) Swingle)	fruits
Limon dulce (*Citrus* L. sp.)	fruits
Macambo (*Theobroma bicolor* Bonpl.)	fruits
Mamey (*Syzygium malaccense* (L.) Merr. & L. M. Perry)	fruits
Orange (*Citrus sinensis* Osbeck)	fruits
Papaya (*Carica papaya* L.)	fruits
Peach palm (*Bactris gasipaes* Kunth)	racemes
Peppers (*Capsicum* L. sp.)	fruits
Pineapple (*Ananas comosus* (L.) Merr.)	fruits
Plantain (*Musa paradisiaca* L.)	racemes
Pona (*Socratea* H. Karst. sp.)	trunks
Sachamangua (*Grias peruviana* Miers)	fruits
Sachapapa (*Dioscorea trifida* L. f.)	plants
Sugar cane (*Saccharum officinarum* L.)	stalks
Huitina (*Xanthosoma* Schott sp.)	plants
Tropical cedar (*Cedrela odorata* L.)	trunks
Umari (*Poraqueiba sericea* Tul.)	fruits
Uvilla (*Pourouma cecropiifolia* Mart.)	racemes
Yarina (*Phytelephas macrocarpa* Ruiz & Pav.)	trunks
Zapote (*Quararibea cordata* (Bonpl.) Vischer)	fruits

Table 12.4 Extractive reserves in Amazonian Brazil (source: IBAMA/CNPT, 1994)

Name	State	Area (ha)	Population	Principal products
Chico Mendes Xapuri	Acre	970,570	7,500	Rubber Brazil nut Copaiba oil
Alto Juruá	Acre	506,186	6,000	Rubber
Rio Cajari	Amapá	481,650	5,000	Brazil nut Rubber Copaiba oil Açaí palm
Rio Ouro Preto	Rondônia	204,583	3,410	Brazil nut Copaiba oil Rubber
Mata Grande	Maranhão	10,480	1,500	Babaçu fruit Fish
Quilomobe de Freixal	Maranhão	9,542	900	Babaçu fruit Fish
N. Tocantins	Tocantins	9,280	2,000	Babaçu fruit Fish
Ciriaco	Maranhão	7,050	1,150	Babaçu fruit

Another disadvantage of extraction is that, because of the diversity of the forest and consequent dilution effect of the trees, it requires a large area to sustain each person. Rubber trees, Brazil nut trees and other products from extraction are widely dispersed in the forest rather than growing in large clusters. For example, the best-known extractive reserve named Chico Mendes at Xapuri in Acre has 7,500 people on 970,570 ha (129 ha per person). The agroforestry systems described above are much more intensive and support greater population densities.

The use of products extracted from the forest often leads to their cultivation. This is not bad if they are then cultivated in sustainable systems such as in agroforestry or in gardens around local dwellings (home gardens). For example, cupuaçú (*Theobroma grandiflorum* (Willd. ex Spreng.) K. Schum.), which used to be much extracted in the forest, is now grown in most home gardens and also in small plantations. Plants that have been tested thoroughly in this way are more likely to succeed than introduced species that require unnatural systems of cultivation that are less integrated with the local ecology and community.

Table 12.5 **Extraction and agroforestry products from native plants which enter the Brazilian economy** (source: IBGE, 1992)

Product	Production (tonnes) 1988	Production (tonnes) 1989	Value 1988 (Cz$)
Colourant:			
Urucu (*Bixa orellana* L.)	793	845	71,497
Fibres:			
Buriti (*Mauritia flexuosa* L. f.)	972	991	34,116
Carnaúba (*Copernicia cerifera* Mart.)	2,544	2,876	96,597
Piaçava (*Leopoldinia* Mart.)	959	1,444	47,326
Other Fibres	2,023	364	139,363
Food and beverage:			
Açai fruit (*Euterpe oleracea* Mart.)	117,119	114,304	16,623,748
Cashew (*Anacardium occidentale* L.)	12,716	8,870	1,775,592
Brazil nut (*Bertholletia excelsa* Bonpl.)	23,391	25,672	1,351,852
Erva mate (*Ilex paraguariensis* A. St-Hil.)	145,064	145,649	20,023,399
Mangaba fruit (*Hancornia speciosa* Gomes)	939	988	100,343
Palmito (*Euterpe* Mart. spp. & other palms)	190,314	202,439	12 230,169
Pinhão fruit	3,118	2,919	313,648
Umbu fruit (*Phytolacca dioica* L.)	19,555	18,999	1,638,624
Latex:			
Caucho (*Castilla ulei* Warb.)	39	42	
Rubber coagulant (*Hevea* Aubl. spp.)	23,035	22,990	171,488
Rubber liquid (*Hevea* spp.)	2,409	1,784	743,454
Mangabeira (*Hancornia speciosa* Gomes)	2	1	55
Balata (*Manilkara bidentata* (A. DC.) A. Chev.)	21	21	4,938
Maçaranduba (*Manilkara huberi* (Ducke) Chevalier)	192	127	40,531
Sorva (*Couma* Aubl. spp.)	1,059	1,106	63,980
Medicines:			
Ipecacuana (*Cephaelis* spp. Sw.)	17	3	3,196
Jaborandi (*Pilocarpus* Vahl spp.)	1,765	1,676	244,793
Miscellaneous:			
Other minor products	2,388	934	70,787
Tanins:			
Angio (*Piptadenia peregrina* (L.) Benth.)	1,557	1,185	59,014
Barbatimão (*Stryphnodendron barbatimam* Mart.)	1,527	1,387	32,508
Others	19	15	1,502
Vegetable oils:			
Pequi (*Caryocar brasiliense* Canbess.)	1,394	1,593	785,221
Tucum (*Astrocaryum tucuma* Mart.)	5,109	5,092	273,490
Babaçu (*Attalea phalerata* Mart. ex Spreng.)	200,031	195,378	14,004,499
Copaiba (*Copaifera* L. spp.)	54	49	14,780
Cumaru (*Dipteryx odorata* (Aubl.) Willd.)	15	9	3,603
Licuri (*Syagrus coronata* (Mart.) Becc.)	3,632	12,421	1, 196,527
Oiticica (*Licania rigida* Benth.)	10,277	15,968	332,641
Andiroba (*Carapa guianensis* Aubl.)	363	–	38,730
Wax:			
Carnaúba wax (*Copernicia cerifera* Mart.)	7,373	7,372	2,730,850
Carnaúba powder (*Copernicia cerifera* Mart.)	10,734	11,011	3,349,219

Plantas do Nordeste

Programme to utilise the plant diversity of North East Brazil

Action taken

For some years the plant taxonomists of the Royal Botanic Gardens, Kew, UK (RBG, Kew) had been working in the arid North East Brazil. Work with local institutions indicated that a lot more than taxonomic surveys of the vegetation were needed if the problems of vegetation destruction by unsustainable practices, and the increase in poverty, were to be addressed. A consortium of institutions in the eight states of this region set up a programme called 'Plantas do Nordeste' (PNE). The slogan of the project explains its purpose: 'local plants for local people'. It is a programme that seeks to integrate the three areas of biodiversity: survey, economic botany and the dissemination of information and training. It is controlled by an Association incorporated in Brazil, the officers of which are representatives of the collaborating institutions. These are a good mixture of local university biology, computer science and chemistry departments, government institutions and non-governmental organisations (NGOs) working in the field with local people. The programme brings together the various elements that are necessary for the successful exploitation of biodiversity: good taxonomic identification, surveys of the vegetation types, studies of plant uses in the region, laboratory work in chemistry and computer science and, most importantly, contact with the local people through organisations carrying out social work among them (See Araujo *et al.*; 1999).

Description of actions

The programme has three areas for action:
1. The biodiversity subprogramme is working with local systematists to collect, identify and name as much of the local flora as possible. This also includes detailed ecological inventories of certain vegetation types such as the 'brejo'. These are the small patches of forest on cloud covered hills that are vital to water supply in an arid region. The use of biodiversity for development projects requires the substrate of sound taxonomy. In the PNE programme we found that one of the factors holding back the development side was the poor knowledge of the flora. Today, in addition to the traditional taxonomy based on the morphology of herbarium specimens, molecular techniques are most useful for development because they are producing more accurate systems of classification and enable the analysis of genetic variation within populations. The more predictive and accurate the classification, the easier it is to search for new useful

products. However, it is not just plant identification that is important; several of the projects in this subprogramme involve surveys of the different types of vegetation in the region to determine their composition and distribution. PNE is gradually building up a taxonomically authenticated checklist of all the species in the region that will be accessible on the Internet.

2. The economic botany subprogramme is examining various useful plant species and setting up projects to develop their use. One project is the 'living pharmacies', which is encouraging medicinal plant use in the poor areas of some of the larger cities and towns in the State of Ceará. The project employs former street children to cultivate the plants and to market the medicines through street theatre. The collaboration with the chemistry department of the local university, gives the project a sound analytic capability to ensure effectiveness of the medicine and to avoid the use of potentially toxic plants. This project is enabling the poor population to obtain cheap cures for many of their common ailments. The economic botany projects also include the study of nectar-producing plants for bee-keeping, the cultivation of plants for dried flower arrangements and the management of goats in corrals through the use of fast growing nitrogen-fixing legumes as fodder.

3. The information subprogramme, which is often a neglected area in such programmes. This subprogramme seeks to disseminate the information generated by the entire programme. This is being carried out in many ways, such as the installation of local databases in the collaborating institutions, and co-operation with various local NGOs that are in contact with the potential users of the information. This subprogramme is largely supported by the British Department for International Development (DFID). Since this project is designed to help the needs of the local people, rather than to develop new industrial products, the informatics side is vitally important and as a result new plant uses are entering into the local economy. A plant information centre (Centro Nordestino de Informações sobre Plantas) has been established in Recife (Pernambuco).

Practical issues and costs

PNE is a large programme that needed extra funding to achieve its goals. All participating institutions have devoted resources such as staff time, transport and laboratory facilities to the programme. Once RBG, Kew had raised £1 million sterling from a private donor, it was easier to set up the co-ordination needed for such a project and to attract other funds from small donors in the UK and in Brazil. Once established, the programme was able to attract an annual contribution of about half a million dollars from the National Research Council of Brazil (Conselho Nacional de Desenvolvimento Cientifico e Technológico) for scholarships to train people in the local universities and for research. This has made the programme truly binational with good support from the host country. After several years of intensive negotiation, the information subprogramme secured support from the UK overseas aid programme of the DFID in 1998. This £1.8 million grant is ensuring that the information generated by the biodiversity and economic botany subprogrammes is being effectively disseminated to the users, through setting up databases in many different government and non-government agencies. There is no doubt that the success of PNE is due to to its ability to attract funding based on the appropriateness of the programme for North East Brazil.

[Information about PNE can be obtained on the Internet from: www.rbgkew.org. uk/ceb/pne or www.cnip.org.br.]

Lessons learnt

One of the most important aspects of this programme was the way in which it brought together a large number of local institutions in the different States that were not previously collaborating. This was synergistic and increased the effectiveness not only of the programme but also of the participating institutions. Lessons learnt were certainly patience and persistence; for example, it took eight years of negotiations to obtain funds for the information subprogramme. The delays were from both countries and each step took a long time before a binational agreement was reached. This could easily have failed if the principal people involved had not been so persistent.

Future steps

The initial focus of the programme, centred on the identification of economic plants in vegetation surveys, is still a problem. Funds are being sought to improve the capacity of the local and regional herbaria to identify plants. The information subprogramme is in its initial phase and during the next four years much will be done to spread information both to the academic institutions involved and to the local user of the plants under study. One of the problems encountered in a programme of this size is the unevenness of funding. This has led to the funding of some projects within the programme and the delay of others; not always in the most logical order because components of the programme selected for funding are those that appeal to a particular donor agency. One of the most stabilising aspects has been regular funding from the Brazilian government.

The PNE programme highlights two factors essential to projects that aim to utilise botanical diversity for development:

1. Such projects will only succeed when the local people are heavily involved. It is essential to gather local knowledge and to involve local people in the project; and

2. The projects that are likely to succeed are those which are of an interdisciplinary nature. PNE involves botanists, agronomists, medical doctors, chemists, computer scientists and experts in other fields. It could not be run without this mixture of expertise.

There is an enormous diversity of tropical vegetation types and tropical plant species and these offer great unrealised potential for future development. However, so much of this is being lost through habitat destruction and species loss. The potential of this diversity for new medicines, agrochemicals, resins, foods, fodder and many other uses should be a strong argument to halt the destruction and conserve tropical diversity for use by future generations.

References

Allegretti, M.H. (1990). Extractive reserves: an alternative for reconciling development and environmental conservation in Amazonia, pp. 252–264. In: A.B. Anderson (ed.). *Alternatives to Deforestation.* Columbia University Press, New York, USA.

Araujo, F.D. de, Prendergast, H.D.V. and Mayo, S.J. (1999). Anais do I Workshop Geral. Royal Botanic Gardens, Kew, UK.

Balée, W. (1986). Análise preliminar de inventário florestal e a etnobotânica Ka'apor Maranhão. *Bol. Mus. Paraense Emílio Goeldi Ser. Botânica* **2**(2): 141–167.

Bennett, B.C. (1992a). Plants and people of the Amazonian rainforests. *BioScience* **42**: 599–607.

Bennett, B.C. (1992b). Use of epiphytes, lianas and parasites by the Shuar people of Amazonian Ecuador. *Selbyana* **13**: 99–114.

Boom, B.M. (1987). Ethnobotany of the Chácobo Indians, Beni, Bolivia. *Advances in Economic Botany* **4**: 1–68.

Boom, B.M. (1989). Uses of plant resources by the Chácobo. *Advances in Economic Botany* **7**: 78–96.

Browder, J.O. (1990). Extractive reserves will not save the tropics. *BioScience* **40**: 626.

Browder, J.O. (1992). The limits of extractivism: tropical forest strategies beyond extractive reserves. *BioScience* **42**: 174–182.

Denevan, W.M. and Padoch, C. (eds.). (1988). Swidden-fallow agroforestry in the Peruvian Amazon. *Advances in Economic Botany* **5**: 1–107.

Denevan, W.M. *et al.* (1984). Indigenous agroforestry in the Bora Indian Management of swidden fallows. *Interciencia* **9**: 346–357.

Faber-Langdoen, D. and Gentry, A.H. (1991). The structure and diversity of rain forests at Bajo Calima, Chocó Region, western Colombia. *Biotropica* **23**: 2–11.

Gentry, A.H. (1988). Tree species richness of upper Amazonian forests. *Proceedings of the US National Academy of Science* **85**: 156–159.

Gentry, A.H. and Terborgh, J. (1990). Composition and dynamics of the Cocha Casha 'mature' floodplain forest, pp. 141–157. In: A.H. Gentry (ed.). *Four Neotropical Rain Forests.* Yale Univiversity Press, New Haven, USA.

Hecht, S.B. and Posey, D.A. (1979). Preliminary results on soil management techniques of the Kayapó Indians. *Advances in Economic Botany* **7**: 174–188.

Homma, A.K.O. (1994). Plant extractivism in the Amazon: limitations and possibilities,. pp. 35–57. In: M. Clusener-Godt and I. Sachs (eds). *Extractivism in the Brazilian Amazon: Perspective on regional development.* MAB Digest No. 18. UNESCO, Paris, France.

IBGE (1992). Anuário estatístico do Brasil. Instituto Brasileiro de Geografia e Estatística, Rio de Janeiro, Brazil.

Padoch, C. (1990). Santa Rosa: the impact of the forest products trade on an Amazonian place and population. *Advances in Economic Botany* **8**: 151–158.

Padoch, C., Chota Inuma, J., de Jong, W. and Unruh, J. (1985). Amazonian agroforestry: a market-oriented system in Peru. *Agroforestry Systems* **3**: 47–58.

Padoch, C. and de Jong, W. (1989). Production and profit in agroforestry: an example from the Peruvian Amazon, pp. 102–112. In: J.O. Browder (ed.). *Fragile Lands of Latin America: strategies for sustainable development.* Westview Press, Boulder, Colorado, USA.

Phillips, O.L., Hall, P., Gentry, A.H., Sawyer, S.A. and Vasquez, R. (1994). Dynamics and species richness of tropical rainforests. Proceedings of the US National Academy of Sciences **91**: 2805–2809.

Posey, D.A. (1985). Indigenous management of tropical forest ecosystems: the case of the Kayapó Indians of the Brazilian Amazon. *Agroforestry Systems* **3**: 139–158.

Prance, G.T. (1989). Economic prospects from tropical rainforest ethnobotany, pp. 61–74. In: J.O. Browder (ed.). *Fragile Lands of Latin America: strategics for sustainable development.* Westview Press, Boulder, Colorado, USA.

Prance, G.T. (1994). Amazonian tree diversity and the potential for supply of non-timber forest products, pp. 7–15. In: R.R.B. Leakey and A.C. Newton (eds). *Tropical trees: the potential for domestication and the rebuilding of forest resources.* HMSO, London, UK

Prance, G.T. (1998). Indigenous non-timber benefits from tropical rain forest, pp. 21–42. In: F.B. Goldsmith (ed.). *Tropical Rain Forest: a wider perspective.* Chapman & Hall, London, UK.

Prance, G.T., Balée, W., Boom, B.M. and Carneiro, R.L. (1987). Quantitative ethnobotany and the case for conservation in Amazonia. *Conservation Biology.* **1**: 296–310.

Proctor, J., Anderson, J.M., Chai, P. and Vallack, H.W. (1983). Ecological studies in four contrasting lowland rain forests in Gunung Mulu National Park. *Sarawak Journal of Ecology* **71**: 237–260.

Reitsma, J.M. (1988). Vegetation forestière du Gabon. Technical series. *Tropenbos* **1**: 1–142.

Schwartzman, S. (1989). Extractive reserves: the rubber tappers' strategy for sustainable use of the Amazonian rainforest, pp. 150–165. In: J.O. Browder (ed.). *Fragile Lands of Latin America: strategies for sustainable development.* Westview Press, Boulder, Colorado, USA.

Schwartzman, S. and Allegretti, M.H. (1987). *Extractive production and the rubber tappers movement.* Environmental Defense Fund, Washington, DC, USA.

Toledo, V.M., Batis, A.I., Becerra, R., Martínez, E. and Ramos, C.H. (1995). La selva util: etnobotánica cuantitativa de los grupos indígenas del trópico húmedo de México. *Interciencia* **20**: 177–187.

Valencia, R., Balslev, H. and Paz y Mino, G. (1994). High tree alpha diversity in Amazonian Ecuador. *Biodiversity and Conservation* **3**: 21–28.

Wright, D.D., Heinrich Jessen, J., Burke, P. and Silva Garza, H.G. da. (1997). Tree and liana enumeration and diversity on a one-hectare plot in Papua New Guinea. *Biotropica* **29**: 250–260.

Chapter 13

Conservation and Management of Forest Plots in Mauritius

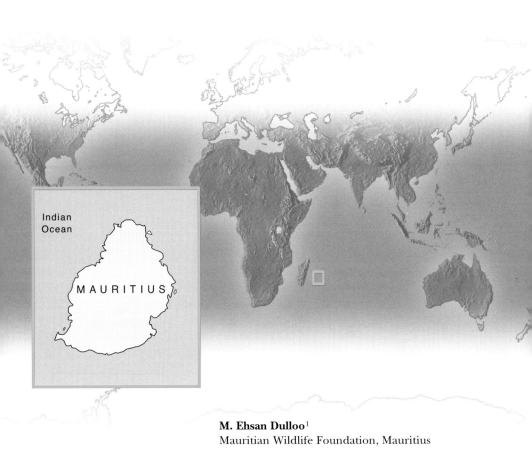

Indian Ocean

MAURITIUS

M. Ehsan Dulloo[1]
Mauritian Wildlife Foundation, Mauritius

[1] Now at: International Plant Genetic Resources
Institute, Nairobi

Introduction

The oceanic island of Mauritius is located in the south-west Indian Ocean, some 900 km east of Madagascar (longitude 57°30'E and latitude 20°20'S). The island is part of the Mascarene group which also includes La Réunion (a French Département d'Outre-Mer) and Rodrigues Island (a territory of Mauritius) (Figure 13.1). They are all volcanic in origin, having emerged from the ocean floor between eight million years (Mauritius) and 1.5 million years (Rodrigues) ago. Mauritius has a total land area of 1,865 km^2 and consists of a central plateau, rising to an altitude of about 550 m giving way to coastal plains on the fringes of the island, more extensively in the north. The highest peak in Mauritius rises to 828 m (Black River Peak) in the south-west of the island. Mauritius was uninhabited until the beginning of the sixteenth century and was then entirely covered with dense native forests containing a rich diversity of species, a high proportion of which were endemic. The natural vegetation was cleared largely for agriculture, and forest now covers about 5% of Mauritius (Safford, 1997; Figure 13.2). Mauritius is now one of the most densely populated countries in the world with a population density of about 565 inhabitants/km^2.

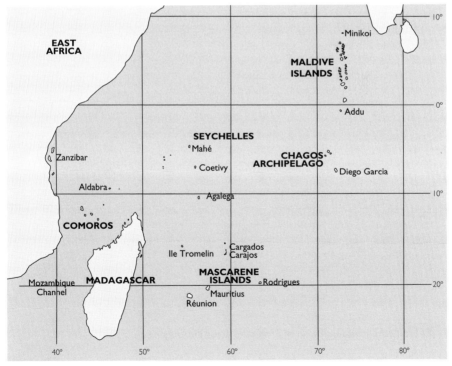

Figure 13.1 Map of the south-west Indian Ocean showing location of Mascarene Islands.

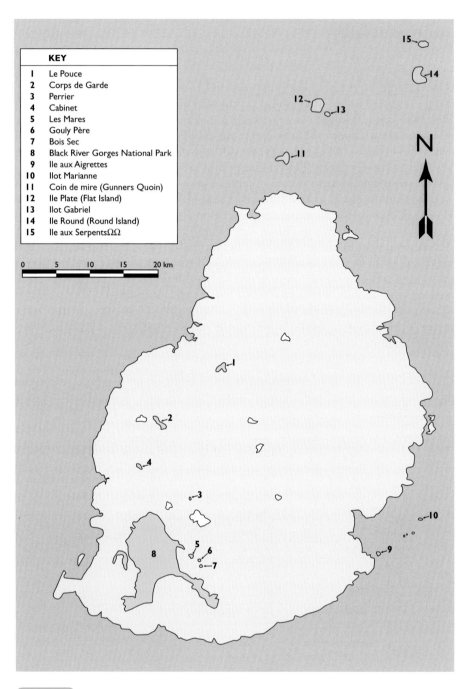

KEY

1	Le Pouce
2	Corps de Garde
3	Perrier
4	Cabinet
5	Les Mares
6	Gouly Père
7	Bois Sec
8	Black River Gorges National Park
9	Ile aux Aigrettes
10	Ilot Marianne
11	Coin de mire (Gunners Quoin)
12	Ile Plate (Flat Island)
13	Ilot Gabriel
14	Ile Round (Round Island)
15	Ile aux SerpentsΩΩ

Figure 13.2   Distribution map of protected areas in Mauritius.

The first ecological studies were undertaken by Vaughan and Wiehe (1937, 1941). They described three broad vegetation associations: a palm community now restricted to the northern offshore islets, lowland dry forest under 365 m altitude and with 1,000–2,500 mm/year rainfall and an upland wet forest over 365 m and with 2,500–5,000 mm/year rainfall. Due to the topography and climate, the upland forest contains a variety of different habitat types and Vaughan and Wiehe have divided them into five intergrading vegetation types. These include a marsh community rich in *Pandanus* Parkinson (Pandanaceae) species (at least 15 species are endemic to Mauritius), a *Philippia* Klotzsch (Ericaceae) heath community, a *Sideroxylon* L. (Sapotaceae) thicket; a *Stillingia* Garden ex L. (Euphorbiaceae) – *Croton* L. (Euphorbiaceae) community and a mossy forest in the wettest areas.

Mauritius possesses a rich diversity of endemic plants, with 750–900 native taxa of which 45% are endemic (Maunder *et al.* (unpublished data); Strahm, 1994, 1996). During the past 300 years, the native vegetation has been greatly altered as a consequence of human colonisation. The indigenous animals were hunted for food (dodo and tortoises). Forests were cleared for growing introduced crops, particularly sugarcane (Figure 13.3). The disturbances to the native flora are not only due to deforestation and direct harvesting, but also as a result of displacement by invasive species (see Appendix 5). Island biotas are notoriously vulnerable to introductions of exotic organisms (Loope and Mueller-Dombois, 1989; Vitousek, 1988). In Mauritius, the introduced Javanese deer (*Cervus timorensis*), wild pigs (*Sus scrofa*) and rats (*Rattus rattus* and *R. norvegicus*) have caused much damage to the indigenous forests. Many of the exotic plants, for

Figure 13.3   The lowland forest communities of Mauritius have been virtually eradicated and replaced with urban developments and monocultures of sugar cane. The surviving upland forests, whilst extensive, are heavily degraded and subject to invasion by exotic plants.

instance Chinese guava (*Psidium cattleianum* Sabine (Myrtaceae)), privet (*Ligustrum robustum* subsp. *walkerii* (Decne.) P.S. Green (Oleaceae)), wild pepper (*Schinus terebinthifolius* Raddi. (Anacardiaceae)) have become invasive in Mauritian forests.

As a result of large-scale deforestation, natural forest only exists in inaccessible places such as river gorges and mountain tops or on marginal lands unsuitable for other uses. The majority of the surviving indigenous flora and fauna are found in these forest patches. However, these remnant natural areas are themselves threatened with further degradation by invasive species. As a result of these factors, the indigenous flora and fauna have declined, both in terms of population and genetic diversity. This paper will review the activities undertaken to conserve the indigenous forest biota in Mauritius. The different approaches to conservation are described, and illustrated with practical examples. The lessons learnt from these activities are analysed and the future steps to ensure the conservation of biodiversity in Mauritius are discussed.

Conservation Approaches

Two fundamental approaches are recognised: *in situ* and *ex situ* conservation. The different approaches to plant conservation have been discussed by various authors (Damania, 1996; Dulloo *et al.*, 1998; Falk, 1990; Maxted *et al.*, 1997). *Ex situ* conservation refers to the conservation outside the natural habitat while *in situ* conservation is the conservation within the natural ecosystem and/or the agro-ecosystem where the target species/germplasm has evolved. No single conservation technique can effectively conserve the full range of genetic diversity of target species (Falk, 1990). The *ex situ* conservation of wild plants can be achieved in a number of ways, depending upon the mode of reproduction of target species, storage behaviour and the amount of resources available. Plants can be conserved as seeds in seed banks, as living tissues in *in vitro* gene banks or as whole living plants in field gene banks, arboreta, botanic gardens, etc. Various technologies are now being developed to conserve their genetic material in DNA libraries.

In situ conservation can be viewed in two different contexts: natural ecosystem conservation and on-farm conservation (see www.ipgri.cgiar.org/themes/in_situ_project/home/insituhome.htm). We shall focus here on natural ecosystem conservation. *In situ* conservation of natural ecosystems refers to the conservation of biological resources in natural habitats. Various methods for protecting natural ecosystems have been developed. Essentially these methods involve the setting up of managed protected areas.

Early conservation work

In Mauritius, various actions, both at the *in situ* and *ex situ* level, have been undertaken to save threatened plants and animals from extinction and to prevent forest degradation. The realisation of the progressive loss of the native flora and fauna and the degradation of the natural ecosystems came shortly after the first ecological studies of vegetation of Mauritius were undertaken in the 1930s (Vaughan and Wiehe, 1937, 1941). This prompted the set up of the first nature reserves in 1951, which were representative of the different successional stages of the vegetation as determined by Vaughan and Wiehe (1937). In the succeeding years, other plant communities on the mainland and small islets became protected, and by 1975 there were 20 nature reserves proclaimed. These were reduced to 15 as several adjacent reserves were merged to form one nature reserve of 3,611 ha in extent, the Macchabe/Bel Ombre Nature Reserve, which was also given the status of Biosphere Reserve in 1979. Later in 1994, the Macchabe/Bel Ombre block and adjacent lands were turned into the first National Park of Mauritius, Black River Gorges National Park, which covers 6,574 ha (3.5% of the island) and includes most of the important areas for wildlife on the mainland. Figure 13.2 gives the distribution of the different protected areas on Mauritius.

Threatened species work

Mauritius has approximately 685 native taxa of flowering plants with 311 endemic to the island and another 150 shared only with the other Mascarene islands (Strahm, 1993). Walter and Gillett (1998) consider that 294 vascular plants of Mauritius are globally threatened. This represents 39.2% of the vascular plants of the island. For 28 species, mostly endemic, less than ten wild individuals are known (Maunder *et al.*, in press). Simply setting aside an area as a reserve, does not always ensure the protection of the resident biota. There may be a range of influences causing a deterioration of the ecosystems, leading to continued loss of biodiversity. In the 1970s, a number of reports on the conservation status of the flora and fauna also highlighted the threats to the local biodiversity and the need to undertake conservation work to prevent the impoverishment of the island's biological resources. The Forestry Service of Mauritius and later the National Parks and Conservation Service have been collecting seeds and other propagating material to raise native plants in *ex situ* nurseries (Dulloo *et al.*, 1996; Owadally *et al.*, 1991). In 1983, with the support of the World Wide Fund for Nature (WWF) and the World Conservation Union (IUCN), a plant programme was set up to rescue critically endangered plants through *ex situ* and *in situ* conservation and to document the endangered plants in the wild. This task was undertaken by Strahm (Strahm, 1993), and a Red Data Book for Rodrigues

Island was published as a result of this work (Strahm, 1989). Many of the critically endangered species have been targeted for propagation *ex situ* and for eventual reintroduction into the wild or management in botanic gardens. For example, *Dombeya mauritiana* F. Friedman (Sterculiaceae), which was reduced to one individual in the wild, was saved *in extremis* from extinction through vegetative propagation (cuttings). The mother tree is dead, but its clones have been re-introduced to a number of reserves in Mauritius (Mondrain, Ile aux Aigrettes) and are in cultivation in botanic gardens (Royal Botanic Gardens Kew, UK; Conservatoire Botanique National de Brest, France). Similarly *Hyophorbe vaughanii* L.H. Bailey (Palmae) with a wild population of four trees, has been propagated and reintroduced back into managed reserves. However, for the majority of species, the reproductive biology is little understood and it has been difficult to establish them in secure cultivation.

Conservation Management Areas

Most of the conservation efforts have concentrated on species rather than on ecosystems. However in the past ten years, there has been increasing integration with habitat protection and restoration. In Mauritius, it became apparent early on that alien species, both animals and plants, were causing irreversible damage to the forest ecosystems (Vaughan, 1968). Vaughan (1968) recommended the weeding of some of the nature reserves. By this he meant the manual uprooting of all alien species. The idea was that by removing the alien species, the native species would no longer be suppressed and would naturally regenerate. For example, Perrier, a small nature reserve of 0.25 ha has been continually 'gardened' by the Forestry Service since 1969.

In 1986, Dulloo and Strahm, started a series of experimental 'sample plots' which later became more intensively managed conservation areas. This work was based on the re-survey of a study quadrat at Macchabe, now referred to as the Vaughan plot, originally established and surveyed by Vaughan and Wiehe in the 1930s (Vaughan and Wiehe, 1941). This plot has been sporadically weeded since the 1940s. In our study, we re-surveyed the Vaughan plot to measure the changes that have occurred over the 49 years since its establishment (Figure 13.4). We also established a control plot adjacent to the Vaughan plot to compare the effect of weeding on the species composition and on forest degradation (Strahm, 1993). It became clear from this survey that natural regeneration within the native forest was heavily reduced by alien species and that weeding was essential to reduce further degradation of the forests. It also showed that even sporadic weeding could retard considerably the degradation process caused by invasive species.

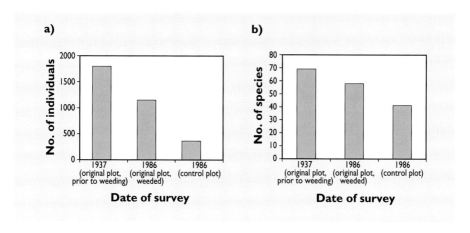

Figure 13.4 Changes in a) number of individuals and b) number of species in the Vaughan plot under weeding management since the 1940s compared with a control plot in the (1,000 m²) Macchabe Forest (Mauritius) over 49 years.

Based on these results, we established other sample plots in the upland forests. To date, some of these plots have been expanded and other new Conservation Management Areas (CMAs), within the protected areas, have been set up, totalling an area of 38 ha (Table 13.1). In addition, a number of private native forest areas on the mainland and small islets are being managed to halt the degradation of the forest ecosystems, e.g. Mondrain, a private nature reserve run by the Royal Society of Arts and Science of Mauritius and managed by the Mauritian Wildlife Foundation (MWF).

Table 13.1 **Conservation Management Areas (CMAs) in Mauritius**	
Name of CMA	**Approx. area (ha)**
Vaughan plot	0.4
Brise Fer	19.3
Mt. Cocotte	0.4
Florin	2.5
Mare Longue	2.4
Le Petrin	6.2
Belloguet (Bel Ombre)	2.5
Fixson (Bel Ombre)	4.3
Total	38.0

Management of forest plots

The CMA now represents the unit of management within the protected areas. Management of these areas involves manual weeding of exotic plants and the control of exotic animals. One of the disadvantages of manual weeding is that it is a very labour intensive and costly operation. Some 315 person-days are on average required for weeding (manual cutting and stump removal) 1 ha of degraded forest. Our experience has shown that it is necessary to repeat weeding at least three times a year. The frequency of maintenance weeding can then be reduced to one weeding per year after a period of four to five years intensive weeding. In Mauritius, the initial establishment of the CMAs has cost the government some US$37,000 and a recurrent sum of US$74,000 for maintenance per year (Bachraz, pers. com.). These figures are based on contract labourers being recruited for the work. It must be noted that the cost of weeding is very variable and depends upon many factors, such as the type of labour used (government labour versus contracted labour), degree of invasion of the forest plot, method used, type of terrain, etc. Thus, such figures should not be overly generalised.

Despite its high cost, manual weeding has been the only option available. However some research has been initiated, under a United Nations Development Program (UNDP)/Global Environmental Facility (GEF) Biodiversity Restoration project, in Mauritius to find alternative methods that are cost effective over longer time periods, such as chemical control. Some trials have been carried out to determine the best herbicide and its optimal concentration to control two of the worst invasive plant species, namely *P. cattleianum* and *L. robustum*. The results showed that Garlon® 4E (active ingredient – Triclopyr Ester) was the most effective herbicide giving 100% mortality at full concentration (refer to Santos *et al.* (1992) for Garlon treatment trials on alien plant species). Other weed control trials using different weeding implements have been done to increase the efficiency of weeding. For example, on Ile aux Aigrettes, an islet nature reserve, MWF has developed a mechanical uprooter to remove *Leucaena leucouphala* Benth. (Leguminosae) stumps and reduced the weeding time by two-thirds (Khedun, pers. com.). An experiment is presently ongoing to compare current manual weeding practices with other weeding techniques, such as using the mechanical uprooter, herbicide, or copper nails. The results of this experiment will hopefully provide more cost effective weeding techniques.

Deer and pigs also cause much damage to vegetation, the former by grazing plants and de-barking trees and the latter by uprooting seedlings. All the established CMAs are fenced to keep these animals out. Fencing is one of the most expensive management operations, and the extent of the area of the CMA has largely been dependent on the financial resources available. In Mauritius, as a rule of thumb, the cost of fencing (using square wire mesh rather than chain-link) is roughly US$50 per linear metre. Rats also predate seeds thereby reducing the regenerative capacity of the forest, as well as on the eggs and young of native birds. Within some of the CMAs (e.g. Brise Fer and Mare Longue), a rat control programme has been implemented by MWF. Control is presently effected using Klerat® poison (active ingredient Brodifacoum 0.4%, an anticoagulent), with poison stations set 50 m apart over a grid. Trapping of other exotic predators is also undertaken. These predators include: feral cats (*Felis catus*) and mongoose (*Herpestes auropunctatus*), both of which take endemic birds and reptiles, and the omnivorous tenrecs (*Tenrec ecaudatus*) which predate invertebrates. Strahm (1993) has shown that rat control and the exclusion of deer and pigs from the forest plots has a beneficial effect on native forest regeneration.

Whilst the weeding of invasive species and control of exotic animal pests are seen to be appropriate methods for reversing forest degradation, the effects of such management on the other components of the forest biodiversity are largely unknown. It is important to emphasise, that when dealing with *in situ* conservation, any management intervention must take into account the different components of the ecosystem to ensure the effectiveness of the management regime. A series of investigations on the effect of weeding on birds (Jones *et al.*, 1992), land snails (Florens, 1996) and butterflies (Mauremootoo *et al.*, in press) have been carried out. The results showed that weeding had beneficial effects on plants, birds and butterflies, but there are some concerns regarding indigenous land snails that seem to be adversely affected after weeding. Florens (1996) showed that during the first eight years of forest regeneration after drastic weeding of exotics, densities of native arboreal species of land snail decreased and densities of exotic snails and slugs more than doubled. He recommended a more gradual removal of invasive plants together with a rapid replanting of nursery-grown native plants in weeded gaps. On the other hand, Mauremootoo *et al.* (in prep.) showed that mean butterfly abundance was 19 times greater in CMAs than in non-managed forest, and attributed this to the effects of weeding on forest structure and microclimate. Thus, the lesson learnt here is that one should carefully evaluate the effect of any management intervention on the forest ecosystem, taking into account its various components of biodiversity (see Appendix 5).

Genetic level

Conservation efforts should also be addressed at the genetic level in addition to the species and ecosystem level. Many plant species in the wild exist as a mosaic of genetically differentiated populations that have evolved in response to local selection pressures arising from environmental heterogeneity. An understanding of the amount of genetic variability and its distribution within and between existing populations is of great importance in the formulation of conservation strategies. Conservation of ecosystems does not necessarily ensure continuing adaptive change unless the genetic base of the species is sufficiently wide. For example, ecogeographic and diversity studies of wild coffee (*Coffea* L. (Rubiaceae)) species in Mauritius has helped to document the extent of genetic variability in their wild population and led to the development of conservation strategies (Dulloo *et al.*, 1998). Such studies can also be used to identify areas of high diversity, which require *in situ* protection and thereby help in the design of the reserve.

Ecological restoration

There is still a lot we do not know about the dynamics of the Mauritian forest ecosystem. What are the interactions between the indigenous flora and fauna? How are the species dispersed and what are the pollinators and dispersers? What are the conditions that will promote the regeneration of target species? How much gene flow is occurring within and between populations to ensure sufficient variability for long-term population viability? Weeding and control of animal pests alone may not be sufficient to retain existing levels of indigenous biodiversity. If a self-sustaining forest ecosystem is the desired goal, then we have to actively manage the conservation areas through habitat restoration techniques.

Reintroduction is an attempt to establish a species in an area which was once part of its historical range, but from which it has been extirpated or become extinct (IUCN, 1998; see Appendix 4). In Mauritius some attempts at plant reintroduction have been undertaken (Owadally *et al.*, 1991). However, it is becoming apparent that many of these introductions have failed owing to an inappropriate choice of species and planting regimes. The Society for Ecological Restoration (SER) defines ecological restoration as 'the process of assisting the recovery and management of ecological integrity. Ecological integrity includes a critical range of variability in biodiversity, ecological processes and structures, regional and historical context, and sustainable cultural practices' (see also Box 14.1). It is also important to establish the associated animal species and microorganisms that will ensure pollination, dispersal and establishment, ensuring ecological integrity. Restoration will only have been successful if the native species become established as viable populations. In Mauritius, such restoration work is being carried out at five different sites, namely: the offshore islands of Ile aux Aigrettes and Round Island (Figure 13.5); degraded forest sites at Brise Fer (Mauritius); Grande Montagne (Rodrigues, inland), and at Anse Quitor (Rodrigues, coastal). Each of these sites represents different forest ecosystems. The projects are being implemented by the MWF in collaboration with the Government of Mauritius, and with the Durrell Wildlife Conservation Trust, Royal Botanic Gardens, Kew and Fauna and Flora International who are under contract from the GEF Biodiversity Project (World Bank and UNDP).

<u>Figure 13.5</u>   Habitat restoration work on Round Island, Mauritius, undertaken after the eradication of introduced goats and rabbits. Here, teams from the National Parks and Conservation Service and the Mauritian Wildlife Foundation are shown surveying a site and replanting it with endemic species.

Future Steps

CMAs have proved to be a successful interim measure for protecting the threatened plant species of Mauritius. However, the future of plant conservation in Mauritius, will be dependent on the following:

- development of cost effective restoration methods integrating weed control and replanting;
- better understanding of forest ecosystem dynamics through research activities particularly on reproductive, pollination and the dispersal biology of key species;

Figure 13.6 Ecological monitoring within a managed conservation plot assessing seed regeneration.

- development of field gene banks, especially for threatened plant species which are no longer capable of reproducing;
- development of novel propagation techniques for the mass propagation of indigenous species for restoration;
- sustaining staff training and development;
- improved and supportive public perceptions on biodiversity.

Box 13.1   Case study: Habitat restoration on Ile aux Aigrettes

Ile aux Aigrettes Nature Reserve, is a 25 ha low coralline islet located off the south-east coast of Mauritius (see Figure 13.2).

This islet contains the last remnant of Mauritian dry coastal ebony forest and many of the plants found there are endemic to Mauritius.

The island is highly degraded as a result of past occupation and extensive illegal wood cutting.

In 1985, MWF initiated a restoration programme to protect and restore the native vegetation of the island (Dulloo et al., 1996). Exotic mammalian predators, such as feral cats and rats, have been eradicated and a programme for eradication of the shrew is ongoing.

The restoration work on Ile aux Aigrettes is a component of the GEF Mauritius Biodiversity Restoration Project. This work followed a number of steps:

- The production of a management plan stating the vision and the conservation objectives and detailing the activities to be undertaken on the island (Dulloo et al., 1997);

- The production of a restoration plan providing a detailed prescription on the restoration process;

- Areas earmarked for restoration were manually weeded. Nursery-grown seedlings of native pioneer species, such as Scaevola taccada (Gaertn.) Roxb. (Goodeniaceae), Dodonaea viscosa Jacq. (Sapindaceae), Hibiscus tiliaceus L. (Malvaceae), were then planted out at high densities (up to 12 plants per m²) to produce a native cover on the ground for suppressing exotic weeds.

  - In parallel to this, native bird species were established.

  - Studies to investigate ecosystem functioning have been initiated, for example, current studies are focusing on the pollination biology of native plant species by endemic reptiles, and the effect on the island's vegetation of re-introducing the land tortoise.

A monitoring programme (Figure 13.6) of species establishment and dynamics has been initiated to provide baseline data and to detect changes in the vegetation communities in relation to habitat management practices.

References

Damania, A.B. (1996). Biodiversity conservation: a review of options complementary to standard *ex situ* methods. *Plant Genetic Resources Newsletter* **107**: 1–18.

Dulloo, M.E., Jones, C.J., Strahm, W. and Mungroo, Y. (1996). Ecological restoration of native plant and animal communities in Mauritius, Indian Ocean, pp. 83–91. In: D.L. Pearson and C.V. Klimas. (eds). *The Role of Restoration in Ecosystem Management*. Society for Ecological Restoration, Seattle, USA.

Dulloo, M.E. *et al.* (1997). *Ile aux Aigrettes Management Plan 1997–2001*. Technical Series, No.1/97. Mauritian Wildlife Foundation, Port Louis, Mauritius.

Dulloo, M.E.*et al.* (1998). Complementary conservation strategies for the genus *Coffea*: a case study of Mascarene *Coffea* species. *Genetic Resources and Crop Evolution* **45**: 565–579.

Falk, D. (1990). Integrated strategies for conserving plant genetic diversity. *Annals Missouri Botanic Garden* **77**: 38–47.

Florens, F.B.V. (1996). *A study on the effects of weeding exotic plants and rat control on the diversity and abundance of land gastropods in the Mauritian wet upland forest*. B.Sc. thesis, School of Science, University of Mauritius, Mauritius.

IUCN (1998). *IUCN Guidelines for Re-introduction*. Prepared by the IUCN/SSC Re-introduction Specialist Group. IUCN, Gland, Switzerland.

Jones, C.G., Swinnerton, K.J., Taylor, C.J. and Mungroo, Y. (1992). The release of captive-bred Pink Pigeons *Columba mayeri* in the native forest on Mauritius. A progress report July 1987 – June 1992. *Dodo Journal of the Wildlife Preservation Trust* **28**: 92–125.

Loope, L.L. and Mueller-Dombois, D. (1989). Characteristics of invaded islands, with special reference to Hawaii, pp. 257–280. In: A. Drake, F. Di Castri, F. Groves, R. Kruger, and H.A. Mooney. (eds). Biological Invasions: a Global Perspective. SCOPE. John Wiley & Sons, Chichester, UK.

Maunder, M. *et al.* (unpublished data). The Status and Management of the Critically Endangered Endemic Flora of Mauritius.

Mauremootoo, J.R., Fowler, S.V., Florens, F.B.V. and Towner, C.V. (in prep). The effect of weeding alien plants and of canopy cover on the butterfly fauna of Mauritian montane forest.

Maxted, N., Ford-Lloyd, B.V. and Hawkes, J.G. (1997). Complementary conservation strategies, pp. 15–39. In: N. Maxted, B.V. Ford-Lloyd and J.G. Hawkes (eds). *Plant Genetic Conservation: the in situ approach*. Chapman and Hall, London, UK.

Owadally, A.W., Dulloo, M.E. and Strahm, W. (1991). Measures that are required to help conserve the flora of Mauritius and Rodrigues in *ex situ* collections, pp. 95–117. In: V.H. Heywood and P.S. Wyse-Jackson (eds). *Tropical Botanic Gardens – Their Role in Conservation and Development*. Academic Press, London, UK.

Safford, R.J. (1997). A survey of the occurrence of native vegetation remnants on Mauritius in 1993. *Biological Conservation* **80**: 81–188.

Santos, G.L., Kageler, D., Gardner, D.E., Cuddihy, L.W. and Stone, C.P. (1992). Herbicidal control of selected alien plant species in Hawaii Volcanoes National Park, pp. 341–375. In: C.P. Stone., C.W. Smith and J.T. Tunison (eds). *Alien Plant Invasions in Native Ecosystems of Hawaii: Management and Research*. University of Hawaii Cooperative National Park Resources Study Unit, Honolulu, USA.

Strahm, W. (1989). *Plant Red Data Book for Rodrigues.* Koeltz Scientific Books, Konigstein, Germany.

Strahm, W. (1993). *The Conservation and Restoration of the Flora of Mauritius and Rodrigues.* PhD. thesis, University of Reading, UK.

Strahm, W. (1994). Regional overview: Indian Ocean islands, pp. 265–292. In: WWF and IUCN (1994–1997). Centres of plant diversity. A guide and strategy for their conservation, 3 volumes. IUCN Publications Unit, Cambridge, UK.

Strahm, W. (1996). The vegetation of the Mascarene islands. *Curtis's Botanical Magazine* **13**: 214–217.

Vaughan, R.E. (1968). Mauritius and Rodrigues. In: I. Hedberg and O. Hedberg (eds). Conservation of Vegetation in Africa South of the Sahara. *Acta Phytogeographica Suecicica* **54**: 265–272.

Vaughan, R.E. and Wiehe, P.O. (1937). Studies on the vegetation of Mauritius. I. A preliminary survey of the plant communities. *Journal of Ecology* **25**: 289–343.

Vaughan, R.E. and Wiehe, P.O. (1941). Studies on the vegetation of Mauritius III. The structure and development of the upland climax forest. *Journal of Ecology* **29**: 127–160.

Vitousek, P.M. (1988). Diversity and biological invasions of oceanic islands, pp. 181–189. In: E.O. Wilson (ed.) *Biodiversity.* National Academy Press, Washington DC, USA.

Walter, K.S. and Gillett, H.J. (eds) (1998). *1997 IUCN Red List of Threatened Plants.* Compiled by the World Conservation Union Monitoring Centre. IUCN – The World Conservation Union, Gland, Switzerland.

Chapter 14

Restoration of a Small Tropical Coastal Forest in Kenya: Gede National Monument Forest Restoration Project

Ann Robertson
Malindi, Kenya

Clare Hankamer
Royal Botanic Gardens, Kew, UK

Mathias Ngonyo
Malindi, Kenya

Introduction

Tropical dry forests are amongst the most threatened of tropical habitats. Indeed, the level of degradation is far more advanced than that of tropical wet forests (Cabin *et al.*, 2000; Janzen, 1988; Mooney *et al.*, 1995). In Africa, dry woodlands and forest are increasingly impacted by human utilisation (Menaut *et al.*, 1995). The East African coastal forests of Kenya and Tanzania have been identified, in a number of key studies setting global priorities for plant conservation, as particularly rich in endemic species and under serious threat as a result of habitat loss (Myers *et al.*, 2000; Olsen and Dinerstein, 1998; WWF and IUCN, 1994–1997). Olsen and Dinerstein (1998) categorise this ecoregion as critically endangered in terms of its ability 'to maintain viable species populations, to sustain ecological processes and to be responsive to short-term and long-term environmental changes'. Myers *et al.* (2000) place these forests as one of the eight 'hottest' global biodiversity hotspots for conservation support (Figure 14.1). The Convention on Biological Diversity (CBD) (Glowka *et al.*, 1994) (Article 8(f); see Appendix 1) requires Contracting Parties to 'rehabilitate and restore degraded ecosystems'. This chapter discusses the initial results of a project in Kenya to restore a degraded forest remnant in this globally important coastal region.

The forests of coastal Kenya are fragmented relics of the once extensive Zanzibar-Inhambane Regional Mosaic (White, 1983). In the mid 1990s, the coastal forest in Kenya was estimated to cover about 660 km^2 (Burgess *et al.*, 1998, Burgess and Clarke, 2000). The surviving forest patches range in size from the extensive Arabuko-Sokoke Forest Reserve of approximately 400 km^2, to minute fragments of a few hectares, which are often the remnants of sacred forest or Kayas (see Githitho, Chapter 8). The high floral diversity of these forest fragments has been well documented (Beentje, 1988; Birch, 1963; Dale, 1939; Gillett, 1979; Hawthorne, 1984, 1993; Lucas, 1968; Polhill, 1989; Robertson and Luke, 1993) with more than 50% of Kenya's rare trees occuring in the Coast Province (Beentje, 1988). This vegetation type is increasingly cleared for hotel and residential developments, and it has only been possible to protect a few small areas. Some are protected as National Forest Reserves, others are National Parks or National Reserves, and those known to be Kayas have been given National Monument Status. They are administered by the National Museums of Kenya (NMK) Coastal Forest Conservation Unit (CFCU) and managed in collaboration with the local communities (Robertson, 1992; Wilson, 1993; see Githitho, Chapter 8).

There have been increasing and ongoing efforts to protect and conserve the forests. Indeed, the Kenya National Environmental Action Plan (1994) calls for immediate action to protect Kenya's coastal forests alongside planning for their sustainable use by the local community. It is likely that all the forests extant today have been affected by centuries of human settlement along the

Figure 14.1   The coastal forests of East Africa are recognised as global priorities for conservation. Any regional conservation strategy should focus on retained forest areas such as the Shimba Hills, Kenya, but also recognise the imperative for restoration.

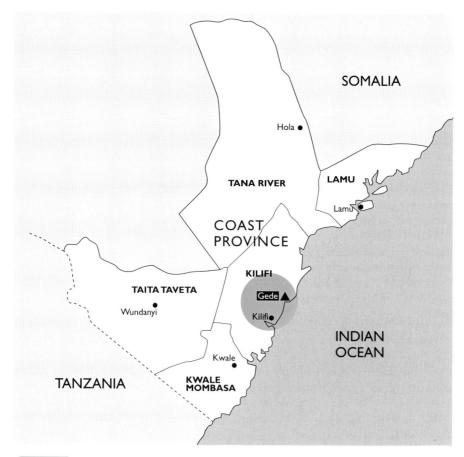

Location of Gede on the Northern Kenya coast.

East African coast (Hamilton, 1981). Forest areas have been repeatedly cleared and grown back following abandonment of plots and settlements. The greatest current threat is from coastal tourism with the clearance of coastal forest for hotel complexes (Luke, 1996). Ironically, income from tourism in Kenya has dropped dramatically over the past few years, partly as a result of environmental degradation (Shah *et al.*, 1997).

Site description and project rationale

Gede is the excavated ruins of a thirteenth–seventeenth century stone city, surrounded by a 44 ha patch of forest, near the tourist resorts of Malindi and Watamu (Figure 14.2). In 1948, Gede was declared a National Park and later became a National Monument (NM). In 1969, following independence in Kenya, the administration was taken over by NMK, and the site, including the surrounding forest, became a popular attraction for both residents and tourists. It is thought that when the city was thriving, most of the area now supporting dense woodland would have been covered with gardens and buildings, with only a few relict trees, such as baobabs, surviving. Since the site was abandoned in the seventeenth century, the forest has grown back, seeded from the extensive forest that separated Gede from the historic town of Malindi to the north.

There is evidence that the forest was revered in local tradition, as particular trees, and the ruins, were considered sacred and were the foci of prayers and offerings until quite recently (Alaussy, pers. com.). Although the Gede site is managed and promoted primarily for tourism, with a new museum complex developed in 1999, the natural forest has also been protected. The site attracts paying visitors and educational tours, and a short illustrated guide (Faden and Faden, 1972) to some of the common trees of the forest was produced with these groups in mind. The administrative unit of the NMK Kipepeo butterfly farming project for the adjacent Arabuko-Sokoke forest community was set up at Gede in 1993, and this also emphasised the importance of the forest patch as a refuge for rare butterfly species (Gordon, 1998).

Surveys of the flora (Faden, unpublished; Gerhardt and Steiner, 1986, and this project) have led to the identification of a new species and a list of over 280 taxa of flowering plants. The present flora is typical of the coastal remnant and re-growth forests (Birch, 1963; Frazier, 1993). They contain a mixture of tall, fast-growing pioneer species, such as *Gyrocarpus americanus* Jacq. (Hernandiaceae), *Sterculia appendiculata* K. Schum. (Sterculiaceae), *Antiaris toxicaria* Lesch. (Moraceae) and *Ficus* L. spp. (Moraceae), and hardwoods, such as *Combretum schumannii* Engl. (Combretaceae), *Cordyla africana* Lour. (Caesalpiniaceae) and *Mimusops obtusifolia* Lam. (Sapotaceae). Numerous baobabs, (*Adansonia digitata* L. (Bombacaceae)), some of enormous girth, are a typical feature of these forests.

Black and white colobus monkeys (*Colobus angolensis*) were reported in Gede during the 1960s (Williams and Arlott, 1981). The fragmented nature of the coastal forests of this region and the particularly small size of the forest patches suggests that mammal populations may become less viable over time. It is unclear also what effect this fragmentation has had on individual plant species possibly reducing reproductive success. *Ficus faulkneriana* C.C. Berg (Moraceae), a Critically Endangered (*sensu* IUCN, 1994) coastal endemic of Kenya and Tanzania, is thought to exist as a few moribund populations on cleared and marginal land (Mwasumbi *et al.*, 1998). The wasp pollinator for this species is thought to be extinct as no viable seed is produced from these individuals (Luke, pers. com.).

Project Aims and Objectives

In the 1980s, as the nearby Gede village expanded and the surrounding areas were cleared for agriculture, it became evident that the forest patch was at risk. The whole site was securely fenced in 1991, and casual cultivation, cutting of poles and collection of firewood within the boundary was stopped. A degraded area in the northern section of the site, extending over about 5 ha, was designated for forest restoration. The Curator planted some indigenous tree seedlings, obtained from the local Forest Department Nursery, in this area in 1991. Unfortunately, these were not all suitable species (and most have since died), but this gave rise to the suggestion to try a more structured restoration project (Njuguna, pers. com.). With the increasing interest in tropical forest conservation, recent experience gained from the three year World Wide Fund for Nature (WWF) NMK Coastal Forest Survey (Robertson and Luke, 1993), and the first author's horticultural expertise and situation (based in nearby Malindi), it was decided to seek funding to carry out a formal restoration project. Ecological restoration has become increasingly prominent as a component of biodiversity conservation (see Box 14.1) and yet there are relatively few case studies published for tropical ecosystems detailing the techniques and results.

With the backing of the NMK Centre for Biodiversity, the first author prepared a project proposal which was submitted to the Royal Netherlands Embassy in Nairobi. The project was carried out in two phases: Phase I from January 1992 to December 1993, with funding from the Netherlands Koningschool Fund; and Phase II from January 1994 to November 1996 as the Gede National Monument Forest Restoration Project, with funding from the Netherlands Ministry of Agriculture, Nature Management and Fisheries.

The objectives of the project were to:

- restore a 5 ha degraded area of forest within the secure, fenced, boundaries of the NM site
- carry out enrichment plantings in other poorly forested sections of the site
- increase awareness of the importance of tropical forest conservation among visitors to the site, and among the adjacent community
- increase knowledge of the propagation and growth potential of selected tree species of this forest type
- train local staff in propagation, planting and monitoring techniques and in knowledge of the flora and fauna of the site

---

Box 14.1   Ecological restoration: its importance and goals

Ecological restoration is integral to the requirements of the Convention on Biological Diversity. Under Article 8(f) of the CBD (see Appendix 1) each Contracting Party is required to 'rehabilitate and restore degraded ecosystems and promote the recovery of threatened species *inter alia*, through the development and implementation of plans or other management strategies'.

The goals for ecological restoration (Atkinson, 2000) may be:

- the reconstruction of an historical ecosystem;
- the conservation of biological diversity at the threatened species level, genetic level or successional community level;
- the restoration of the evolutionary context of the species in an ecosystem.

Project Implementation: Phase I

The project was initiated in November 1991, and was marked by a tree planting ceremony by HRH Prince Bernhard of the Netherlands. Work on the Gede Koningschool Forest Project began early in 1992.

1. Design of restoration plots

The restoration site was designed and set up on the basis of priority areas for restoration and ease of maintenance and monitoring of plots over the restoration period.

- A species inventory of the site was initiated at the start of the project and updated throughout the project.
- Priority was given to open, unvegetated areas for restoration planting.
- The main restoration site of 5 ha was surveyed and divided into a grid of adjacent plots of 20 × 20 m. The area had varied cover, from some almost cleared patches, with differing soil fertility, to areas with quite thick cover, including a few large trees (Figure 14.3).
- 32 plots were set up, each with 1 m paths around the perimeter to act as fire breaks.
- 40 trees were planted per 400 m² plot. This planting density was adopted so that good tree cover would be obtained even if several trees died per plot.

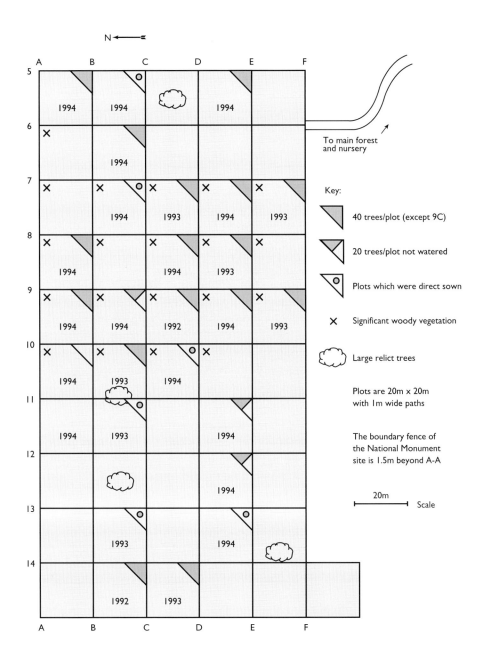

Figure 14.3 Layout of restoration plots at Gede National Monument showing dates when plots were planted.

- Planting was carried out around existing relief trees. This 'copse' planting was felt to be more appropriate than systematic planting over the whole plot area.
- Trees were planted over a four year period (project implementation phases I and II).
- Other areas were identified as priority for restoration within the forest. Planting in these areas was not done on a grid design.
- 75 species were planted, with an average of 24 species per plot.

Control plot

A control plot of 40 trees was set up in 1993, where plants were not watered, to monitor what effect this had on seedling establishment. Plants were measured in the same way as for the other plots.

2. Collection of plant material for the restoration site

Prior to the start of the project the following material was obtained:

- Wild seedlings (wildlings) of trees and shrubs were collected from the existing Gede NM forest by resident staff;
- Other suitable seedlings were donated from the WWF/NMK Coast Forest Survey project (Robertson and Luke, 1993);
- For the rest of the project all propagating material (seeds, seedlings and cuttings) were obtained from within Gede forest;
- All source material (apart from seed used for the direct seeded plots) was grown on in the nursery until ready for planting:
  - 55% of the plants used for restoration were from seed source (84% of this was sourced from Gede NM forest);
  - 45% of the plants used for restoration were wildlings, grown on in the nursery (93% of this was sourced from Gede NM forest);
- In the initial planting in 1992, material was used that had been held in the nursery for some time (up to seven years). Younger material was used in subsequent years to meet planting demand.

3. Nursery management of seedlings and wildings

A tree nursery was established, a small office renovated, simple hand tools, black bags and other equipment purchased, and water storage tanks installed. Young seedlings or wildings were planted into 9 cm diameter black polythene bags initially, and then transplanted into 16 cm diameter bags. They were transplanted again to 27 cm diameter bags if necessary.

4. Criteria for selecting plants for plots

Species were chosen for planting in the restoration plots using a combination of factors:

- known to commonly occur at Gede; information on species indigenous to Gede was obtained from the species inventory. As this inventory was started at the beginning of the project, information on what species to select for the restoration plot improved with time;

- found at other collection sites, but thought likely to occur historically at Gede;

- ease of propagation and establishment, taking into account:
  - availability of propagation material;
  - ability to germinate;
  - transplantation success of wildlings;
  - speed of growth and health in the nursery, e.g. *Erythrina* L. (Leguminosae) spp. was avoided as it is prone to attack by tip gallwasps, causing high levels of mortality;

- attractive to frugivores: birds, bats and monkeys, e.g. *Ficus* L. spp., to encourage natural seed dispersal within the restoration area.

5. Planting, plot management and monitoring

In the first wet season, in May 1992, 104 seedlings of trees and shrubs were planted out (93 in plots and 10 in other areas), and in April and May 1993, 351 seedlings were planted (241 in plots and 110 in other areas).

i) Planting regime

- Seedlings were planted during the wet season to help ensure successful establishment.

- Through tree planting experience gained before the project, it was known that tree seedlings with a sizeable root ball, which usually correlated with a height of about 0.5 m, would establish better than smaller seedlings. The erratic rainfall and moisture availability in coastal Kenya had to be considered;

- Planting holes of about 25 cm diameter and depth (to allow space for compost and the large root ball), and 1 m apart from each other, were prepared in the selected plot;

- No seedlings were planted under the canopy of large remnant trees, as roots usually obstructed the digging of planting holes;

- Compost (see below) was placed in the bottom of the hole and seedlings were removed from their bag, placed in their planting positions and firmed in using the remaining soil.
- All the planted trees in the plots were staked (using poles cut from naturalised neem (*Azadirachta indica* A. Juss. (Meliaceae)), and numbered and mapped for future reference.

The restoration of the main degraded area was achieved by staggered planting over four years (Figure 14.3). Plots to be planted each year were chosen using a combination of factors. The barest (most recently cultivated, poorest soil) were planted first, but in 1992 only a limited number of suitable seedlings were available. Also in 1992, quite a few shrubs were planted, but because of limitations of funds, it was decided to concentrate on planting trees in the following years. As rainfall was very unreliable, it was felt necessary to ensure aftercare in the field with watering and weeding. This plot management was kept to a minimum to reduce costs.

ii) Plot maintenance

Basic plot maintenance was carried out in the following ways:

- Compost was produced for use in the potting mix and for the tree planting holes using leaf sweepings and manure;
- The newly planted trees were kept clear of competing weeds and climbers;
- The fire break around the main planted area, and the paths between the tree plots, were kept cleared;
- Other tasks included pruning natural re-growth of coppiced trees and removal of woody weeds, such as *Lantana camara* L. (Verbenaceae) (see Appendix 5) but little of this was done due to shortage of staff;
- The boundary fence was also checked regularly to prevent goat-herders forcing entry for goats searching for food during the dry seasons.
- In dry periods, irrigation of planted trees was the main task, this was done for up to a year after planting to take the plants through to the following long rains. This entailed pushing a cart with eight 20 litre (l) water cans to the two main planted areas, the furthest trees being over 500 m from the water source at the nursery. A hand pushed water bowser (44 gallon (gal) drum on wheels) was commissioned and purchased to make watering easier. Trees at a dry site received about 7 l of water twice a week.

Those in better soil received 7 l of water once a week. Staff willing to work overtime ensured that watering was done in the early morning and late afternoon for maximum efficiency. The mains water supply to the site was erratic, and on a number of occasions failed completely for several weeks. Tankers of water had to be purchased from Malindi when the limited amount of stored water was used up. When the Kipepeo Butterfly Farm was established at Gede the cost of obtaining tankers of water was shared, and the water supply improved in 1996. Watering stopped at the onset of the long rains in April. Rainfall records were obtained from nearby sites (Figure 14.4), and from a rain gauge installed in the nursery, and were used to make decisions on when to water the trees.

iii) Plot monitoring

- Sufficient staff (4–6 per day) were employed to care for the nursery and the planted out seedlings. A supervisor was trained to take day-to-day charge of the project and to monitor the growth of the planted trees.
- Natural regeneration of seedlings in the plots was recorded.
- The species inventory for the whole forest patch was updated as more species were identified.
- Herbarium voucher specimens were collected.
- The following data were collected for each tree planted:
  - height,
  - diameter at breast height (dbh) for each tree with > 1 cm (centimetre) dbh.
- Data were recorded at the end of the long rains (July/August) in each year after planting (up to 1996).
- Measurements were also made in 1997, after the end of the project with collaboration from the site Curator.
- Information on the project and general forest conservation was displayed to the public.
- Members of the Wildlife Clubs of Kenya from a nearby school were involved in the project in 1993 when they were asked to plant, care for and monitor 40 seedlings in a specially set up school project plot.

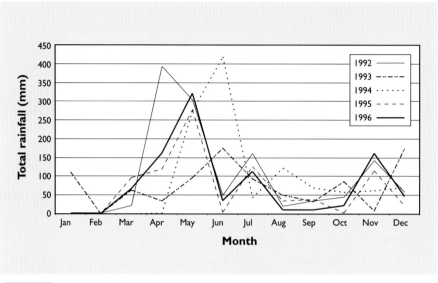

Figure 14.4   Rainfall data for the Gede National Monument site.

Project Implementation: Phase II

In Phase II of the project the nursery was maintained, additional equipment was purchased and the numbers and range of species was increased. In April 1994, 400 trees were planted in the main restoration site. In May 1995, 420 trees were planted in the plots and 125 trees were planted in other degraded areas. Over the whole project about 75 species, mostly trees, but including some shrubs, were planted in the main restoration area. Dead trees were replaced in some of the previously planted plots. Mr. Ngonyo continued searching the site for new species for the nursery and for the species inventory, which by early 1997 listed 284 species.

Seeded plots

A number of plots were selected for restoration by direct sowing. These characteristically had bushy re-growth of stumps of older trees that had been coppiced or fire damaged for agricultural clearance in the past. Due to lack of funding and resources, these remnant trees were not mapped nor monitored, but the main species present were from the genera: *Xylopia* L.

Table 14.1 **Species attaining a mean relative gain in height of 0.2 m or greater per year**

Species with a mean relative gain in height of >0.2m	Mean relative gain in height (m/yr)	Standard deviation
For all plots planted in 1995		
Afzelia quanzensis Welw.	0.26	0.23
Grewia plagiophylla K. Schum.	0.39	0.11
Haplocoelum inoploeum Radlk.	0.40	0.06
Milletia usaramensis Taub.	0.26	0.09
For all plots planted in 1994		
A. quanzensis	0.32	0.18
Balanites wilsoniana Dawe & Sprague	0.34	0.33
Berchemia discolor (Klotzsch) Hemsl.	0.22	0.11
Bourreria petiolaris (Lam.) Thulin	0.72	0.28
Carpodiptera africana Mast.	0.39	0.11
Combretum schumannii Engl.	0.25	0.18
Commiphora zanzibarica (Baill.) Engl.	0.40	0.34
Cordyla africana Lour.	0.21	0.07
Ficus bussei Warb. ex Mildbr. & Burret	0.25	0.16
Ficus sansibarica Warb.	0.56	0.32
G. plagiophylla	0.71	0.10
Kigelia africana (Lam.) Benth.	0.62	0.27
Lepisanthes senegalensis (Juss. Ex Poir.) Leenh.	0.23	0.15
M. usaramensis	0.20	0.09
Mimusops obtusifolia Lam.	0.28	0.03
Sideroxylon inerme L.	0.49	0.15
Sorindeia madagascariensis Thouars ex DC.	0.30	0.17
Suregada zanzibariensis Baill.	0.49	0.22
Terminalia spinosa Engl.	0.37	0.26
Trichilia emetica Vahl	0.30	0.23
Ziziphus mucronata Willd.	0.35	0.14
For all plots planted in 1993		
M. usaramensis	0.77	0.20
S. zanzibariensis	0.51	0.26
T. emetica	0.39	0.18

Note: Other species are not listed where standard deviation exceeded mean relative gain in height.

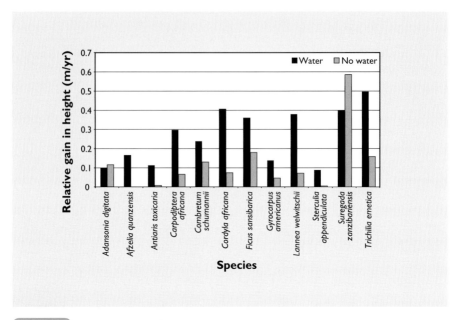

Figure 14.5   Comparison of two plots planted in 1993, to show the effect of watering on plant growth rate.

(Annonaceae), *Trichilia* P. Browne (Meliaceae), *Dalbergia* L. f. (Leguminosae) and *Azadirachta* A. Juss (Meliaceae). As an alternative strategy to planting in these plots, a mixture of suitable seeds was scattered in two plots in 1993 and in five plots in 1994. These seeds had been collected earlier in each year. Where possible, more than 50 seeds of each species were used for each plot, but not enough seeds were available for some species. This method of restoration proved to be difficult to monitor. As more natural regeneration took place, it was difficult to accurately identify the seedlings growing from the seeds scattered by the project.

Initial observations on the restoration process

A reliable source of fresh water was essential for the maintenance of the nursery and for watering the seedlings for at least two dry seasons, after planting in the long rains (April/May/June). Usually in one year there are two dry seasons, separated by the long and short rains, but there is great variability around this norm. In the event, only 14% of the 1,165 plants planted subsequently died (Of this 14%, 59% were collected as wildlings and 41% as seeds prior to growing on in the nursery). This is a fairly high success rate for establishment where one would normally expect high mortality (Milton *et al.*, 1999). This success was due to the quality of the horticultural aftercare. Such skills proved to be important in planning to bulk up enough stock for subsequent plantings. Lack of suitable stock was a problem early on in the project but this was alleviated as knowledge about the cultivation of each species developed. Practical plant husbandry and observational skills are vital to understand why some species fail to establish (Atkinson, 1990).

Where no watering was given, most species grew more slowly, but very few species died (Figure 14.5). Interestingly, *A. digitata* and *Suregada zanzibariensis* Baill. (Euphorbiaceae) grew faster in the control (unwatered) plot. As the remnant woody vegetation in the plots re-grew, the number of trees planted per plot was reduced from 40 to 20–25 where necessary. Unfortunately, it was not possible to monitor the natural re-growth because of limitations of funding.

The diversity of species had also been partially restored, but was limited by the availability of species in the nursery. Table 14.1 shows the species that attained a relative growth rate of >20 cm per year. These fast growing species were identified as colonisers. Surprisingly, *G. americanus, S. appendiculata, A. toxicaria* described as fast-growing pioneers in earlier studies (Robertson and Luke, 1993) did not demonstrate such a rapid annual growth rate. Further work on the demography and ecology of these species compared with species such as *Millettia usaramensis* Taub. (Papilionaceae) and *Trichilia emetica* Vahl (Meliaceae) would prove useful. Atkinson (2001) claims that such single species studies can very effectively improve our understanding and management of habitat restoration processes.

Tropical restoration research has suggested that the successful establishment of indigenous tree species is helped by the use of exotic species as a nurse crop (Ray and Brown, 1995), which are harvested at a later stage (Chapman and Chapman, 1996; 1999; Parrotta, 1995). However, if indigenous cover species can be identified using these methods then this serves the same purpose.

During the life of the project, vigorous re-growth of cut stumps, and root suckers was observed. There were many coppiced neem trees (*A. indica*), a naturalised tree along the Kenya coast, and thick growth of the woody weed *L. camara*. It had been planned to prune out these invaders, but there was not sufficient project staff to do this. It is hoped that the fast growth of the planted trees will crowd out the woody weeds.

Because of the fence and continuous monitoring by project staff, illegal cutting in the restoration area by people from Gede village was kept to a minimum. Over time, it became obvious that prevention of fire, and stopping cultivation and pole cutting was also leading to spontaneous forest re-growth. Restoration is a very long-term process, and a relatively short-term project should only aim to accelerate a natural process of forest re-growth. However, with some of the first trees planted measuring between 4–6 m by July 1997, and with over 80% survival, there were, by mid-1997, enough saplings established to ensure future forest cover. This is an important factor, which has to be taken into account when examining the success of restoration projects. It was pointed out that if the plot had just been left alone, natural re-growth, and eventual natural re-seeding from the extant forest, would have achieved the same objective. However, without project supervision of the site, and the interest and concern of the staff in ensuring their hard work in planting and watering achieved results, it is likely that the natural re-growth would have been retarded. The establishment of numerous trees and a diversity of species would have taken a lot longer, maybe decades.

For the direct sown plots, general trends in species establishment only were observed. *Tamarindus indica* L. (Leguminosae) attained the greatest height in 2-3 years, followed by *Afzelia quanzensis* Welw. (Melastomataceae). It is interesting to note that both species are very slow to grow if planted as seedlings. Some species did not germinate at all. Growth potential varied according to whether the seeds were scattered widely or slightly buried in a clump. Some of those not buried were noted to have been eaten by mammals or insects. Those scattered in the open did not do so well as those in shade, and those in partial shade did better than those in dense shade. Of those in buried clumps, the strongest seedlings (often *A. quanzensis*) survived at the expense of the weaker seedlings. The planting methods utilised here such as 'copse' planting within existing vegetation, and planting of seedlings appear to show greater establishment success than direct sowing methods. This trend is reflected in other studies on translocations of coastal species in South Africa (Milton *et al.*, 1999) and in Caribbean dry forest restoration (Ray and Brown, 1995).

Project Funding and Capacity Building

1. **Broader utilisation of nursery-grown stock**

   i) **Public sales.** During the project, excess plants from the nursery were sold to the public and other institutions for approximately US$1 each.

   ii) **Public awareness.** The indigenous tree nursery was continued after the end of the project at the request of the Curator, as a service to the community and to maintain awareness of the importance of planting indigenous trees. It was recognised that this would not be a revenue earner for NMK, but indigenous tree seedlings could be grown to order if suitable propagating material could be obtained locally or from elsewhere at the coast.

   iii) **Support to other local restoration projects.** The bulk of plants were made to the Bamburi Nature Trail in Mombasa. This project has been responsible for the restoration of areas quarried for coral for the Bamburi Cement factory.

   iv) **Support to small-scale business.** Seedlings of food plants for the Kipepeo butterfly farming project were also propagated.

2. Linking with regional coastal forest conservation activities

It is intended that the long-term role of the Gede restoration nursery should complement the role of the CFCU indigenous nursery at Ukunda. The CFCU nursery has focused on plant sales to the local community, but in particular to the hotel industry. The first author worked with local hotel landscapers and horticulturists to consider the importance and efficacy of incorporating indigenous species into hotel landscapes and planting. The CFCU is also involved in species recovery work of a number of threatened species (see Chapter 8 and Chapter 15).

3. Project staff and administration

The project was administered by the Project Executant, Mrs. S.A. Robertson, Research Scientist and Coastal Biodiversity Adviser with NMK, resident in Malindi, on average two days a month (approximately 120 person days, January 1992 to November 1996). The four regular staff were employed full-time on casual terms, five days a week (approximately 5,000 person days); one being the higher paid Nursery Supervisor, Mr. Mathias Ngonyo. Extra workers were taken on as necessary. These are very rough estimates of time invested as both the Project Executant and the Nursery Supervisor carried out plant identification, seed and plant collection, and general monitoring in their spare time. The field staff often worked overtime to do the watering in the dry seasons. A major difficulty has been retaining skilled staff over the long term as their work has been funded on a series of short-term contracts. Links were developed with the Conservation Projects Development Unit (CPDU), Royal Botanic Gardens, Kew through a number of workshops, notably the Biodiversity Restoration Workshop as part of a regional capacity building project: the NMK-Darwin Plant Conservation Techniques Course for East Africa.

4. Training and public awareness

The Project Executant and the Nursery Supervisor participated in National Tree Planting Days in the area, and trees were donated from the nursery for many other tree planting ceremonies. Through a generous grant from the Kenya Museum Society, a slide projector and screen were purchased for the project, and this enabled talks to be given. Display boards were maintained at the tree nursery to explain the project to visitors and students, and to promote environmental awareness and the importance of tree planting. The Nature Trail set up by Gerhardt and Steiner was maintained, with name boards replaced where necessary, and a map of the Nature Trail displayed on a board at the entrance to the site. It proved difficult to involve schools directly in the project. It had been hoped to further involve the Wildlife Club from Gede Secondary School, but the Club went through a difficult time so no additional collaboration was possible.

5. Project budget and expenditure

The project expenditure is given in Table 14.2. The two initial grants, for US$7,352 and US$9,868, were increased by US$550 and US$2,723 by interest accrued on fixed deposits, by tree sales, by one donation, and by the sale of the water bowser at the end of the project. The sum remaining from Phase I, US$91, was carried over to Phase II. The travel and transport costs for Phase II were minimal due to use of a vehicle from another project, for which NMK provided basic office and nursery space, water (in Phase II) and electricity. The total expenditure for both phases of the project came to US$20,000 at November 1998 rates for restoration of 5 ha.

Follow-on Monitoring

The management regimes used here are valuable in giving an indication of how succession can be manipulated to support the still relatively small amount of tropical ecosystem restoration research. Numbering and mapping the plots and trees in the main area ensured that the height measurements could be continued and would yield reliable data on growth rates. Unfortunately, despite interest in continued monitoring shown by permanent staff at the Centre for Biodiversity at NMK and by the Kenya Forestry Research Institute, no action has been taken. The last measurements were done in 1997.

A species inventory and specimen collection is essential to build the long-term capacity for such restoration sites. At the end of the project there were 284 taxa on the plant list. There are undoubtedly more taxa to be collected, including many herbs, grasses and sedges, but some of those already noted will probably die out as the forest is restored. Some of the names need to be

Table 14.2 **Breakdown of project expenditure for Phases I & II from January 1992–March 1997**

Budget lines	Phase I (US$)	Phase II (US$)
Salaries and allowances	4,696	10,498
Travel and transport	1,070	170
Capital expenditure	474	470
Running expenses	1,571	1,054
Total	**7,811**	**12,192**

Figure 14.6   *Ziziphus robertsoniana* Beentje (Rhamnaceae) planted in the grounds of the Nomad Beach Hotel, Diani, Kenya as part of a scheme promoting the planting of indigenous coastal forest species in hotel grounds. This species is Endangered (*sensu* IUCN, 1994).

checked, and voucher specimens collected where none exist. The objective of updating the plant list provided by Gerhardt and Steiner (1986) was to try to ensure that only species already known to be growing in the Gede NM forest were used in the restoration work. There were a few inconsistencies. *M. usaramensis* and *Ficus natalensis* Hochst. (Moraceae) seedlings were planted, because they were available, and came from very similar coastal coral rag forest, but were never identified as growing naturally in the Gede NM forest.

Restoration also presents the possibility of restoring species that may have once been part of the flora of that site, but have since been lost. This project did not attempt to replace lost species, in part because there was no record of what was there until recently, but the site could give an opportunity to provide a refuge for threatened and rare plants known from this site (*Balanites wilsoniana* Dawe and Sprague (Zygophyllaceae)) and from other similar sites, such as *Grevea eggelingii* var. *keniensis* (Verdc.) Verdc. (Montiniaceae), *Lasiodiscus pervillei* Baill. (Rhamnaceae), *Ziziphus robertsoniana* Beentje (Rhamnaceae) (Figure 14.6), *Croton megalocarpoides* Friis & Gilbert (Euphorbiaceae) and *Flabellariopsis acuminata* (Engler) Wilczek (Malpighiaceae).

Lessons Learnt

- Given the chosen plant husbandry techniques, seedling survival rate was very high and this should remain so, assuming that current practices continue, including adequate site management, removal of goats and maintenance of fire breaks. This project demonstrates how much can be achieved in restoration using good horticultural skills (Affolter, 1997).

- Financial limitations dictated the number of staff employed and it was estimated that limiting planting to around 400 trees per year would be most effective in terms of time and money spent, and in the labour required for watering. Availability of water was also a limiting factor, and need for water must be considered when planning restoration projects.

- Investment in training and dissemination to the community of Gede village was important to ensure appreciation of the local value of the restoration project. Future activities should involve the community in the planning process.

- Availability of plant inventories of the Gede NM site facilitated the choice of species for restoration and highlights the importance of taxonomic support for restoration work.

- Consideration of statistical design for data analysis in the early planning stages of the site would have facilitated data analysis at the end of the project and also use of the site for ongoing research.

- Lack of long-term funding for such work is common, but the lack of available information on the success of restoration projects in the tropics calls for commitment from funders for long-term support to monitoring and evaluation.

Future Steps and Conclusions

Laying aside considerations of human interference in the restoration process, the goal of any restoration project has to be subjective as it depends on the natural direction of vegetative succession (Brown and Lugo, 1994). When attempting to restore tropical forests, is the combination of species planted representative of a stable or dynamic forest type? There was no information available on the distribution of species in the Gede NM forest fragment prior to restoration, but it is assumed that there was no forest at the site about two or three hundred years ago. Is the Gede NM forest changing naturally from an association of fast-growing pioneers to some more stable association? One study of forest recovery in Kibale National Park, Uganda, showed that despite planting seedlings and direct sowing, natural succession to canopy vegetation was deflected to vegetation dominated by two weedy species, which formed a dense understorey, preventing growth of seedlings through this layer (Chapman and Chapman, 1999). The restoration biologist therefore has to be prepared for a number of unpredictable outcomes. It is best not to be too exacting when planning restoration goals (See Box 14.3), but to emphasise the importance of restoration wherever and whenever possible as a strategy alongside forest conservation.

There is a need to evaluate the restoration project to see how effective it was. Questions for consideration in the long-term monitoring of these plots are:

- How rapidly are viable populations of individual species establishing?

- What are the keystone species in the restoration site? Are they the same as those for the reference ecosystem?

- Are invasive species, such as *L. camara*, being outcompeted by potential canopy species? If so, how and over what period of time?

- How is natural regeneration occurring? Is it via visiting mammals and birds from neighbouring sites?

- Are planted species gradually being replaced by naturally regenerating species? What are the factors affecting this process?

The project was extremely successful in restoring a diversity of trees in degraded areas of a threatened coastal forest within the fenced site. The funding was limited, and the supervision minimal, but the enthusiasm of the staff, and the ability of the Nursery Supervisor, played a large part in the project's success. Also, the project contributed significantly to staff training, and raised awareness of forest conservation and restoration among school children and the general public.This community support will be essential in determining the future use of the site for education and as a source of material for further larger-scale restoration (Atkinson, 2001; Cabin *et al.*, 2000).

Box 14.3 Suggestions for Preparation and planning for degraded tropical forest restoration

Preparation

- Select a 'reference ecosystem' to provide essential information on species and associations. This will provide the framework for producing a potentially similar ecosystem structure at the restoration site. A reference ecosystem can be defined as one that most closely resembles the original undisturbed restoration site in terms of habitat type, species composition and ecosystem structure.

  Examples of desirable information from the reference ecosystem, bearing in mind that much of this information is not available for most tropical ecosystems, include:

  - What are keystone species and pioneers/canopy species?
    Janzen (1994) highlights the problem of the lack of information when deciding what action to take for the conservation and recovery of degraded tropical ecosystems, defining a keystone species as: 'you happen to know enough about it that you can recognise the impact of removing it from its associates'.
  - What species make up the understorey?
  - What frugivores/pollinators operate here?

- Conduct literature searches of previous descriptions of the vegetation of the site prior to degradation, inventories or surveys. These will aid decision-making.

- Remove threats that brought about the original degradation, e.g. control invasive alien species and fire and secure land tenure to prevent illegal extraction of forest products.

- Discuss and obtain support from local community stakeholders on the aims of the restoration work.

- Secure taxonomic support and training for project staff. Aim to produce an inventory of the site over the life of the project, allowing for natural regeneration of species. Inventories of the reference ecosystem can be used to compare species richness over time.

Planning

- Have a flexible approach i.e. one which does not specify the ultimate goal.
- Where possible, use management regimes tailored to support a number of purposes (e.g. use sewage effluent for irrigation).
- Plant around remnant vegetation and select plant species that attract frugivores, to act as dispersal sites and to create 'nuclei of activity' within the restored area.
- Manipulate the soil as needed, e.g. introduce indigenous mycorrhizae.
- The topsoil must be kept moist, cool and shaded. Consider using of mulches.
- Be aware that nutrient cycling strategies may change during the rehabilitation process, e.g. presence or absence of termites is an important factor in African dry tropical forest (Mando *et al.*, 1999; Menaut *et al.*, 1995).
- Aim to maximise vegetation cover, using existing vegetation before substituting new individuals.
- Use a range of seeding techniques when in doubt as to what to plant.

Box 14.3 continued

- Allow the process of natural selection to decide the successional pathway of the restoration site.
- Plant species mixtures according to what is known (if anything) about their ecological associations in the reference ecosystem.
- Use effective control of invasive exotic species throughout restoration process.
- Employ exotic species *only where necessary* to act as a 'nurse crop' for native species. Do not use this method if the chosen exotic species has the potential to become invasive (see Appendix 5).
- Invest in staff with good horticultural skills to manage the site.
- Use nursery facilities to ensure good planting stock and accurate record-keeping of the restoration process.
- If planning to use site for research, use a design that will allow easy statistical analysis of the data. Decide in advance the minimum amount of data required that will adequately answer the research hypotheses. Ensure adequate site monitoring.

(Adapted from: Atkinson, 2001; Brown and Lugo, 1994; Society for Ecological Restoration, 2000)

Acknowledgements

As main author, I would like to thank my first co-author, Dr Clare Hankamer, for all her encouragement and assistance to complete this report.

The project would never have succeeded without the enthusiasm and capability of my second co-author. Mathias Ngonyo proved a steadfast worker and maintained a high level of enthusiasm for the restoration concept. He is looking forward to the time when he can show his grandchildren the trees he planted. My grateful thanks go to Mr. Nico Visser, of the Royal Netherlands Embassy, Nairobi for obtaining the funding for both phases of the project and for his interest and involvement in environmental protection at the coast. I also thank HRH Prince Bernhard of the Netherlands for initiating the project, and for making the first tranch of funding available.

I would also like to thank the Director and staff of NMK, Nairobi, particularly Professor S. Njuguna, for the original idea, and the Curator and staff at Gede NM for starting the tree planting, for office space, sharing facilities and day to day supervision. I thank Dr Clare FitzGibbon and Frankfurt Zoological Society for permission to use the vehicle left in my charge for trips to Gede from Malindi, which made the funds go a lot further.

The staff of the Conservation Projects Development Unit, Royal Botanic Gardens, Kew, UK, particularly Dr Mike Maunder, were very supportive of the objectives of the project and gave me a wider appreciation of the importance of tropical forest restoration and conservation.

Finally I thank my husband, Mr Ian Robertson for continued financial, practical and moral support.

References

Affolter, J.M. (1997). Essential role of horticulture in rare plant conservation. *HortScience* **32**(1): 29–34.

Atkinson, I.A.E. (2001). Introduced mammals and models for restoration. *Biological Conservation* **99**(1): 81–96.

Atkinson, I.A.E. (1990). Ecological restoration on islands: prerequisites for success, pp. 73–90. In: D.R. Towns, C.H. Daugherty, and I.A.E. Atkinson (eds). *Ecological Restoration of New Zealand Islands.* Conservation Sciences Publication No. 2. Department of Conservation, Wellington, New Zealand.

Beentje, H.J. (1988). Atlas of the rare trees of Kenya. *Utafiti* **1**(3): 71–123.

Birch, W.R. (1963). Observations on the littoral and coral vegetation of the Kenya Coast. *Journal of Ecology* **51**(3): 603–615.

Brown, S. and Lugo, A.E. (1994). Rehabilitation of tropical lands: a key to sustaining development. *Restoration Ecology* **2**(2): 97–111.

Burgess, N.D. and Clarke, G.P. (eds) (2000). *Coastal Forests of Eastern Africa.* IUCN, Gland, Switzerland.

Burgess, N.D., Clarke, G.P. and Rodgers, W.A. (1998). Coastal forests of eastern Africa: status, endemism patterns and their potential causes. *Biological Journal of the Linnean Society* **64**: 337–367.

Cabin, R.J. *et al.* (2000). Effects of long-term ungulate exclusion and recent alien species control on the preservation and restoration of a Hawaiian tropical dry forest. *Conservation Biology* **14**(2): 439–453.

Chapman, C.A. and Chapman, L.J. (1996). Exotic tree plantations and the regeneration of natural forests in Kibale National Park, Uganda. *Biological Conservation* **76**: 253–257.

Chapman, C.A. and Chapman, L.J. (1999). Forest restoration in abandoned agricultural land: a case study from East Africa. *Conservation Biology* **13**(6): 1301–1311.

Dale, I.R. (1939). *The woody vegetation of the Coast Province of Kenya.* Imperial Forestry Institute, University of Oxford, Oxford, UK.

Faden, R.B. and Faden, A.J. (1972). *Some Common Trees of Gedi.* National Museums of Kenya, Nairobi, Kenya.

Frazier, J.G. (1993). Dry coastal ecosystems of Kenya and Tanzania, pp. 129–150. In: E. van der Maarel (ed.). *Ecosystems of the World 2B. Dry Coastal Ecosystems, Africa, America, Asia and Oceania.* Elsevier, Amsterdam, The Netherlands.

Gerhardt, K. and Steiner, M. (1986). *An inventory of a coastal forest in Kenya at Gedi National Monument including a check-list and a Nature Trail.* Report from a minor field study. Swedish University of Agricultural Sciences International Development Centre Working Paper 36. Uppsala, Sweden.

Gillett, J.B. (1979). Kenya, pp. 93–99. In: I. Hedberg (ed.). *Systematic Botany, Plant Utilisation and Biosphere Conservation.* Institute of Systematic Botany, Uppsala, Sweden.

Glowka, L., Burhenne-Guilmin, F., Synge, H., McNeely, J.A. and Gündling, L. (1994). *A Guide to the Convention on Biological Diversity.* Environmental Policy and Law paper no. 30. IUCN Gland, Switzerland.

Gordon, I. (1998). Kipepeo, Conservation with a butterfly touch. National Museums of Kenya. *Horizons* **3**: 9–13, July 1998.

Guerrant, E.O. (1996). Designing populations: demographic, genetic and horticultural dimensions, pp.171–208. In: D.A. Falk, C.I. Millar and M. Olwell (eds). *Restoring Diversity: strategies for reintroduction of endangered plants.* Center for Plant Conservation, Missouri Botanic Garden, Island Press, Washington, DC, USA.

Hamilton, A.C. (1981). The Quaternary history of African forests; its relevance to conservation. *African Journal of Ecology* **19**: 1–16.

Hawthorne, W.D. (1984). *Ecological and biogeographical patterns in the coastal forests of East Africa.* PhD. Thesis, University of Oxford, UK.

Hawthorne, W.D. (1993). East African coastal forest botany, pp. 57–99. In: J.C.Lovett and S.K.Wasser (eds). *Biogeography and Ecology of the Rainforests of Eastern Africa.* Cambridge University Press, Cambridge, UK.

Janzen, D.H. (1988). Tropical dry forests, the most endangered major tropical ecosystem, pp. 130–137. In: E.O. Wilson (ed.) *Biodiversity.* National Academy Press, Washington, USA.

Janzen, D.H. (1994). Priorities in Tropical Biology. Trends in Ecology and Evolution **9**(10).

Kenya National Environment Action Plan (1994). Ministry of Environment and Natural Resources, Nairobi, Kenya.

Lucas, G. (1968). Kenya. In: Conservation of vegetation in Africa south of the Sahara. *Acta Phytogeographica Suecica* **54**: 152–159.

Luke, Q. (ed.) (1996) *Conservation Assessment and Management Planning (CAMP) Training Workshop for Selected Species of the Kenya Coastal Forests,* 20–25 November 1996, Coastal Forest Conservation Unit, Kenya. Royal Botanic Gardens, Kew, UK.

Mando, A., Brussaard, L. and Stroosnijder, L. (1999). Termite- and mulch- mediated rehabilitation of vegetation on crusted soil in West Africa. *Restoration Ecology* **7**(1): 33–41.

Menaut, J.C., Lepage, M. and Abbadie, L. (1995). Savannahs, woodlands and dry forest in Africa, pp. 64–92. In: S.H. Bullock, H.A. Mooney and E. Medina (eds). *Seasonally Dry Tropical Forests.* Cambridge University Press, Cambridge, UK.

Milton, S.J. *et al.* (1999). A protocol for plant conservation by translocation in threatened lowland fynbos. *Conservation Biology* **13**(4): 735–743.

Mooney, H.A., Bullock, S.H. and Medina E. (1995). Introduction, pp. 1–8. In: S.H. Bullock, H.A. Mooney and E. Medina (eds). *Seasonally Dry Tropical Forests.* Cambridge University Press, Cambridge, UK.

Mwasumbi, L., Johansson, S., Luke, Q., Simiyu, S., Mashauri, S., Ipulet, P., Maunder, M., Clubbe, C. and Hankamer, C. (1998). *Conservation Assessment and Management Planning (CAMP) Training Workshop for Selected Plant Species of the East Usambara Mountains, Tanzania.* 2–6 March 1998, East Usambara Catchment Forest Project, Tanzania. Royal Botanic Gardens, Kew, UK.

Myers, N., Mittermeier, R.A., Mittermeier, C.G., da Fonseca, G.A.B. and Kent, J. (2000). Biodiversity hotspots for conservation priorities. *Nature* **403**: 853–858.

Olsen, D.M. and Dinerstein, E. (1998). The Global 200: A representation approach to conserving the Earth's most biologically valuable ecoregions. *Conservation Biology* **12**(3): 502–515.

Parrotta, J.A. (1995). Influence of overstory composition on understory colonization by native species in plantations on a degraded tropical site. *Journal of Vegetation Science* **6**: 627–636.

Polhill, R.M. (1989). East Africa, pp. 219–225. In: D.G. Campbell and H.D. Hammond (eds). *Floristic Inventory of Tropical Countries*. New York Botanical Garden, New York, USA.

Ray, G.J. and Brown, B.J. (1995). Restoring Caribbean Dry Forests: evaluation of tree propagation techniques. *Restoration Ecology* **3**(2): 86–94.

Robertson, S.A. (1987). *Preliminary Floristic Survey of the Kaya Forests of Coastal Kenya*. Report to World Wide Fund for Nature and National Museums of Kenya, Nairobi, Kenya.

Robertson, S.A. (1992). Kaya forests of the Kenyan coast, p. 152. In: J.A. Sayer, C.S. Harcourt and N.M. Collins (eds). *The Conservation Atlas of Tropical Forests*. IUCN. Macmillan Publishers Ltd., UK.

Robertson, S.A. and Luke, W.R.Q. (1993). *Kenya Coastal Forests*. The report of the NMK/WWF Coast Forest Survey. WWF Project 3256: Kenya, Coast Forest Status, Conservation and Management. Report to World Wide Fund for Nature and National Museums of Kenya, Nairobi, Kenya.

Shah, N.J., Linden, O., Lundin, C.G. and Johnstone, R. (1997). Coastal management in Eastern Africa: status and future. *Ambio* **26**(4): 227–234.

Society for Ecological Restoration (2000). Internet www. ser.org >(27/3/2000).

White, F. (1983). *The Vegetation of Africa; a descriptive memoir to accompany the UNESCO/AETFAT/UNO Vegetation map of Africa*. UNESCO, Paris, France.

Williams, J.G. and Arlott, N. (1981). Gedi National Park, pp. 65–66. In: *A Field Guide to the National Parks of East Africa*. Collins, UK.

Wilson, A. (1993). Sacred forests and the elders, pp. 244–248. In: E. Kemf (ed.). *The Law of the Mother: protecting indigenous peoples in protected areas*. Sierra Club Books, San Francisco, USA.

WWF and IUCN (1994–1997). *Centres of plant diversity. A guide and strategy for their conservation*. 3 volumes. IUCN Publications Unit, Cambridge, UK.

Chapter 15

Capacity Building for Plant Conservation in East Africa:

a case study of the National Museums of Kenya-Darwin Plant Conservation Techniques Course

Clare Hankamer
Royal Botanic Gardens, Kew, UK

Perpetua Ipulet
East African Herbarium, National Museums of Kenya, Nairobi, Kenya

Colin Clubbe
Royal Botanic Gardens, Kew, UK

Mike Maunder[1]
Royal Botanic Gardens, Kew, UK[1]

[1] Now at: The National Tropical Botanical Garden, Kauai, Hawaii, USA

Introduction

Sub-Saharan Africa contains four of the world's 25 biodiversity hotspots (*sensu* Myers *et al.*, 2000). These four areas – the Eastern Arc Mountains and Coastal Forests of East Africa, the West African Forests, the Cape Floristic Province and the Succulent Karoo – collectively hold 11,370 endemic plant species. In addition, 11 of the world's 50 most botanically diverse countries are African; three of which are the East African countries of Tanzania with 10,000 vascular plant species, Kenya with 6,000 and Uganda with 5,000 species (Mittermeier and Werner, 1990; Polhill *et al.*, 1952–). Collectively, the East African region contains 20 Centres of Plant Diversity (*sensu* WWF and IUCN, 1994); these include the Taita Hills, Kenya, the East Usambara Mountains, Tanzania and Bwindi Forest, Uganda.

At the national level, Kenya alone has five Centres of Plant Diversity (*sensu* WWF and IUCN, 1994):

- the limestone bush/woodland of the Somali-Masai Regional Centre of Endemism
- the Shimba Hills within the Indian Ocean Coastal Belt
- three sites of Afro-montane vegetation:
  - mount Kenya
  - mount Elgon
  - the Taita Hills

Some of these areas are highly threatened with destruction.

It is projected that Kenya's population will increase to 63,360,000 by 2025 from an estimated 6,265,000 in 1950 (World Resources Institute, 1996). The increase in human population and associated rates of habitat conversion will continue to erode important habitat areas (Stuart *et al.*, 1990; UNESCO, 1992; WWF and IUCN, 1994). Habitat loss continues as a serious problem with, for example, less than 2,000 km^2 of the Eastern Arc and Coastal Forest hotspot remaining from an original area of 30,000 km^2 (Myers *et al.*, 2000). In Kenya, for example, the 60 surviving sacred Kaya forest groves cover less than 2,000 ha and yet, in combination with other forest patches in the Coast Province, contain more than 50% of Kenya's threatened trees (Beentje, 1988; Robertson and Luke, 1995). Less than 3% of the original forest cover survives in the Taita Hills, Kenya (WWF and IUCN, 1994). This is home to the Taita Hills African violet (*Saintpaulia teitensis* B.L. Burtt (Gesneriaceae)), a highly threatened endemic that survives as only six small colonies in a forest patch of less than 10 ha (Simiyu, pers. com.). In the East African region as a whole, relatively little is known of the ecology or conservation status of threatened species, and although some lists and schedules are available, no completed national or regional Red Lists exist.

Species conservation problems in East Africa can be divided into three categories:

- extinction of localised endemics e.g. *Aloe murina* L.E. Newton (Aloaceae), *Euphorbia tanaensis* Bally (Euphorbiaceae), *Saintpaulia* Wendland species, through habitat loss and degradation

- potential loss of national populations of widespread but uncommon East African species, e.g. *Camptolepis ramiflora* (Taub.) Radlk. (Sapindaceae) in Kenya, as a result of habitat loss and degradation

- over-harvesting of economically important species, e.g. *Dalbergia melanoxylon* Guill. & Perr. (Leguminosae), and medicinal plants, e.g. *Prunus africana* (Hook. f.) Kalkman (Rosaceae)

Overall, botanical conservation, in terms of botanical inventory, habitat protection and capacity for species management is relatively poorly resourced within the region (Maunder and Göhler, 1995). Unfortunately, most herbaria and botanic gardens on the continent (Table 15.1) have insufficient adequately trained staff and are effectively isolated from other institutions in the country or region (Baijnath, 1995). This has been largely as a result of an historical focus on forest management, with the primary concerns of forest productivity and protection of water catchments; and on national parks, established for large mammal conservation (Rodgers, 1995; Sayer *et al.*, 1992).

Table 15.1 **Taxonomic and botanic garden resources in East Africa** (Adapted from Heywood (1995) and Walter and Gillett (1998))

Country	Plant and fungal biosystematic collections		No. of botanic Gardens	Threatened species/total number of vascular plant species	National herbaria
	No. of collections	No. of specimens			
Kenya	2	550,000	5	240/6500	East African Herbarium, National Museums of Kenya
Tanzania	4	111,750	3	436/10,000	National Herbarium of Tanzania, Arusha
Uganda	4	70,967	2	15/5400	Makerere University, Kampala

Identification of the Need for Training

In a strategic review of plant conservation networking needs for Kenya (Maunder and Göhler, 1995) it was demonstrated that Kenya requires an effective network of collaborating organisations actively supporting plant conservation. It was recommended that the proposed network would promote:

- an integrated approach to plant conservation, utilising and promoting the available professional skills of Kenya;
- developing further collaborative relationships with the protected area network (for instance, the Kenya Wildlife Service and Forest Department);
- collaboration between state, parastatal, commercial, non-governmental organisation (NGO) and amateur groups;
- Development of in-country taxonomic expertise;
- the initiation of a national system for identifying plant and habitat conservation priorities based on information gathering and monitoring;
- institutional strengthening through professional training;
- a network of public display facilities and research/conservation facilities in the different vegetation zones.

Capacity building (*sensu* Berg, 1993), through training and institutional strengthening, has been promoted as a priority to enable the sustainable management of East African natural resources (Williams, 1998). Calls have been made to focus these training activities on taxonomic groups other than megafauna, and to review the success of current conservation procedures as a preliminary to the identification and implementation of new procedures (USAID, 1993). This has been particularly recognised in the forestry sector with support given to Forestry Action Plans through the Tropical Forestry Action Programme. Foresters are typical of many government sector workers in natural resource management, in receiving traditional training that is inadequate for the conservation management of indigenous forest species (Williams, 1998). The Protected Area Conservation Strategy for East, Central and Southern Africa highlighted that protected area managers identified the need for training particularly in intervention programmes, such as vegetation management (Pitkin, 1995). A number of national and regional studies have identified the need for training in plant conservation, including the Kenyan National Environmental Action Plan (and subsequent Country Study (Republic of Kenya, 1998). Similarly, the Ugandan Country Study identified capacity building as a priority activity (Republic of Uganda, 1998).

Training is recognised as the cornerstone to implement capacity building and networking activities. Both the National Museums of Kenya (NMK) and the Royal Botanic Gardens, Kew (RBG Kew) are committed to implementing the United Nations Convention on Biological Diversity (CBD) (Glowka *et al.*, 1994; see Appendix 1). RBG Kew has run an active training programme in Plant Conservation Techniques since 1993 to address the need for training plant conservationists in integrated plant conservation techniques (*sensu* Falk *et al.*, 1996).

The NMK, with its reputation as a leader in biodiversity research, was keen to develop its role in regional training. Staff from NMK's Plant Conservation Programme (East African Herbarium) had received training on the International Plant Conservation Techniques Course at RBG Kew. During this period plans for a collaborative capacity building exercise were developed. The Plant Conservation Programme had approached local donors with a proposal for a training course and had received an initial commitment from these organisations. Further discussions were then held in 1996, between RBG Kew and the East African Herbarium (NMK). It was recognised by both institutions that, as part of their commitment to the CBD under Article 12 (Research and Training), priority should be placed on building capacity for plant conservation in the region by putting together a collaborative East African regional training course. Article 12 identifies specific training needs, including the need for training conservation biologists and *in situ* and *ex situ* conservation managers within developing countries. It was agreed that this transfer of skills should contribute to expanding the skills base of individuals as well as to strengthening of participating institutes.

A funding proposal to support a regional course was submitted to the Darwin Initiative for the Survival of Species, initiated in 1992 as part of the UK's response to the Rio Earth Summit (DETR, 1998). The aim of the Darwin Initiative is to support biodiversity conservation in developing countries by building long-term capacity for institutions involved in conservation through to supporting the implementation of the CBD, training, research, environmental education and awareness.

Establishing the Course

A Darwin grant was awarded to the RBG Kew, for a three-year project in collaboration with staff from the East African Herbarium, NMK. The project incorporated three annual courses to train staff involved in plant resources management in key regional institutions.

The project team felt that 'parachuting' an externally designed course into East Africa would result in failure. Consequently, the curriculum was developed in full collaboration with in-country partners to best meet the local, documented, training needs. The training was designed to emphasise applied techniques and strategies for plant conservation, focusing almost entirely on regional plant conservation issues, with input on relevant international conventions and policy. The formal taught sessions were provided by regional experts with tuition and workshops supported by the Kew team and selected external facilitators. It was agreed that the regional training project must have a clear role and serve the needs of a clearly defined user group. The training programme for East African plant conservationists adopted the following points of principal:

- the need to develop open communication lines between currently isolated plant conservation agencies and individuals

- to demonstrate soon after establishment a commitment to strengthening all participating organisations

- to maintain a transparent organisational structure where all organisations and individuals can effectively participate

- to identify and support the professional needs of the 'customers', course content influenced by the needs of partners not administrators

- in contrast with European or North American networks the training programme will be working with relatively under-funded and under-resourced collaborators

The Plant Conservation Techniques Course (PCTC) for East Africa was conceived as a three-year programme with the goal:

'To enable participants to develop appropriate skills in conservation planning and practices in order to facilitate effective implementation of plant conservation initiatives in East Africa.'

The goal of the programme and the nature of the partnership between the organising institutions matched the priorities of both collaborating institutions as well as the main funding agencies.

Course objectives

The course objectives were:

- to review plant conservation issues in Eastern Africa and outline mitigating measures

- to introduce the theoretical background to plant conservation practice

- to develop appropriate skills in plant conservation planning and management

- to enhance knowledge in project proposal writing and fund-raising techniques

- to promote cross-disciplinary action and regional networking

The training programme comprised five linked components:

- a formal taught component based at NMK, composed of lectures on a range of subjects pertinent to plant diversity management in the region (Table 15.2)
- course workshops, building on the formal sessions, to encourage group working skills and a cross-disciplinary approach to plant conservation
- a field-based component, designed to allow participants to apply the lecture and workshop materials: such as the World Conservation Union (IUCN) Categories of Threat (see Appendix 2), inventory and monitoring techniques and species recovery planning, to local plant conservation project priorities (Figure 15.1)
- short applied projects
- three Specialist Workshops on applied conservation issues held outside of the main taught courses

The three Specialist Workshops were planned and run between the three main courses. These were designed to 'train the trainers' and to support the establishment of a regional IUCN Species Survival Commission (SSC) plant conservation network.

Course Implementation

1. Course co-ordination

An advisory committee carried out the initial planning and preparation. This contained representatives from the National Museums of Kenya, related NGOs and international agencies. This committee acted as an advisory body to the course management team. A full-time co-ordinator was employed, whose role was to co-ordinate the courses and workshops, providing important links with regional institutions and tutorial and teaching support to participants during the course, as well as enlisting financial support.

2. Selection of participants

The target number of participants was 15 per year, a large enough group to generate a sense of cohesion whilst small enough to enable individual tuition and mentoring. The selection of participants was guided by their current work activities e.g. whether they were directly involved in plant conservation work. The selection process attempted to ensure even coverage between the three countries and representation from all the major relevant institutions (Table 15.3).

Table 15.2 **Subjects covered in Plant Conservation Techniques Courses for East Africa**

- Biodiversity issues for East Africa
- Biogeography/vegetation of Africa
- Dryland ecosystems
- Conservation of wildlife resources in East Africa
- Conservation of forest resources in East Africa
- Approaches to plant conservation
- Invasive species (see Appendix 5) – impacts and responses including field case study
- Species conservation – orchids, African violets
- Role of mycorrhizae
- The role of herbaria in conservation
- Systematics and conservation
- Conservation of genetic resources/Global Plan of Action
- Seed banks
- The role of the Nairobi Botanic Garden and Arboretum in conservation
- Plant quarantine and legislation
- Vegetative propagation including tissue culture
- Conservation through domestication
- IUCN activities in East Africa
- Education for conservation
- Community approaches to conservation
- Indigenous knowledge/ethnobotany surveys
- Role of mapping and remote sensing in conservation
- Conservation information management
- International policies and conventions
- Biodiversity policy issues in East Africa
- Biodiversity Action Plans
- World Conservation Union (IUCN) Red List categories
- Survey and monitoring techniques
- Recovery planning and reintroduction techniques
- Conservation Assessment and Management Plans
- Habitat and species assessment techniques
- Protected area management strategies

Figure 15.1 Investigating forest plots within Amani Nature Reserve as possible training locations for the National Museums of Kenya – Darwin Plant Conservation Techniques Course. Left to right: Perpetua Ipulet, Clare Hankamer, Patrick Muthoka (NMK), Shedrack Mashauri (East Usambara Conservation Area Management Programme, Tanzania), Ahmed Mndolwa, (Tanzanian Forest Research Institute).

3. Management of donor support for scholarships

Local and regional collaboration was recognised as a fundamental component of the course. Whilst the Darwin Initiative provided significant funding it was felt that this should be matched by a local commitment. The Darwin Initiative grant was subject to the receipt of matching funding. About 50% of funding was gained from local donors, mainly through initial public relations work carried out by the NMK Darwin Project Manager. This raised the profile of the course, particularly in Nairobi where many regional donors are based. For many donors, this support was committed throughout the three years of the project, with interest in continuation of funding beyond the life of the project. Much time was invested in meeting with individual donors to update them and to report on the course, but this paid dividends in terms of continued or increased support.

4. Field training

Field training was incorporated into a seven to ten day residential course after the formal teaching and course workshops held at the National Museums of Kenya (Figure 15.2). This training was undertaken in collaboration with local plant conservation projects.

Table 15.3 Country and institutional representation for the Plant Conservation Techniques Course for East Africa

Course	No. of participants	No. of institutions represented	Country representation		Ratio of women:men	Projects submitted (late/invited to resubmit)	Proportion of participants who passed
Plant Conservation Techniques 1996	14	11	Uganda Tanzania Kenya	3 2 9	3:11	5 (3)	42%
Plant Conservation Techniques 1998	14	12	Uganda Tanzania Kenya	4 5 5	2:12	5 (5)	57%
Strategies for Plant Conservation 1999	14	10	Uganda Tanzania Kenya Madagascar	3 4 6 1	4:10	14 (0)	85%
Summary	**42**	**21**	**Uganda Tanzania Kenya Madagascar**	**10 11 20 1**	**9:33**	**24 (8)**	**62%**

Note: Annual target number of participants = 15.

Figure 15.2   Left to right: Shedrack Mashuari (East Usambara Conservation Area Programme, Tanzania), Mike Maunder, and Mollel Neduvoto (National Herbarium, Tanzania), prepare a field herbarium to support identifications during fieldwork. Labelled specimens, with full or provisional identifications, are used for reference during field inventory work.

This resulted in a number of benefits:

- it ensured that participants were exposed to real plant conservation scenarios of local and regional relevance
- it encouraged participants to test their professional understanding and personal philosophies against real problems
- it exposed the host project staff to a period of ideas and information exchange, and informal field training
- it exposed laboratory or office-based participants to the joys (and otherwise) of field work
- it established a relaxed and informal working atmosphere that is sometimes difficult to generate in the formal surroundings of a major urban institution

The course team worked closely with the host project staff, using their knowledge of the site and species, to design the field exercises. These were designed to provide an intellectual and practical challenge for the course participants and to directly respond to selected management or scientific issues facing the host institution. Host project staff were invited to facilitate in all aspects of the field training.

Course Assessment

Course projects

Participants were invited to select a project in consultation with their managers; this was to ensure that the project matched institutional interests. Course participants were asked to present a written project proposal. Following a review of the proposal, a formal presentation of this proposal was made by the participants to their fellow participants and course lecturers. Participants were assessed on the quality of the presentation, its practical relevance to institutional and national objectives, and on the completed project document. Requirements for project acceptance were that they were based on the participants' existing work remit and demonstrated the application of plant conservation techniques to address a particular issue (Box 15.1). An independent moderator from the International Plant Genetic Resources Institute (IPGRI) based in Nairobi, plus course tutors, assessed the project proposals and final projects. Comments on the proposal were discussed with each participant prior to their departure back to their home institution. Where projects did not meet the required standards,

---

Box 15.1 An example of course project proposal – survey of *Mimusops somaliensis* Chiov. (Sapotaceae) (Mududu, 1998)

Goal: To survey the protection and conservation status of *M. somaliensis*. Vernacular name: mgama (duruma and digo); endemic to coastal Kenya and Somalia, and harvested for timber.

Project objectives:

- to establish the conservation status of this species using the World Conservation Union (IUCN) Categories of Threat
- to identify the current local and commercial uses of this species
- to assess options for the conservation of germplasm (*ex situ* conservation) for enrichment planting
- to develop a monitoring plan for propagation and conservation of this species
- to establish partnerships with the local communities, Forest Department, National Museums of Kenya (NMK) and other organisations in the conservation of this species

Work plan:

- species inventory
- socio-economic survey
- assessment of level of *in situ* propagation/regeneration
- development of *ex situ* propagation techniques to raise plants for enrichment planting
- raising local community awareness
- community mobilisation for enrichment planting

Project Executant: Hamisi Juma Mududu, Coastal Forest Conservation Unit, NMK.

---

participants were given the opportunity to resubmit their project within a given deadline based on further comments from the course tutors.

The theme of the final course in 1999 was modified to encompass the application of strategy. The 1999 course, Plant Conservation Strategies for East Africa, was aimed at training middle management and researchers. It was felt important to train this level of management in the same key institutions so that they would provide support to the two cohorts of participants from the 1996 and 1998 courses. The course structure prompted more discussion of concepts. Projects were completed during course time and demonstrated a synthesis of the course in developing a strategy for action for the institution or within the participants' existing research remit.

Course certification

Participants were assessed by formal examination and by submission of a project. The examination expected a working understanding of the CBD, national strategy and the fundamentals of applied plant conservation biology. Successful candidates were awarded an International Certificate in Plant Conservation Techniques, or Strategies, for East Africa.

Training in Conservation Assessment

Setting priorities for threatened species management: the Conservation Assessment and Management Planning (CAMP) for protected areas

Conservationists in the tropics face the fundamental challenge of where to invest time and facilities most effectively. Currently, there is a paucity of data on the conservation status of wild plant species and as a result the management of threatened plant species is not widely practised in the region. The Darwin Course provided a review of the structures and processes for species recovery. Importantly, a significant amount of time was given to the identification of both conservation priorities and appropriate target species for recovery planning. It was felt important that established and effective protocols were examined and tested for local applicability during the training. Accordingly, IUCN/SSC protocols were tested as a potential mechanism for local teams to employ for data gathering and priority setting. The Conservation Assessment and Management Plan (CAMP) protocol was developed by the Conservation Breeding Specialist Group of the IUCN/SSC (Ellis and Seal, 1995; see Appendix 3). The protocol was developed to provide a participatory approach to conservation planning; using a broad range of available information to provide a rapid and effective means of assessing conservation status using the IUCN Categories of Threat (see Appendix 2). Importantly the process enables species to be prioritised for conservation management (Figure 15.3).

Figure 15.3   Leonard Mwasumbi of the University of Dar es Salaam leads a workshop developing field identification skills in the East Usambara Mountains, Tanzania. Plant identification in the field is often difficult as manuals and Floras often require reference to flowering and fruiting material that may only be seen seasonally. Here Leonard is offering some practical tips on the field identification of a *Cynometra* L. (Leguminosae) species using leaf characteristics.

The aim of incorporating a CAMP training workshop into the Plant Conservation Techniques Course for East Africa was to equip participants with practical tools to assess the status of threatened species. The IUCN Categories of Threat are an important tool for conservation planning and hence, sustainable management. It was therefore important for participants to be comfortable in their use and application. In addition, this process was designed to encourage participants to use their own field observations, and to critically analyse technical and non-technical sources of information including unpublished or oral sources of information on plant diversity. The CAMP training workshops were applied in two Centres of Plant Diversity: the Kenya coastal forests and the East Usambara forests (*sensu* WWF and IUCN, 1994). This allowed participants to contribute to an actual CAMP, which was subsequently used by the host project as a contribution to its own activities. The 1996 course was hosted by the Coastal Forest Conservation Unit (CFCU), Kenya, looking at the threatened species of the coastal Kayas (Luke, 1996; Figure 15.4). The 1998 course was hosted by the East Usambara Catchment Management Project (EUCAMP) (see Box 15.2), looking at threatened species in the East Usambaras, Tanzania (Mwasumbi *et al.*, 1998).

For each course, an initial list of 10–15 species was compiled. For ease of data gathering, endemic species with known range were chosen where possible. The four-day CAMP exercise followed a set programme:

- Preliminary lectures and a briefing session prior to departure, established the working protocols for the week, including field safety issues;

- On arrival, introductory lectures and walks established the principles of landscape change and the biodiversity response to that changing landscape;

- The opening of the CAMP started with half a day of discussions and briefings from a wide range of the relevant community, and scientific and legislative stakeholders. This allowed participants to get to grips with threats to biodiversity and to explore how different stakeholder groups viewed and/or valued plant conservation. Major economic, social and biological trends were identified through presentations and group work;

- Selected field trips to local sites, of two half days during the three to four days of activities, allowed participants to see the species *in situ*, assess local threats, and to hear briefings from stakeholders and host project personnel. Participants worked during this stage in small groups and were encouraged to collect, collate and assess field data, anecdotes and scientific reports (Figure 15.5);

- IUCN Categories of Threat were assigned based on consensus within each working group. All the information on a given species was recorded on a taxon data sheet. Management recommendations were then made based on this pooled information;

- Plenary sessions each evening allowed the facilitators to discuss taxon data sheets with the whole group and to highlight any problems with assumptions made over data;

- Participants were also asked to prepare and present two short papers on regional priorities for plant conservation and training. Topics were identified and ranked by workshop participants;

- The findings of the workshop were then written up as a report. The findings are compiled and edited by the workshop facilitators, with the lead botanical resource person acting as lead editor. This report was made available to the host project, participants and to IUCN.

Facilitation by the host institution's staff proved very fruitful, leading to involvement in subsequent CAMP workshops. For instance, some of the species selected for assessment in the 1996 workshop in Kenya were re-assessed during the 1998 workshop in Tanzania. This allowed a comparison in available data and allowed professionals studying the Kenyan and Tanzanian populations to meet and exchange information. For instance, the Coastal

---

Box 15.2 The role of the Conservation and Management Planning (CAMP) training workshop to build capacity for a species recovery programme

Cola usambarensis **Engl. (Sterculiaceae)**

• Endemic understorey tree of moist upland forest of the East Usambara Mountains, used for building poles and for medicinal practice.

Sequence of training activities in relation to development of recovery plan:

1996

• 1996 course – Shedrack Mashauri (Amani Nature Reserve Officer) supported by East Usambaras Conservation Area Management Programme (EUCAMP) attended course. Submitted project entitled 'Assessment of IUCN Categories of Threat for three endemic tree species of the East Usambaras', which incorporated an assessment of *C. usambarensis* (Mashauri, 1996).

1997

• *C. usambarensis* selected by host project (EUCAMP) as candidate species for the 1998 course CAMP training workshop.

1998

• CAMP workshop – *C. usambarensis* categorised as Endangered. Management recommendations made: carry out conservation research to determine distribution and autecology of this species.

• Discussion with Royal Botanic Gardens, Kew (RBG Kew) and EUCAMP on the feasibility of drafting a recovery plan for this species.

• A Masters study was undertaken in collaboration with University College, London, RBG Kew and EUCAMP to determine factors affecting distribution of *C. usambarensis* and to determine the current level of harvesting (Muir, 1998).

1999

• Data used for a training workshop on recovery planning for 1999 course – findings to be incorporated into final recovery plan.

• Further monitoring of populations, setting up of permanent plots to measure population demography in relation to harvesting pressure.

• Preparation of draft recovery plan.

Forest Conservation Unit (CFCU) (Kenya) team was invited, with the agreement of the Tanzanian hosts, to assess the Tanzanian status of three threatened coastal forest species (*Ficus faulkneriana* C.C. Berg (Moraceae), *Cephalosphaeria usambarensis* (Warb.) Warb. (Myristicaceae) and *Gigasiphon macrosiphon* (Harms) Brenan (Leguminosae)). The workshop report therefore provided useful information for CFCU in assessing the overall regional status of these species and in promoting regional collaboration. Likewise, the workshop enabled EUCAMP, which was in the process of finalising a management plan for the Amani Nature Reserve, to consider single species management for the first time and to incorporate priorities into the plan.

Figure 15.4   Quentin Luke of the Coastal Forest Conservation Unit, Kenya leading a field visit to the Kaya forests of the Kenya coast.

The 1999 field course structure was altered to put an emphasis on two related scientific themes: (1) single species demography and (2) habitat quality assessment techniques. This was held at the Mwaluganje Elephant Sanctuary, Shimba Hills, Kenya, and looked at the effects of an enclosed and increasing wild elephant population on coastal woodland structure and populations of an indigenous cycad *Encephalartos hildebrandtii* Braun & Bouché (Zamiaceae). Staff from RBG Kew, CFCU and EUCAMP jointly facilitated the fieldwork with the support of the Kenya Wildlife Service (Figure 15.6).

Post-course evaluation

Each course was evaluated by the course management team and by the participants. The latter were asked to fill in an evaluation form, which was then reviewed by the course management team and, where appropriate, recommendations were incorporated into structure of the next course.

Figure 15.5   Colin Clubbe (RBG Kew) works with Maud Kamatenesi (Makerere University, Uganda) and Steevern Shunda (Institute of Traditional Medicine, Huhimbili University College of Health Sciences, Tanzania) on the analysis of survey data from the day's fieldwork.

Figure 15.6   Participants in the 1999 NMK Darwin Plant Conservation Strategies Course for East Africa. The field component was held in the Mwaluganje Elephant Sanctuary, Kenya to review conservation issues for the coastal forest ecosystems and the impact of elephant populations on these habitats.

Table 15.4 **Country and institutional representation for the three Darwin Specialist Workshops**

Course	No. of participants	No. of institutions represented	Country representation		Ratio of women: men
Seed Conservation, 1997	23	8	Kenya Uganda Tanzania UK	21 0 0 2	3:20
Botanic Garden Conservation Role, 1998	14	11	Kenya Uganda Tanzania UK	7 3 3 1	3:11
Biodiversity Restoration, 1998	28	16	Kenya Uganda Tanzania UK	25 1 0 2	10:18
Summary	65	35	Kenya Uganda Tanzania UK	53 4 3 5	16:49

Specialist workshops

Three specialist workshops were held, that ran for two to five days (Table 15.4). Venues for the specialist workshops were selected as appropriate for the subject, and the Course Co-ordinator invited participants from key regional institutions. Staff from RBG Kew facilitated the three workshops.

In each case, participants were encouraged to form a regional Specialist Working Group that would meet beyond the workshop to take forward the objectives laid out during the workshop. Thus, the Seed Conservation Workshop discussions fed into the preparation of a strategy for plant genetic resources conservation for Kenya. The Botanic Garden Workshop has resulted in the formation of the East African Botanic Garden Working Group; an incipient group of interested botanic gardens leading to a possible regional network. The Biodiversity Restoration Workshop has resulted in the development of the Habitat Restoration Group meeting in Kenya. The criteria from this workshop are outlined in Box 15.3.

Box 15.3 Specialist Workshop: habitat restoration

Habitat restoration

The workshop was held in Voi, Kenya. A three day workshop consisting of one day in the field visiting forest reserves and potential restoration sites and two days as a workshop. The workshop was facilitated by Royal Botanic Gardens, Kew and the Reintroduction Specialist Group of the World Conservation Union (IUCN) Species Survival Commission.

Workshop goals

To introduce participants to current thinking on biodiversity restoration (species reintroduction and habitat restoration) as applicable to East Africa. The strategic goal of the workshop was to assess options to halt habitat loss in the Taita Hills and to restore degraded habitats.

Key issues

- What is habitat restoration? Can restoration be undertaken at different spatial scales serving different functions, for instance 1–10 ha micro reserves for localised endemics or habitat fragments, 10–100 ha 'habitat' restoration areas, and 1,000–10,000 ha ecosystem restoration plots.

- How can habitat restoration as a tool for biodiversity management contribute to social and development issues? International case studies were presented and definitions established (Society for Ecological Restoration and Reintroduction).

- At what point do we move from reducing biodiversity loss towards reversing biodiversity loss? Do some critical areas, e.g. the Taita Hills, demand urgent action to consolidate surviving areas and to initiate linkages between fragments?

- In 10–20 years will restoration, including the translocation of animal populations and the restoration of vegetation, be core components of protected area management?

Current activities in the region were reviewed including species reintroductions (Hirola antelope and timber tree species), habitat restoration (coastal forests) and repatriation of crop germplasm to farming communities in the region by the International Plant Genetic Resources Institute (IPGRI).

Lessons Learnt

Three years experience in running a complicated and extensive training initiative taught all parties a number of important lessons. The following points were taken from course reviews carried out by the course management team, from a final Darwin project review as part of the Darwin grant management process and from feedback from post-course evaluations completed by the course participants.

Selection of participants

Whilst care was taken to select participants who would most benefit from the training, a number of the participants were nominated by their institutions or supporters. These individuals were not actively involved in conservation, and accordingly, found it extremely difficult to fully participate in the course. This was a particular problem during project preparation, where participants were asked to focus their projects on a component of their existing work. In some cases this resulted in projects not being submitted.

Regional representation

The course was designed to serve the needs of the East African region, however in reality, both the location and activities favoured participants from Kenya. Better representation from the region by facilitators and participants would have strengthened the regional value of the course. This was partly a result of difficulties in regional communications between NMK and some institutions. We held field courses in Tanzania and Kenya, and it was hoped to host the third year fieldwork in Kibale National Park, Uganda. However, a decision was made to move the venue due to concerns about security in this area.

Course timetable structure

We had intended to provide a good balance between formal taught sessions and informal tutorials and independent study periods. However, the 1996 timetable was overloaded with formal lectures and left little time for discussion (unless it was specifically requested by lecturers or participants). We found that lecturers would extend their lectures right up to the end of the allocated time slot (two hours) rather than lecturing for part of the time and using the rest for discussions. This created some frustration among participants who needed time between lecture slots to visit the library or to arrange meetings with course tutors. Free time for personal study was allocated in the fifth week, which in hindsight was both too little and too late for participants to adequately prepare for projects and the field trip. After discussion with the Course Co-ordinator, participants and tutors, and using feedback from course assessments, a decision was made to remove some of the lower priority topics from the timetable. This freed up time for private study in subsequent courses.

Allocation of time to non-formal teaching

Word processing and research skills varied among participants from course to course. In the final two courses, more time was allocated for informal tutorials to help take participants right through the project preparation process, including training in literature searches, word processing skills or feedback on project proposals and content.

Course projects

For the 1996 and 1998 courses, participants were asked to produce a project proposal that related directly to their institutional goals and current work activities. The project itself was then implemented over a five-month period, after the course, with the understanding from their line management that this should be done as part of their existing remit and in work time. The Course Co-ordinator ensured that participants were made aware of the need to discuss potential projects with their line management prior to attending the course so as to enlist adequate institutional support. This was not effectively communicated within the institution and was compounded by the mistaken view that the NMK-Darwin course would provide funding to support this work. As a consequence, much time was spent by the course management team to re-work proposals and to write to line managers to ensure projects were feasible and completed on time. The Project submission rate for the 1996 and 1998 courses was relatively poor (42% and 57% respectively). It was decided therefore that for the 1999 course that projects should be completed within the formal NMK-based component of the course.

Field training

Fieldwork experience naturally varied within any course group, but there was a common weakness in field observation skills. In particular, in perceiving indicators of threat (other than the more obvious land clearance activities). This had not been fully anticipated by course tutors. Therefore, care was taken at field sites to highlight the importance of this for the CAMP process through formal and informal data collection. This also applied to published biodiversity inventory data. Participants were nervous about assigning a Category of Threat to a species for which there was little verified field data. Effective facilitation of discussion groups enabled participants to build confidence in this over time. There was also a lack of confidence in producing this material for official use. This was a trade-off in attempting to run a legitimate CAMP workshop at the same time as a training exercise. This process was generally thought to have been worthwhile, however, it relied very heavily on good taxonomic support. Outside of the training forum, discussions were held with facilitators from the host projects on the use of such conservation management techniques. It is interesting to note that even though conservation planning is recognised as a 'strongly emerging field of specialisation' it has not been adopted into undergraduate course curricula in either Africa or Europe (Noss, 1997).

The initial CAMP taxon datasheet was developed for use on medicinal plants. Consequently, a number of data fields had to be modified to suit broader use. Modifications were passed on to the Conservation Breeding Specialist Group to be incorporated into their new guidelines for CAMP workshops.

Course development

As a vocational course, the participants came from a wide range of institutes and represented an equally wide range of environmental ethics and philosophies. A large proportion of the early courses was needed to refine a consensual definition of plant conservation specific to East Africa. This definition was used to direct workshops and project development.

Future Steps

Collaborative training events, if held in-country and specifically designed to match regional and local needs, can make a cost-effective contribution to the development of regional capacity. During three years, 107 professionals from 35 institutions were trained. It represented the first custom-designed course in the conservation of wild plant resources; accordingly, professionals were exposed to training that was designed for their needs. In addition, they received professional training that they would not otherwise have received. The course has resulted in new collaborative relationships and new projects have been established. In particular, new networks, formal and informal, were established and supported.

The success should be countered against recurrent issues that face any vocational course serving a wide range of clients. Course managers should give very close attention to maintaining flexibility in both materials and teaching styles. The timetable should allow sufficient time for the establishment of common skills required by all participants. In addition, when using a range of resource personnel great care should be taken to ensure that teaching styles do not clash; indeed, if there is a dramatic variation in teaching styles these should be presented as compartmentalised sectors in the timetable.

The strong links built up with local donors will help to ensure valuable funding support for such a follow-on activities. Capacity for regional collaboration has been built at the institutional level through the Specialist Workshops. It was felt that this was an effective way of presenting and discussing current thinking on specialist areas that are viewed as regional priorities. At the same time, it targeted staff at management level, in key institutions, who are in a position to disseminate and implement this information more broadly through in-house training and management of plant conservation activities.

The next stage in professional development for the region's plant conservationists will be through applying training to research and management projects. After all, the cataloguing of diversity will only be an academic/historical exercise, unless that diversity can be retained and managed in the real world.

Useful Websites

- Darwin Initiative at
  www.defra.gov.uk/environment/darwin/index.html
- National Museums of Kenya at
  www.museums.or.ke/research.html
- United Nations Convention on Biological Diversity (CBD) at
  www.biodiv.org
- RBG training at
  www.rbgkew.org.uk/education/index.html
- Conservation Breeding Specialist Group at
  www.cbsg.org/
- East Usambara Conservation Area Management Programme
  (EUCAMP) (formerly EUCFP) at
  www.usambara.com/

Acknowledgements

The authors wish to thank all those mentioned in the Acknowledgements at the beginning of this publication for their assistance, support and funding. We would also like to thank:

Darwin Initiative for the Survival of Species, Department of the Environment, Transport and the Regions (DETR), UK; UK Department For International Development (DIFD); British Council, Nairobi; African Wildlife Foundation, Cambridge, UK; United Nations Education, Science and Cultural Organisation (UNESCO); International Plant Genetic Resources Institute (IPGRI), Rome, Italy; Royal Botanic Gardens, Kew, UK; East Usambara Conservation Area Management Programme (formerly EUCFP), Tanzania; Royal Netherlands Embassy, Nairobi; Kenya Wildlife Service; National Museums of Kenya; IUCN Regional Office, Nairobi; British High Commission, Nairobi; Boulevard Hotel, Nairobi.

References

Baijnath, H. (1995). Networking African herbaria for biodiversity conservation, pp. 193–198. In: L.A. Bennun, R.A. Aman and S.A. Crafter (eds). *Conservation of Biodiversity in Africa: local intiatives and institutional roles.* National Museums of Kenya, Nairobi, Kenya.

Beentje, H.J. (1988). Atlas of the rare trees of Kenya. *Utafiti* **1**(3): 71–123.

Berg, E.J. (1993). *Rethinking Technical Co-operation: reforms for capacity building in Africa.* United Nations Development Programme, New York, USA.

DETR (1998). *First Darwin Report: Darwin Initiative for the Survival of Species.* Department of the Environment, Transport and the Regions, London, UK.

Ellis, S. and Seal, U.S. (1995). Tools of the trade to aid decision making for species survival. *Biodiversity and Conservation* **4**(6): 553–72.

Falk, D.A., Millar, C.I., Olwell, M. (eds). (1996). *Restoring Diversity: strategies for reintroduction of endangered plants.* Island Press. Washington DC, USA.

Flint, M. (1991). *Biological Diversity and Developing Countries: issues and options.* Overseas Development Administration, London, UK.

Glowka, L., Burhenne-Guilman, F., Synge, H., McNeely, J.A., and Gündling, L. (1994). *A Guide to the Convention on Biological Diversity.* Environmental Policy and Law paper no. 30. IUCN, Gland, Switzerland.

Heywood, V.H. (1995). *Global Biodiversity Assessment.* Cambridge University Press, Cambridge, UK.

Luke, W.R.Q. (ed.) (1996). *Conservation Assessment and Management Planning (CAMP) Training Workshop for Selected Species of the Kenya Coastal Forests,* 20–25 November 1996, Coastal Forest Conservation Unit, Kenya. Royal Botanic Gardens, Kew, UK.

Maunder, M. and Göhler, C. (1995). *Feasibility Study for the Development of a National Plant Conservation Network.* Report undertaken for the National Museums of Kenya under assignment from the Overseas Development Administration. The Royal Botanic Gardens, Kew, UK.

Mittermeier, R.A. and Werner, T.B. (1990). Wealth of plants and animals unites megadiversity countries. *Tropicos* **4**: 4–5.

Muir, C. (1998). *A Study to Investigate the Factors Affecting the Distribution of Cola usambarensis, an Endangered Endemic Tree of the East Usambara Mountains, Tanzania.* M.Sc. Dissertation in Conservation, University College, London, UK.

Mwasumbi, L., Johansson, S., Luke, Q., Simiyu, S., Mashauri, S., Ipulet, P., Maunder, M., Clubbe, C. and Hankamer, C. (1998). *Conservation Assessment and Management Planning (CAMP) Training Workshop for Selected Plant Species of the East Usambara Mountains, Tanzania.* 2–6 March 1998, East Usambara Catchment Forest Project, Tanzania. Royal Botanic Gardens, Kew, UK.

Myers, N., Mittermeier, R.A., Mittermeier, C.G., Da Fonseca, G.A.B. and Kent, J. (2000). Biodiversity hotspots for conservation priorities. *Nature* **403**: 853–858.

National Biodiversity Unit (1992). *The Costs and Unmet Needs of Biological Diversity Conservation in Kenya.* The National Biodiversity Unit (National Museums of Kenya) and Metroeconomica Ltd., Nairobi, Kenya.

Noss, R.F. (1997). The failure of universities to produce conservation biologists. *Conservation Biology* **11**(6): 1267–1269.

Pitkin, B. (1995). Protected Area Conservation Strategy (PARCS): training needs and opportunities among protected area managers in Eastern, Central and Southern Africa. Biodiversity Support Program, Washington DC, USA.

Polhill, R., Milne-Redhead, E., Hubbard, C.E., Turrill W.B., Beentje H.J., Whitehouse, C.M. and Smith, S. (eds). (1952–). *Flora of Tropical East Africa.* Royal Botanic Gardens, Kew, UK.

Republic of Kenya (1998). Draft First National Report to the Conference of the Parties (COP). Available from World Wide Web: www.biodiv.org/doc/world/ke/ke-nr-01-en.pdf>

Republic of Uganda (1998). First National Report on the Conservation Biodiversity in Uganda. Available from World Wide Web: www.biodiv.org/doc/world/ug/ug-nr-01-en.pdf>

Robertson, S.A. and Luke, W.R.Q. (1995). The coastal forest survey of the National Museum of Kenya and the World Wide Fund for Nature. In: L.A. Bennun, R.A. Aman and S.A. Crafter (eds), p. 89. *Conservation of Biodiversity in Africa: Local intiatives and institutional roles.* National Museums of Kenya, Nairobi, Kenya.

Rodgers, W.A. (1995). The conservation of biodiversity in East Africa: approaches of forestry and wildlife sectors compared. In: L.A. Bennun, R.A. Aman and S.A. Crafter (eds), pp. 217–230. *Conservation of Biodiversity in Africa: Local initiatives and institutional roles.* National Museums of Kenya, Nairobi, Kenya.

Sayer, J.A., Harcourt, C.S. and Collins, N.M. (eds). (1992). *The Conservation Atlas of Tropical Forests. Africa.* Macmillan, UK.

Stuart, S.N., Adams, R.J. and Jenkins, M.D. (1990). *Biodiversity in Sub-Saharan Africa and its islands: conservation, management and sustainable use.* IUCN, Cambridge, UK.

UNESCO (1992). *Managing Protected Areas in Africa.* UNESCO World Heritage Fund, Paris, France.

USAID (1993). *African Biodiversity: foundation for the future. a framework for integrating biodiversity conservation and sustainable development.* Biodiversity Support Program, Washington DC, USA.

Walter, K.S. and Gillett, H.J. (eds). (1998). *The 1997 IUCN Red List of Threatened Plants.* Compiled by the World Conservation Monitoring Centre. IUCN – The World Conservation Union, Gland, Switzerland and Cambridge, UK.

Williams, P.J. (1998). Building capacity for sustainable management of natural forests in East Africa, pp. 71–107. In: P. Veit (ed.). *Africa's Valuable Assets: a reader in natural resource management.* World Resources Institute, Baltimore, USA.

World Resources Institute (1996). *World Resources 1996–97.* Oxford University Press, Oxford, UK.

WWF and IUCN (1994). *Centres of Plant Diversity: A strategy for their conservation. Vol.1 Europe, Africa, South West Asia and the Middle East.* IUCN Publications Unit, Cambridge, UK.

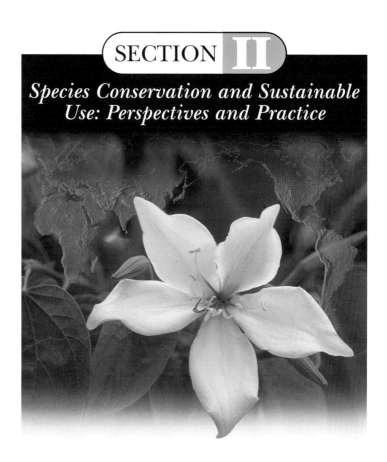

SECTION **II**

*Species Conservation and Sustainable
Use: Perspectives and Practice*

Chapter 16

Section Overview:
Species Conservation
Issues in The Tropics

Mike Maunder [1]
Royal Botanic Gardens, Kew, UK

Colin Clubbe
Royal Botanic Gardens, Kew, UK

[1] Now at: The National Tropical Botanical Garden, Kauai, Hawaii, USA

Introduction

"What we do (or do not do) within the next few decades will determine the long-term future of a vital feature of the biosphere, its abundance and diversity of species"

(Myers *et al.*, 2000)

As tropical habitats continue to decline in both area and quality, an increasing number of plant species will require conservation management to ensure their survival (Figure 16.1). The species is a discrete and readily recognisable unit of conservation management, and often the unit of national and international legislation e.g. Convention on International Trade in Endangered Species of Wild Fauna and Flora (CITES). Species are the traditional focus for *ex situ* agencies, such as botanic gardens and gene banks. The compositional elements

Figure 16.1   A member of the 'living dead', *Hyophorbe amaricaulis* Martius (Palmae), known from a single tree surviving in the Curepipe Botanic Garden, Mauritius. Whilst such examples are valid targets for conservation management, the priority should always focus on wild populations and habitats.

of species, namely populations, individuals and genes, are increasingly becoming the foci of targeted management actions. As populations of threatened species become more isolated and fragmented there is an increasing need to manipulate both the demographic and genetic dynamics.

The majority of the world's species can be retained through the 'coarse filter' approach of habitat conservation that can potentially conserve all levels in the biodiversity hierarchy (Hunter *et al.*, 1988). However, an increasing number of protected areas will require conservation management because of external influences impacting on ecological processes and promoting changes in both community structure and composition, leading to species loss. Accordingly, a 'fine filter' approach is required to retain those species not secured through the priority action of habitat conservation. Protected areas have been established with the assumption that environmental conditions and community patterns/composition have been stable for long periods in the past and will continue to be stable into the future. There is, however, increasing evidence that habitats and associated species assemblages are loosely organised collections of species whose coexistence is dependent on their individual ecological tolerances and subsequent distribution along environmental gradients (Hunter *et al.*, 1988).

Plant extinction trends

Increasing concern is being expressed about the extinction of plant species. The avoidance of extinctions, and the loss of biological capital, is driving many national and local conservation initiatives such as national biodiversity action plans, a requirement of the Conservation on Biological Diversity (CBD) (Glowka *et al.*, 1994; see Appendix 1). Extinctions are being recorded at three different levels:

- global extinction – where the last individuals of the taxon have died. This may be preceded by a period where individuals persist in *ex situ* conditions (Extinct in the Wild *sensu* IUCN; see Appendix 2) (Figures 16.2 and 16.3).

- national extinction – where politically defined sections of the global range become extinct.

- local extinction – where local populations are lost from the national populations.

The World Conservation Union (IUCN) maintains the global list of threatened and recently extinct plant species. The latest global review of plant species decline and extinction (Walter and Gillett, 1998) lists 33,798 threatened vascular plant species, representing 12.5% of the world's 270,000 described species. The same study identifies 380 species of plant that are Extinct. The 380 IUCN recorded extinctions are from 59 countries, with five countries accounting for about 70% of the global records: Australia (71), Republic of South Africa (53), Mauritius (47), USA (22) and India (19). In

Figure 16.2

Euphorbia taruensis S. Carter (Euphorbiaceae) (scandent succulent – shown from bottom right to top left of figure) in the last known site for this species in southern Kenya (1996). The whole site has since been destroyed by quarrying activities and this species is now thought to be Extinct in the Wild. The future of this species will be dependent upon good management until a re-introduction can be undertaken.

numerical terms the list is dominated by oceanic islands: Mauritius (47), Cuba (23), St. Helena (11), French Polynesia (12), Réunion (6), New Caledonia (5), New Zealand (7), Bermuda (3), Norfolk Island (2); accounting for 30% of all recorded plant extinctions. Do these records suggest that these countries are subject to unusually high levels of habitat loss and associated species extinction or are these records from countries with active botanists and sufficient historical records to judge species loss?

The distribution of recorded threatened species is not uniform in terms of taxonomy. Ten plant families account for around 35% (12,689/33,798) of the recorded threatened plant species: Compositae (2,553), Orchidaceae (1,779), Leguminosae (2,206), Rubiaceae (1,120), Scrophulariaceae (969), Euphorbiaceae (933), Arecaceae (859), Myrtaceae (747), Cruciferae (747) and Gramineae (776). This family distribution indicates that a high proportion of the IUCN records are from the Euro-Mediterranean regions,

Figure 16.3　　One of two surviving cultivated populations of *Euphorbia taruensis* S. Carter (Euphorbiaceae). The species survives in the nursery of the Coastal Forest Conservation Unit at Ukunda shown here with nursery manager Mbinda Jeremiah, on the Kenya coast, and the Nairobi Botanic Garden at the National Museums of Kenya.

with well studied diverse floras, with high diversities of Scrophulariaceae, Cruciferae, Euphorbiaceae, and Leguminosae. Of the ten families, only five are predominantly tropical in terms of species richness, namely: Orchidaceae, Leguminosae, Rubiaceae, Myrtaceae and Arecaceae.

Ten countries hold approximately half of the world's recorded threatened species (approximately 17,000 species), namely: USA (4,669), Australia (2,245), Turkey (1,876), Republic of South Africa (2,215), Brazil (1,358), Mexico (1,593), Panama (1,302), India (1,236) and Spain (985). Australia and the USA alone hold over 20% of the recorded globally threatened species. These countries share a number of characteristics: all are species rich with a high diversity of habitats and ecosystems, and the majority have an established taxonomic infrastructure with ongoing botanical exploration and study. Two countries are Mediterranean with modern floristic accounts: Spain and Turkey; four are highly diverse 'continental' countries encompassing temperate, mediterranean and tropical regions: USA (including Hawaii), Republic of South Africa, India and Australia; and three are tropical Latin American nations: Brazil, Panama and Mexico. This group does not closely match with either recorded plant diversity or recorded habitat loss; for instance, the megadiverse nations of the Philippines, Malaysia and Indonesia do not feature highly in the list of recorded extinctions. Tropical Africa, an area with a poorly supported taxonomic infrastructure and historically poor levels of collection (Morat and Lowry, 1997), reveals apparently low levels of threat, with few countries reporting more than 2% of their flora as threatened. It would be difficult to use these figures to generate political support for the 'biodiversity crisis', a 2% loss from a poorly defined resource over a 20–50 year time scale would appear reasonable to some! It can be argued that current assessments of global plant loss reflect the extent of conservation and taxonomic investment rather than the realities of species and habitat decline.

Existing assessments of threatened species and estimates of tropical extinctions are based upon an imperfect understanding of taxonomic diversity, the distribution of that diversity and the response of that diversity to the various types and intensities of habitat modification. For the largest part of the planet there is no clear consensus on the rate of species and population loss. For instance, there is a dramatic lack of congruity between the IUCN historical records (Walter and Gillett, 1998) and the estimates of tropical extinctions made by Koopowitz *et al.* (1993, 1994) based on levels of local species endemism and recorded habitat destruction. This is particularly notable for Brazil where IUCN records only five extinctions and Koopowitz estimates a loss since 1950 of 2,261 species. Using documented habitat loss may prove a more accurate gauge of species loss than the traditional approach documenting extinctions through observations of single species. The lack of field survey work and the rapidity of habitat loss means that species extinction will be identified only in retrospect. It is evident that the information pertaining to plant extinctions is not sufficient to identify in advance which tropical species are at greatest risk of extinction.

The plant conservationist in the tropics faces the challenge of matching available, and often declining resources, with the demands of dealing with increasingly catastrophic habitat loss and associated species decline. However, conservationists rarely have sufficient data to effectively deploy their time and resources. At present, it is very difficult for a tropical agency to effectively identify targets for conservation at the species level. Only a few attempts have been made to identify extinction prone tropical plant species in advance (Martini *et al.*, 1994). However, there are basic steps that can be taken to ensure a better understanding of prevailing environmental trends and the likely candidate species requiring conservation management. These steps are dependent upon working with collaborating agencies and individuals to gather information across a broad range of disciplines.

Using habitat loss and endemism may allow more realistic assessments of threat. Myers *et al.* (2000) have identified 25 global 'hotspots' where extraordinary concentrations of endemic species are subject to catastrophic habitat loss (see Table 1 pages xxiv–xxv). These 25 enclaves contain as many as 44% of the world's vascular plants on only 1.4% of the Earth's land surface. The 25 hotspots have in total lost 88% of their primary vegetation, for instance Madagascar has only 10% of its primary vegetation surviving. In contrast, 8,000 of Madagascar's estimated 10,000 vascular plants are endemic, with only 306 species listed by the IUCN as globally threatened (approximately 3%). Clearly, for hot spot areas such as Madagascar there is an urgent need to survey the flora, assess taxonomic and geographical priorities, and establish effective protected areas (Prance *et al.*, 2000).

Species Conservation Assessments

Increasingly conservation and protected area agencies are charged with responsibilities for managing and recovering threatened plant species. How can these species be identified (Box 16.1)? Threatened species, those species threatened with global extinction, can be defined as those species characterised by one or more of the following features (See IUCN Categories of Threat, Appendix 2):

- declining range,
- restricted and declining areas of occupancy (e.g. the specific habitat within the range),
- fragmented populations or population concentrated into vulnerable single locations,
- fluctuating range and occupancy areas, locations and numbers of individuals,

- declining numbers of locations and individuals,
- low numbers of individuals.

Can provisional lists for careful review be drawn up without investing in major and prolonged national surveys of biodiversity (see Bytebier and Pearce, Chapter 4; Hernández and Gómez-Hinostrosa, Chapter 18)? Where are the available information sources that can be used to initiate such a review and guide agencies in their identification of candidate species?

Suggested steps for species conservation assessments

1. **Define geographical boundaries of survey.** An agency or institution may have a defined geographical remit defined by national or local political boundaries, or defined by the boundaries of a national park. Establishing such boundaries will focus the exercise within a defined management and legal context.

2. **Establish project context.** Prior to initiating detailed research the context for the study should be clearly established with regard to the legal mandate and responsibilities for the work. If the exercise is being conducted at the national level the project executors should seek briefing from the national agencies responsible for biodiversity. If the exercise is focused on a single protected area or restricted land area, an initial short site visit should be undertaken to identify:

- key collaborators and stakeholders,
- relevant information sources,
- prevailing biodiversity issues and trends.

3. **Literature survey.** A considerable amount of information on threatened plants may lie in published sources, and this will be of varying degrees of usefulness and should always be subject to review and verification:

   i) **International sources:** The IUCN and World Conservation Monitoring Centre (WCMC) maintain a global list of threatened plant species – which is available to all interested parties. Obtain national lists and those from neighbouring states/provinces that share contiguous habitats and ecosystems. A number of international groups can provide data on nationally and globally threatened plant taxa. These include the IUCN Species Survival Commission (SSC) and its Specialist Groups, the International Institute for Plant Genetic Resources (IPGRI) and publications from the IUCN/Worldwide Fund for Nature (WWF) study, Centres of Plant Diversity (WWF and

IUCN, 1994–1997). Increasingly these data sources will be available on-line. For example the IUCN Red List is searchable on its own website at www.redlist.org/.

ii) **National sources:** It is likely that some information will exist on the conservation status of species in your region. National listings of threatened species may be held by government agencies and ministries; indeed, different ministries may hold different lists! Lists can be published in CBD Biodiversity Action Plans, forestry and conservation strategies. Highly relevant information is often located in the 'grey literature' – unpublished and uncatalogued. Reports and documents filed in forestry offices, museums, non-governmental organisations, herbaria and universities can all shed light on conservation issues. In addition to species level information, data should also be collected from government development documents (agriculture, forestry, transport, energy and sustainable development) on specific habitats or regions threatened by development or destruction e.g. hydro-electric or agricultural relocation schemes.

iii) **Scientific sources:** a number of scientific resources will be invaluable for the review. Discussions with the national or regional herbarium should identify the standard floristic work for the region providing the taxonomic and nomenclatural reference for the work, and sources of specialist botanical information, including field workers. The herbarium may be able to help with data supply and it may maintain a database that can be interrogated for conservation information or a manual survey can be undertaken e.g. for the study area how many species are represented by five or fewer collected specimens? Any restricted endemics with no recent records? The herbarium should be able to provide information on local or national endemic plants. In addition to the herbarium, other valuable sources include: national and overseas universities, that may be conducting relevant conservation biology studies, national plant conservation networks and amateur natural history or botany groups.

4. **Human resources.** A vast amount of invaluable information, often field-based, is lodged with individuals and will remain inaccessible and unpublished unless specifically sought out. To obtain a more accurate understanding of species conservation trends, it is imperative that time is given to discussions with key

people. These include: field botanists/taxonomists, foresters, protected area managers and wardens, natural history society members and amateur botanists, traditional users and village groups, and commercial dealers. This process must be handled carefully with all sources of guidance and information scrupulously credited in any reports.

5. **Initial candidate list.** The above steps should allow the compilation of a draft species list that can be subsequently refined with field data. This list will include:

- nationally and internationally listed taxa from IUCN and national listings;
- taxa listed as threatened by adjacent nations or states;
- national or local endemics;
- taxa, particularly national endemics, restricted to vulnerable and heavily disturbed habitats;
- taxa thought to be susceptible to over-harvesting and over-utilisation;
- those taxa identified by experienced contacts that have declined or continue to decline for a wide variety of reasons.

6. **Assessing and sorting the list.** This list may well be long and need taxonomic and nomenclatural editing. The assessment of conservation priorities is a multi-disciplinary task that can only be undertaken through active dialogue between different agencies. Questions need to be asked such as:

- Do a large proportion of your candidate taxa come from a particular province, mountain area or protected area?
- Do particular taxonomic groups dominate your listing?
- Does this represent reality or the fact that certain plant groups are better studied than others?
- Are the candidate listings dominated by utility groups such as medicinals or timber species?

At this stage it may be useful to establish regional working groups to further review the listing. A Conservation Assessment and Management Planning (CAMP) (Ellis and Seal, 1995) workshop or series of workshops will effectively review a candidate list and ensure that all relevant parties can contribute to the assessment, to the priority setting and establishing institutional responsibilities and contributions (see Hankamer *et al.*; Chapter 15 and Appendix 3). Priority setting can be guided by a number of factors:

i) **Level of threat.** This is obviously an important guiding factor. However, whilst this approach allows investment in those species most threatened with extinction it can result in a poorly co-ordinated response that is species-, rather than habitat- or ecosystem- led. It may result in conservation agencies investing in 'high cost – high risk' species. This can be justified if these species are used as flagships species to generate support for protected areas that serve the wider retention of biodiversity.

ii) **A regional habitat based approach to conserve hotspots.** This approach promotes conservation investment in habitats or ecosystems containing concentrations of threatened species. It is probably the most cost-effective response for high diversity areas.

iii) **An evolutionary perspective.** Selection may be based on the recognition that a national portfolio of biodiversity may contain unique evolutionary units such as endemic families, genera and species. In addition, countries may well contain unique and important infraspecific diversity.

iv) **Utility and economic importance.** Species conservation will be increasingly guided by this factor. Plant resources such as wild species that provide forage, timber, medicinals, tropical fruit, etc., or wild relatives of crops, need to be conserved as wild populations to continue as both vital rural resources and to support crop development.

v) **'Leverage' or 'flagship' species that may or may not fall into the above categories.** A number of spectacular threatened species can be used as symbols to lever support for the conservation of tropical habitats and species. Examples include groups with international donor interest (often horticulturally driven) such as orchids, cacti and succulents, palms, carnivorous plants and African violets (*Saintpaulia* H. Wendl. species (Gesneriaceae)).

Species Recovery

Species recovery encompasses the management activities required to halt or reverse the decline in a threatened species population. It can be best achieved through cross-disciplinary recovery teams applying the recommendations of a Species Recovery Plan. Such a programme must have clear numerically based objectives using, where appropriate, information from ecological, taxonomic, genetic, economic and social studies.

The effective conservation management of threatened plants is supported by a five-part process:

1. **Inventory:** the geographically based assessment of taxa that documents their occurrence in mapped political and geographical units; undertaken at a local level (single mountain range or protected area) or nationally. Such information (as discussed above) can be used to contribute towards an initial threatened plant list.

2. **Survey:** the ecologically based assessment of located populations in the field with respect to habitat and threat factors; to gather detailed information on population/demographic status (Figures 16.4 and 16.5).

3. **Monitoring:** repetitive measurements of populations and individuals over time to document numerical and qualitative changes over time (Figure 16.6).

4. **Habitat protection:** the assignment of land for conservation purposes with restrictions on land use and harvesting when appropriate. This may be after the initial survey or a result of information from monitoring.

5. **Recovery:** the management of a threatened taxa or population to decrease the chances of extinction and further secure viable wild populations.

Single species management for threatened species can include a number of actions:

- protection from over-harvesting (illegal and legal);
- protection from invasive organisms and pathogens;
- habitat modification and management, e.g. prescribed burning or clearance of invasive species;
- reintroduction or translocation;
- assisted reproduction e.g. artificial pollination;
- *ex situ* propagation and storage, preferably in-country and directly linked with recovery activities (Figure 16.7).

> **Box 16.1 A method for assessing the conservation status of a species at a national or local level**
>
> ### 1. Inventory
>
> What is known about the location/distribution of the threatened species? What evidence exists to suggest population decline? Assemble all known information. Possible sources of information include:
>
> - herbarium records,
> - published literature,
> - previous surveys,
> - local knowledge, including amateur sources,
> - threatened species list.
>
> Analysis of the gathered data should ideally be undertaken using a database management system e.g. Botanical Research and Herbarium Management System (BRAHMS) (see Pearce and Bytebier, Chapter 4 and Hernández and Gomez-Hinostrosa, Chapter 18). Information to be collated from the above sources can include: location of herbarium specimens, names of collectors and collection numbers, phenological data, collection data (latitude and longitude, altitude), soil type, habitat, vegetation type, slope and aspect, land use, vernacular names and any recorded use. This will provide a list of potential sites with associated ecological characteristics.
>
> ### 2. Field Survey
>
> Starting point depends on whether there are any known sites for the species under investigation. It is useful to supply investigators with a good 'search image' of the species e.g. photograph, drawing or cibachrome, particularly if they are unfamiliar with the species.
>
> | i) Verify status of target species in known field sites | ii) Explore to find new sites for target species |
> |---|---|
> | ↓ | ↓ |
> | Re-locate documented site(s) | Locate suitable habitat where species is likely to occur (is it known to be |
> | ↓ | restricted to a particular soil type, biome, |
> | Locate individuals/populations | land use, geology, season or particular ecological niche?) |
> | | ↓ |
> | | Explore site |
> | | This can be done systematically or randomly: |
> | | • Belt transects can be established across the site and walked systematically to look for species, ensuring coverage of whole site. |
> | | • Timed, random meanders can be undertaken, although these may not cover the whole site. |

Box 16.1 continued

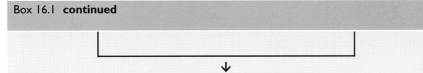

↓

Individuals/population located.
Georeference site using a Global Positioning System (GPS)
↓

Undertake demographic survey of located individuals.
How many of the recorded populations have been lost? How many new sites have been located? Does historical documentation suggest the population structure has changed? Adopt suitable measures to document age structure and reproductive status:

• Is the population dominated by old individuals with few seedlings or immature plants?

• Are there individuals of different ages, seedlings, flowering/fruiting adults?

↓
Describe habitat

3. Monitoring

• Devise monitoring protocol which will document the performance of marked individuals of the species under investigation by making repeated measurements over a period of time e.g. height, girth, reproduction, germination, seedling growth.

• Monitoring is a long-term, resource-intensive commitment and so must be designed to identify problems in the demographics of the species and to provide specific management recommendations.

Five questions to ask when designing a monitoring programme:

  i)   Purpose: what is the aim of monitoring?

  ii) Method: how can this aim be achieved?

  iii) Analysis: how are the data to be handled?

  iv) Interpretation: what do the data mean?

  v)   Fulfilment: when will the aim have been achieved?

References: Bullock, 1996; Given, 1994; Kent and Coker, 1994; Maxted *et al.*, 1995; Mueller-Dombois and Ellenberg, 1974.

Figure 16.4 Steevern Shunda, Institute of Traditional Medicine, Muhimbili University College of Health Sciences, Tanzania and Patrick Mucunguzi of the Makerere University, Uganda, working on a belt transect through cycad populations in the Mwaluganje Elephant Sanctuary, Kenya.

Species as the compositional units of a community or ecosystem are a convenient and discrete unit of management, particularly where that taxon is threatened and requires species specific management. A Species Recovery Plan provides a forum to bring the required expertise together to ensure an effective and integrated approach to species conservation. An integrated approach to plant conservation is being adopted utilising the complementary resources of differing agencies and stakeholders (Bowles and Whelan, 1994; Falk *et al.*, 1996). No one management body or mechanism will be sufficient to deal with the complexities of species recovery.

Species programmes, dealing with single species issues or clusters of threatened species, can be used very effectively to promote habitat conservation. As mentioned earlier, species can be used as flagship (a symbol for conservation), or promoted as keystone (providing a key ecological function) and umbrella species (species requiring large areas of intact habitat) to help secure the conservation of viable habitat reserves. *Ex situ* species displays, such as botanic gardens, can play a fundamental role in public education and fund-raising. Indeed, until accurate listings and assessments of threatened plants are available to guide botanic garden conservation investments it is suggested that

Figure 16.5   David Okebiro, National Museums of Kenya with Shadrack Sanganyi, National Gene Bank, Kenya, using binoculars to survey for *Saintpaulia* H. Wendl. (Gesneriaceae) populations along a river bank, Amani Nature Reserve, Tanzania.

education on species and ecosystem conservation should play a primary role for botanic gardens (see Maunder *et al.*, Chapter 7). In some cases the development of a single species programme has subsequently led to the development of habitat programmes (see Dulloo, Chapter 13). However, poorly planned single species management can result in damaging changes in species abundance and can be interpreted as undermining the value of habitat conservation. For instance, the management for dense concentrations of valued game animals can degrade a habitat.

A recovery plan should utilise the most efficient and most appropriate management responses to ensure species survival. The long-term conservation of threatened species is dependent on the sustained collaboration between a range of agencies responsible for habitat conservation and single species management, both *in situ* and *ex situ*.

Figure 16.6   Shadrack Sanganyi, National Gene Bank, Kenya, assessing wild populations of African Violet (*Saintpaulia* H. Wendl. sp. (Gesneriaceae)) in the Amani Nature Reserve, Tanzania.

Designing and undertaking a Species Recovery Plan

i) The recovery of plant populations

Successful recovery of threatened plant populations is dependent upon:

- a functioning legal and administrative mechanism for funding and implementing recovery activities;
- identification of a valid unit of conservation management. Is the taxon taxonomically valid and does it represent a single phylogenetic clade requiring conservation management?
- sufficient habitat being retained to ensure the persistence of viable wild populations over the long term;
- demographic information being available to identify which stages of the life history are critical to survival and persistence of a population;
- population structures and genetic variation of the managed populations being sufficiently well understood to allow effective population management.

Figure 16.7   The Nairobi Botanic Garden, Kenya holds extensive collections of wild collected African violets (*Saintpaulia* H. Wendl. (Gesneriaceae)) as both a support to taxonomic and phylogenetic research and as a direct conservation investment. A number of Kenyan species and populations are heavily threatened as a result of agricultural expansion into forests (e.g. the Taita Hills) and quarrying of limestone for building and construction. It is likely that at least one taxon is now Extinct in the Wild.

ii) Selection and identification of target species

A Species Recovery Plan should aim to co-ordinate and provide the management activities required to halt, or reverse the decline of a named threatened species. Ideally, it should target a species identified in national biodiversity action plans and take place within a national or governmental framework. Where appropriate, recovery plans can be drafted for clusters of species facing shared risks or sharing a restricted habitat area. The selection of a target species should be based upon reliable information indicating a declining population and an increased risk of extinction, either at a global or regional level. A CAMP workshop offers the opportunity to review a number of species from a geographical area or taxonomic group and can identify potential taxa requiring recovery activities (see Appendix 3 and Hankamer et al.; Chapter 15, Box 15.2).

Box 16.2 A hypothetical recovery plan for an idealised plant species, developed from practical training sessions during the Darwin – National Museums of Kenya Plant Conservation Techniques Course for East Africa

Recovery plan for *Euphorbia darwinensis* (Clubbe and Maunder) Myth (Euphorbiaceae)

Implementing agencies: Plant Conservation Programme of the National Museums of Kenya as part of the Kenyan Plant Conservation Network in collaboration with Kenya Wildlife Service and Kenya Forest Department.

Project summary: Current species status: *E. darwinensis* is restricted to one forest reserve in the Shimba Hills of Kenya and has been categorised as Critically Endangered (*sensu* IUCN). There are only two known populations, one with eight mature individuals, and one of 100 mature individuals. *E. darwinensis* is found at altitudes between 800 and 1000 metres in degraded forest areas within Forestry Department land. Unconfirmed reports describe a wild population in Tanzania. Both Kenyan populations are threatened by proposed extension of plantations, ongoing elephant damage, and invasive *Lantana camara* L. (Verbenaceae) scrub. No regeneration has been observed and all the individuals are mature, with good seed production. The plant is harvested for medicinal purposes. Old herbarium records (1920s) at the National Museums of Kenya suggest a previously wider distribution with specimens recorded from areas now cleared for agriculture.

Recovery objectives: To secure regenerating viable wild populations of *E. darwinensis*. To revise Category of Threat to Vulnerable or Conservation Dependent, within ten years.

Recovery criteria: Establish reproducing and regenerating populations at the natural sites and establish two new populations each of 500 individuals on protected and managed sites.

Actions needed:

1.   Establish legal mandate and regulations for recovery planning.
2.   Assemble recovery project team and confirm working protocols including budget for drafting the recovery plan.
3.   Undertake data search and obtain all relevant papers, reports, etc.
4.   Establish taxonomic identity of the species and undertake a review of distribution and eco-geographic data.
5.   Assess demography and breeding biology. Undertake ecological study of known populations, including impact of invasive weeds and pests, and harvesting.
6.   Protect existing populations from destruction by plantations and elephants.
7.   Survey for new populations, including putative Tanzanian population.
8.   Develop propagation protocol (seed and vegetative) for species.
9.   Locate suitable sites for reintroduction/translocation and undertake feasibility study.
10.   Agree site management procedures.
11.   Validate recovery proposal and draw up implementation schedule. Draw up and agree budget for implementation.
12.   Review on yearly basis for life of project (ten years).

Figure 16.8   Large proportions of the world's temperate and arid land flora can be stored effectively in seedbanks, such as the Millenium Seed Bank at the Royal Botanic Gardens, Kew. However, this is an interim measure. The long term and evolutionary future of these resources is dependent upon retaining wild populations and habitats.

iii) Institutional responsibilities and networking

The most effective recovery projects use a range of professionals from a range of institutions. It is increasingly evident that species recovery requires effective collaboration between a wide variety of institutions e.g. universities, herbaria, protected areas and *ex situ* facilities (Figure 16.8).

iv) Designing a recovery plan

The process of designing a recovery plan can be broken down into a number of stages (see example in Box 16.2):

- **Collect historical and available data on target taxon.** This can include data from herbarium collections (indicating current distribution and historical changes in distribution), field workers, historical accounts, users, unpublished or grey literature, and scientific literature.

- **Undertake a multi-disciplinary review of available information to identify gaps in information.** Collate information from habitat managers, population geneticists, *ex situ* managers, ecologists, community groups and educational specialists.

- **Design initial research phase and identify key players.** Key issues at this stage include identifying the factors causing the species decline and assessing the status and distribution of wild populations. For instance, are all populations declining and how fast? Research should focus on reproductive, demographic and ecological aspects of the target species. Who has the institutional responsibility (legal and otherwise) for the recovery of the species?

- **Establish initial goals and objectives.** Translate recovery plans into objectives. The objectives should be specific, realistic, attainable, measurable (in both numerical terms and with stated time frames), and politically or socially acceptable. Initial priorities should focus on retaining existing populations. Other issues include: increasing the number of individuals or changing the demographic profile of the population, assessing the distribution of genetic diversity, and changing the legal status of species.

- **Design and cost work plan, agree length of project.** It is likely that the recovery activities will be undertaken with a limited budget, probably on a contract basis from a national or international agency. It is vital at this stage to carefully plan and cost the activities: Who will do the work? How long will it take? Will the funding cover the whole of the recovery project or only part of the project? What equipment and resources will be required?

- **Circulate draft recovery plan for review.** The Recovery Group should circulate a draft report to all involved parties (stakeholders) for the purpose of receiving feedback. All comments received should be reviewed by the group and fully acknowledged.

- **Publish recovery plan.** The published recovery plan should be sent to all stakeholders including field workers involved with the target species. Consideration should be given to the languages used. The report should include an executive summary. A Recovery Plan is not the project target, the recovery of the species is the target.

  Many national authorities are publishing their recovery plans on the Internet, for example:

  USA – US Fish and Wildlife Service at endangered.fws.gov/recovery/recplans

  Australia – Wildlife Australia at www.biodiversity.environment.gov.au/wildlife/plans/recovery/index.html

- **Undertake regular review of activities and maintain communication within the group.** As work develops, it is imperative that the team is closely co-ordinated and that all pertinent information is circulated. Regular meetings should be held and regular updates provided to the management and financing authorities.

The management of threatened plant species is a relatively new challenge for conservation. In the tropics, the plant conservationist is practising the management of evolutionary lineages amidst shifting ecological, economic and political contexts. The objective is to stay realistic and to focus on the management of those species that serve a demonstrated social or ecological role, or can act as a lever for retaining wild habitats and wilderness.

References

Bowles, M.L. and Whelan, C.J. (1994). *Restoration of Endangered Species: conceptual issues, planning and implementation.* Cambridge University Press, Cambridge, UK.

Bullock, J. (1996). Plants, pp. 111–138. In: W. Sutherland (ed.), *Ecological Census Techniques.* Cambridge University Press, Cambridge, UK.

Ellis, S. and Seal, U.S. (1995). Tools of the trade to aid decision making for species survival. *Biodiversity and Conservation* **4**(6): 553–572.

Falk, D., Millar C.I. and Olwell, M. (eds) (1996). *Restoring Diversity: strategies for reintroduction of endangered plants.* Island Press, Washington DC, USA.

Given, D.R. (1994). *Principles and Practice of Plant Conservation.* Chapman and Hall, London, UK.

Glowka, L., Burhenne-Guilmin, F., Synge, H., McNeely, J.A. and Gündling, L. (1994). *A Guide to the Convention on Biological Diversity.* Environment Policy and Law paper no. 30. IUCN, Gland, Switzerland.

Hunter, M.L., Jacobsen, G.L. and Webb, T. (1988). Palaeoecology and the coarse filter approach to maintaining biological diversity. *Conservation Biology* **2**: 375–385.

Kent, M. and Coker, P. (1994). *Vegetation Description and Analysis.* John Wiley, London, UK.

Koopowitz, H., Thornhill, A. and Anderson, M. (1993). Species distribution profiles of the neotropical orchids *Masdevallia* and *Dracula* (Pleurothallidinae, Orchidaceae): implications for conservation. *Biodiversity and Conservation* **2**: 681–690.

Koopowitz, H., Thornhill, A. and Anderson, M. (1994). A general model for the prediction of biodiversity losses based on habitat conversion. *Conservation Biology* **8**: 425–438.

Martini, A.M.Z., Rosa, N.A. and Uhl, C. (1994). An attempt to predict which Amazonian tree species may be threatened by logging activities. *Environmental Conservation* **21**: 152–162.

Maxted, N., Van Slageren, M.W. and Rihan, J.R. (1995). Ecogeographic surveys, pp. 255–286. In: L. Guarino, V. Ramanatha Rao and J.R. Reid (eds). *Collecting Plant Genetic Diversity: technical guidelines.* CAB International, Wallingford, UK.

Morat, P. and Lowry, P.P. (1997). Floristic richness in the Africa-Madagascar region: a brief history and prospective. *Adansonia* **19**: 101–115.

Mueller-Dombois, D. and Ellenberg, H. (1974). *Aims and Methods in Vegetation Ecology.* John Wiley, London, UK.

Myers, N., Mittermeier, R.A., Mittermeier, C.G., da Fonseca, G.A.B. and Kent, J. (2000). Biodiversity hotspots for conservation priorities. *Nature* **403**: 853–858.

Prance, G.T., Beentje, H., Dransfield, J. and Johns, R. (2000). The tropical flora remains undercollected. *Annals of the Missouri Botanical Garden* **87**: 67–71.

Walter, K.S. and Gillett, H.J. (eds) (1998). *1997 IUCN Red List of Threatened Plants*. Compiled by the World Conservation Monitoring Centre. IUCN – The World Conservation Union, Gland, Switzerland amd Cambridg, UK.

WWF and IUCN (1994–1997). *Centres of Plant Diversity. A guide and strategy for their conservation.* 3 volumes. IUCN Publications Unit, Cambridge, UK.

Chapter 17

Threatened Species Management in an Ocean Archipelago:

the Galápagos Islands

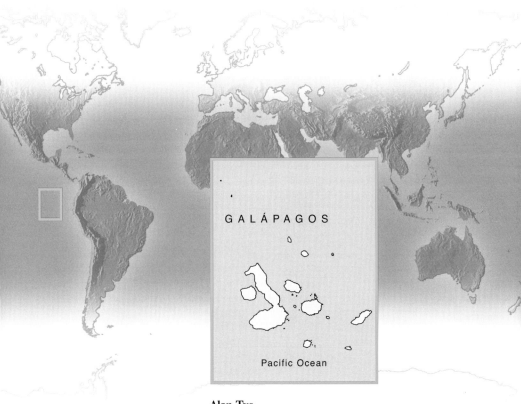

GALÁPAGOS

Pacific Ocean

Alan Tye
Charles Darwin Research Station, Galápagos,
Ecuador

Introduction

Galápagos is an isolated oceanic archipelago of over 130 volcanic islands, islets and rocks, straddling the equator 1,000 km west of Ecuador. They have never been connected to the mainland and are geologically very young, ranging in age from 1 to 3.3 million years (Simkin, 1984). The archipelago is famous for its endemic species, especially for Darwin's finches, which, along with other organisms from the islands, contributed to Charles Darwin's formulation of his theory of evolution by natural selection. Less well known are the endemic plants of Galápagos. The islands support a indigenous vascular flora of about 560 species, of which about 180 (32%) are endemic (Lawesson *et al.*, 1987). Some 53 families of vascular plants have evolved one or more endemic species, among which are seven endemic genera: *Darwiniothamnus* Harling, *Lecocarpus* Decne., *Macraea* Lindl. and *Scalesia* Arn. ex Lindl. (all in the Asteraceae (Compositae)), *Brachycereus* Britton & Rose and *Jasminocereus* Britton & Rose (Cactaceae) and *Sicyocaulis* Wiggins (Cucurbitaceae). Significant radiation has occurred in 11 families and 19 genera of Galápagos plants. Many of the endemic plant species have extremely limited distributions, such as only one island, or part of an island. Of the indigenous plants, the endemic species are of the greatest conservation interest, because their future depends entirely on their conservation for their continued existence in the islands.

The archipelago has a semi-arid, subtropical climate, due to the prevailing cold Humboldt current from the Antarctic and prevailing winds that also come from the south for most of the year. These factors lead to a well-defined vegetation zonation according to altitude and aspect (Wiggins and Porter, 1971). Progressing from lowlands to highlands, the zones are: Littoral (mangroves, dune vegetation and other coastal communities); 'Arid' (actually semi-arid: scrub and light woodland dominated by cacti); Transition (more or less closed woodland); and Humid (broken into a number of sub-zones that vary between islands, including *Scalesia* zone forest, *Miconia* Ruiz & Pavón (Melastomataceae) zone dense scrub, and fern-sedge zone highlands). Most islands, which are low, have no Humid zone; accordingly, biodiversity is highest on the larger and climatically diverse islands.

The first documented discovery of the archipelago was in 1535 by Tomás de Berlanga (Hickman, 1985). Most early visitors were pirates, whalers and sealers, exploiting the islands as a refuge and source of animals. The first settler did not arrive until about 1807. Floreana has been permanently inhabited since 1929, San Cristóbal was settled permanently in 1869, Isabela in 1893 and Santa Cruz in the 1920s. This short history of occupancy, in comparison with other oceanic archipelagoes, has resulted in the Galápagos remaining relatively pristine, with very few extinctions in historical times (only two species of vascular plant). However, the Galápagos Islands now face serious problems, which require rapid and determined action to prevent many more extinctions in the near future.

Threats to the flora

Even before settlement, the pirates and whalers began altering the Galápagos ecosystem. They initiated habitat clearance and direct exploitation of indigenous species, and deliberately or accidentally introduced a number of alien species, including goats, rats and, probably, insects and plants. Even before permanent settlement, the island of Floreana contained large areas dominated by introduced plants such as *Citrus* L. spp. (Rutaceae) (Hamann, 1984; Slevin, 1959). The first settlers introduced plants (pumpkins and potatoes) to Floreana by about 1807 (Porter, 1822). In recent years, the tourism industry has led to a comparatively high standard of living on the islands, attracting more settlers and further increasing the problems. The rapidly increasing settled population of Galápagos, growing at 8% per year in the 1990s through immigration and on-island birthrate, has been paralleled by an enormous number of new introductions of alien plants and animals (Mauchamp, 1997).

i) Direct exploitation
Very few Galápagos indigenous plants are used by the human inhabitants, and direct exploitation is largely limited to logging of timber trees, including endemic species. For instance, the endemic *Lippia salicifolia* Andersson (Verbenaceae) has become very rare partly as a result of logging (Lawesson, 1990; Mauchamp *et al.*, 1998). The very slow-growing *Piscidia carthagenensis* Jacq. (Leguminosae) has been over-exploited for its extremely durable timber, resulting in virtually all large specimens having been cut. Other species, such as the endemic *Psidium galapageium* Hook. f. (Myrtaceae), have been less severely affected, but illegal harvesting combined with a licensing system based on inadequate biological data may be causing long-term damage to the populations.

ii) Habitat clearance
At present, the Galápagos National Park covers 96.7% of the land area of the archipelago. Inhabited areas (urban and agricultural zones, military bases and airports) make up the rest. There are currently five inhabited islands: four with civilian populations (Floreana, Isabela, San Cristóbal, Santa Cruz) and a fifth, Baltra, which is a military base and civil airport. The effect of the relatively small inhabited area is out of all proportion to its size. The military bases and main urban zones are in the extensive lowland Arid and Littoral vegetation zones. However, the agricultural areas are mainly in the humid zones. The Humid zone is very limited in Galápagos (approx. 1,100 square kilometres (km^2) or 14% of the total land area), and agricultural development has destroyed some 60% of the original Humid zone vegetation

in the inhabited islands; indeed, over 90% of the zone has been converted on San Cristóbal (Table 17.1). Some subdivisions of the Humid zone are more badly affected than others; for example, over 80% of the *Scalesia* zone on Santa Cruz has been destroyed. The remaining areas of native Humid zone vegetation are small, and mostly adjacent to agricultural areas that serve as sources of invasion by alien species (Figure 17.1). These factors combine to render the tiny amount of remaining humid zone vegetation, and all of the organisms endemic to it extremely vulnerable to further loss. Despite the serious past effects of habitat clearance, continued clearance is restricted to quarrying. This is not a major threat to the vegetation, although some organisms with tiny ranges, such as the endemic Bulimulidae snails, may be severely affected by it.

Table 17.1 **Area of each vegetation zone destroyed by agricultural development on the four islands carrying a civilian population (in the case of Isabela Island, the figures refer to the inhabited volcano, Sierra Negra)**

Island	Vegetation zone	Total area of habitat type km²	Area of habitat dedicated to agriculture km² (%)
Santa Cruz	Humid	118	90 (76)
	Transition	126	28 (23)
Floreana	Humid	31	5 (15)
	Transition	37	1 (2)
San Cristóbal	Humid	84	78 (93)
	Transition	42	3 (7)
Sierra Negra (Isabela)	Humid	248	53 (21)

iii) Introduced species

The principal current threat to the terrestrial ecosystem of Galápagos is introduced species (Loope *et al.*, 1988). Galápagos is a typical oceanic archipelago in that the indigenous flora is depauperate and lacks many components of more diverse continental systems. A given island area contains many fewer species than a similar area on the mainland. The vegetation lacks many components of more saturated continental systems, such as trees in some habitats, and is highly susceptible to invasion by more competitive exotic species (Loope *et al.*, 1988). Introduced predators, such as cats and dogs, prey on the endemic fauna. Introduced herbivores, especially goats and donkeys, overgraze indigenous vegetation and cause physical damage even to adult trees, to the point of killing them. Introduced invertebrates and small mammals are probably causing as yet hardly-studied effects on seed production

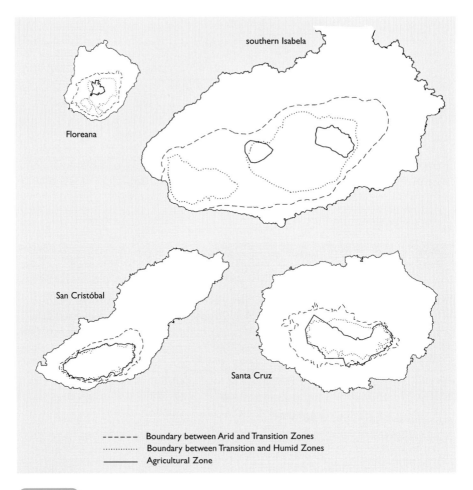

- - - - - Boundary between Arid and Transition Zones
············· Boundary between Transition and Humid Zones
———— Agricultural Zone

Figure 17.1 Map showing the location of the agricultural zones in relation to the major vegetation zones, on the four islands with civilian populations. Note that the islands are not depicted to the same scale nor in their correct relative positions

and dispersal. The effects of introduced herbivores on the vegetation are visually immediate and dramatic. All the indigenous Galápagos herbivores are reptiles (giant tortoises (*Geochelone elephantopus*) and land iguanas (*Canolophus subcristatus*)), accordingly, the vegetation is not adapted to grazing by mammalian herbivores (see Hamann 1979 and 1993). Goats on the Galápagos quickly reach extremely high densities, and graze the herbaceous vegetation very closely, ultimately reducing species diversity as the most sensitive species are eliminated. Goats also browse trees and rub the bark from their trunks, eventually killing them. Goats, and especially donkeys, will eat cactus, and can even fell large *Opuntia* Mill. (Cactaceae) trees by biting into the trunks.

The effects of alien plants are less well known than those of the introduced animals. However, in the long term, the threat from invasive plants is probably

Figure 17.2　The endemic *Scalesia atractyloides* Arn. was reduced to four inviduals in the wild until the recent discovery of a large population in a remote volcanic crater. (Photo: Alan Tye)

greater than that from introduced animals, but at present, the most severely threatened Galápagos plant species have been brought to the brink mainly by overgrazing by introduced herbivores, or habitat degradation, or a combination of the two. For example, until recently (see page 345) the most threatened Galápagos endemic, *Scalesia atractyloides* Arn., had been reduced to four plants in the wild, primarily through damage by goats (Figure 17.2). The settled areas are the major source of invasion into the Galápagos National Park. Most (around 75%) of the alien plant species were introduced deliberately, for their ornamental, agricultural, medicinal or timber value. Some 45% of introduced plant species have become naturalised (Tye, in press). The threat from introduced plants is greatest in the humid zones.

These zones are susceptible to invasion by tree and shrub species from neighbouring agricultural developments that overtop and shade out the low-growing indigenous vegetation. Disturbance is often considered to favour invasion, so that islands or vegetation zones that suffer more disturbance by man are doubly threatened in that both introduction rates and susceptibility to invasion are higher in inhabited areas.

About 30 introduced plant species have so far invaded large areas and/or appear to be causing adverse ecological effects (e.g. altering community composition or threatening individual species). An additional 70 or so already-naturalised species and a further 100 species present in cultivation in the islands, which have not yet escaped, are known to be extremely serious invasives in other parts of the world (Tye, in press).

There have been few studies of the effects of plant invasions. However, where detailed studies have been made, dramatic community changes have been revealed (Jäger, 1999). Some species have caused drastic habitat changes, forming monospecific stands, shading out or otherwise replacing native communities. *Psidium guajava* L. dominates more than 40,000 ha, including most of the highlands of San Cristóbal and huge areas of the two southern volcanoes of Isabela (Cerro Azul and Sierra Negra). Some indigenous species and communities are more susceptible to invasion than others. The shrub *Miconia robinsoniana* Cogn. and the tree *Scalesia pedunculata* Hook. f., which are both endemics and dominant components of the highland *Miconia* and *Scalesia* zones, appear to undergo natural cycles of death and recovery, related to El Niño cycles (Itow and Mueller-Dombois, 1988; Lawesson, 1988). These zones are susceptible to invasion by *Cinchona pubescens* Vahl (Rubiaceae) and *P. guajava*, which are not adversely affected by the rain and drought cycles and which successfully outcompete with indigenous species during the die-back periods. The worst effects seem to be caused by woody species, especially *P. guajava*, *Chronolaena odorata* and *C. pubescens*, and bushes that form impenetrable thickets, such as *Lantana camara* L. (Verbenaceae) and *Rubus* L. (Rosaceae) spp. The main effect of some herbaceous plants is to dominate and replace the natural shrub/herb layer; with some, such as grasses, forming a dense carpet that may also interfere with regeneration of the shrub and tree layers. Some scramblers and climbers form dense mats during the wet season over shrubs, herbs and even trees, and probably restrict the growth of indigenous species. Apart from physical disruption, other invasives may have more insidious effects. The cultivated tomato *Lycopersicon esculentum* Mill. (Solanaceae) appears to have hybridised with the endemic *L. cheesmanii* Riley so as to seriously threaten the latter's existence on some islands.

Conservation Strategy

Two institutions share the major responsibility for Galápagos conservation. The Charles Darwin Research Station (CDRS) is responsible for the research required to determine conservation priorities, and to develop the means of implementing them. The CDRS is the operative arm of the Charles Darwin Foundation for the Galápagos Islands (CDF), an international non-governmental organisation. The Galápagos National Park Service (GNPS) is the Ecuadorian Government agency responsible for managing protected areas in the archipelago. Both the National Park and CDF were established in 1959 and have jointly planned Galápagos conservation ever since. They continue to collaborate in the development of their respective management plans and annual strategy documents, as well as in detail on specific programmes and projects. The GNPS is funded by the Ecuadorian Government, largely through Galápagos tourism revenues, while CDRS raises its funds mainly through proposals submitted to international donors, as well as from interested individuals. Fund-raising for expensive conservation actions is co-ordinated between the two organisations. This conservation partnership has proved remarkably successful.

The Department of Plant and Invertebrate Sciences at CDRS consists of laboratories and offices equipped to undertake this research, with cultivation facilities, an herbarium and a complement of 30 staff, students and volunteers (as of early 2000). The work of the department also benefits from a healthy visiting scientists' programme. Ecuadorian students are financed to work as research assistants, and thereby gain experience valuable to conservation in Galápagos and Ecuador, while international volunteers often contribute expertise unavailable within the permanent staff complement. A programme of student support allows Ecuadorian students to do their thesis research on projects that contribute to CDRS's goals, and international thesis students are also welcomed to work on aspects of CDRS's research priorities. Key elements of the CDRS research strategy are:

- Determination of threat status for endemic species
- Field research on population status and distribution
- Research on the biology of threatened species
- Long-term community and population monitoring
- Monitoring of the threat factors themselves
- Research into treatment of the threats
- Restoration ecology research

Combined, these research elements permit prioritisation of problems, targets and action, and lead to the development of conservation solutions.

Key conservation management elements performed by CDRS, GNPS and other collaborating institutions include:

- short-term protection from threats
- longer-term control or eradication of the identified threat factors
- quarantine
- restoration
- development of management plans and legal frameworks

There will never be sufficient human and financial resources to protect Galápagos fully and restore it to a completely natural state, so effective prioritisation is crucial in determining the most serious threats, their worst effects, and the most urgent and important conservation action to be taken. Combined, the above elements lead to implementation of the conservation solutions devised by the research strategy.

Research and Prioritisation

1. What are the most threatened species and communities?

During the last 40 years, research in Galápagos has generated much information on the indigenous flora, although remaining gaps in our knowledge prevent a complete analysis of status and distribution. In order to prioritise research and conservation action, we need to identify these gaps, and at the same time bring up-to-date our assessment of threat status, especially for the endemic species and subspecies. The last full revision of the status and distribution of the flora of Galápagos was by Lawesson *et al.* (1987). A database of the Flora of Galápagos has since been established, and a revision of the threat status of the endemic taxa, according to the World Conservation Union (IUCN) Categories of Threat criteria (IUCN, 1994) (see Appendix 2), was begun in 1996. As part of this process, distribution maps of the endemics are prepared from the available relevant information, including the CDRS Herbarium, information from publications and unpublished reports, and consultation with knowledgeable individuals. In addition to identifying the most threatened taxa, the classification process also reveals gaps in our knowledge and aids the efficient planning of further field research.

A preliminary analysis of taxa evaluated up to early 1999, revealed that 72% of the extant species (81% of the extant taxa) are threatened (Table 17.2; Tye, 1999), a similar figure of 68% was given in the analysis carried out by Lawesson *et al.* (1987). This high figure reflects two major factors:

Table 17.2 **Numbers (and percentages) of taxa in each threat category** (from Tye, 1999)

	Total taxa evaluated	IUCN Category of Threat[1]					
		EX	CR	EN	VU	LRnt	LRlc
Species	54	1 (2%)	3 (6%)	6 (11%)	29 (54%)	5 (9%)	10 (18%)
All taxa[2]	79	1 (1%)	6 (8%)	12 (15%)	45 (57%)	5 (6%)	10 (13%)

[1] *sensu* IUCN (1994).
[2] Excludes species with infraspecific taxa in order to avoid inflating numbers artificially.

a) Many Galápagos endemics have extremely small ranges, which leads to their being automatically classified as at least Vulnerable, even if they are not currently declining. This is a valid assessment of their status, because they are indeed susceptible to chance events and local disasters, such as the introduction of a competitor, herbivore or disease.

b) Many species have been placed in the higher categories (e.g. Endangered, Critically Endangered) highlighting the proportion of species in serious decline.

i) Research on status and distribution

Field surveys are updating our knowledge of the status and distribution of the indigenous flora. This will allow a more complete and accurate status re-evaluation. The status of some taxa has already been revised: for example, *S. atractyloides* was initially classified as Extinct, until rediscovered by the project in November 1998. The survey work aims to visit known sites of endemic species to re-evaluate their status, as well as surveying all likely sites for additional populations. Vegetation types and species distributions are recorded, and estimates of population sizes made for the rarest species. This ambitious work includes surveys of areas that have apparently never before been visited by botanists, and has already led to many important discoveries, including new populations of some of the rarest and most threatened species. At the same time, valuable information is collected on the threats affecting these species.

ii) Long-term monitoring of threatened populations and communities

Long-term projects have been established to monitor plant communities subject or susceptible to immediate damage, and to support the design and implementation of management plans. Examples include the monitoring of community change on Alcedo Volcano (Isabela island), where goats and donkeys have brought about dramatic changes in the vegetation since the

1970s, reducing the forests near the summit to grassland and remnants of woodland, and on Santiago island, where goats have been present for almost 200 years. In addition, vegetation monitoring continues on Española and Santa Fé islands, from which goats were eradicated in the 1970s, in order to follow the regeneration of indigenous vegetation and to guide proposed habitat restoration. On Pinta, where goats were at last eradicated in late 1999, permanent plots provided data on vegetation regeneration following the reduction of goat populations, and will continue to reveal the process of recuperation following the definitive eradication.

On Alcedo and Santiago, the monitoring takes slightly different forms. On Santiago, goat damage has been so severe that a series of fenced exclosures has been constructed, beginning in 1973. Permanent plots inside and outside each fence permit comparison of protected with unprotected vegetation at 14 sites in a variety of vegetation communities. Plots are monitored at 18-month intervals so as to obtain alternately wet- and dry-season results with less effort than six-monthly monitoring. This project permits evaluation of the effects of the herbivores on the vegetation, determination of the worst-affected areas and species, and planning of further protection. On Alcedo, only two exclosures have so far been constructed, because the management strategy aims at eradication of introduced herbivores from the volcano in the near future. Instead, 14 permanent plots were established in 1995, mainly in the worst-affected humid zone, to assess changes in a variety of plant communities and to monitor recovery after eradication of the introduced animals. On Alcedo, the plots are monitored six-monthly, although this frequency will be reduced to 18 months after having obtained base data for an initial five year period. On both Alcedo and Santiago, vegetation monitoring consists of a complete species inventory of each plot, plus percentage cover measurements using line transects, and measurements of tree growth and mortality.

On Española, Santa Fé and Pinta, plots established about the time of the goat eradication or attempted eradication have been monitored only at intervals of several years but have provided valuable information on the ability of Galápagos vegetation to recover from serious damage caused by introduced herbivores (Hamann, 1979; 1993).

Plots have also been established to monitor community changes associated with plant invasions and natural phenomena, principally the El Niño phenomenon. Populations of selected endemics, including *Scalesia helleri* B.L. Rob. of the Arid Zone and *S. pedunculata* and *M. robinsoniana* of the Humid Zone, are monitored for mortality and recruitment. Some *Scalesia* species experience dramatic population fluctuations associated with El Niño, which may render them more vulnerable to damage by introduced herbivores or to invasion by introduced plants.

iii) Research on the biology of threatened species

In some cases, we need specific information on the biology of a threatened species in order to plan its recovery. An example of a biological research project, whose objective is effective recovery planning, is the study of the phenology, reproductive biology and threats to Galápagos' two rarest plants: *Linum cratericola* Eliasson (Linaceae) and *S. atractyloides*. The aim is to obtain more detailed information about the threats affecting the species (*L. cratericola* L. may still be in decline despite having been protected from grazing by large herbivores), and to collect information required for restoration (cultivation and population augmentation), such as seed production season and viability. The plants are monitored in the field, and results will contribute to planning conservation action.

Research has also been undertaken into the biology of exploited species, especially the timber tree *Piscidia carthagenensis*, in order to determine size structure and recruitment in the various populations in differing climatic and edaphic conditions. Results contribute to determining permitted levels of use. Further work needs to be done on plants subject to direct exploitation, including other species such as *Psidium galapageium*.

More generally, the genetics, systematics and evolutionary relationships of the endemics and other indigenous species are also being examined, which can also contribute towards effective conservation planning. In some cases, uncertainty about the genetic basis of phenotypic variability prevents an assessment of the distribution and threat status of distinct populations. Molecular genetics may solve some of these puzzles and permit better conservation planning.

iv) Research and monitoring of invasives

Surprisingly little research has been done on the effects of introduced animals on indigenous plants, other than the long-term monitoring programmes examining the effects of ungulate grazing. Introduced animals that may be having less obvious effects, such as other insects, molluscs and small mammals, are still relatively neglected by research and much more could be done in this area.

In cases where adverse effects are obvious, research focuses on developing means of eradication or control. Goats have been eradicated from many small islands in Galápagos, and research is now contributing to the design of a large-scale project to remove them from northern Isabela, which will be the biggest goat eradication attempt made anywhere in the world to date. The research element involves studies of goat movements and habitat use, and monitoring of pilot control programmes.

The cottony cushion scale insect *Icerya purchasi* Maskell was accidentally introduced to Galápagos in about 1982 and has rapidly spread to many of the islands. It has so far been recorded as attacking over 50 plant species, including several endemics, and current research is revealing that the insect contributes to increased mortality of these species. *I. purchasii* has been the subject of successful biological control in many parts of the world, so it was chosen as the target of the first biocontrol programme in Galápagos. A research project is examining the potential for use in Galápagos of well-studied control agents, including the Coccinellid beetle *Rodolia cardinalis* Mulsant. The project is investigating specificity of the control agent, to ensure that it will not attack indigenous scale insects, before undertaking controlled release experiments. At the same time, the biology of the scale on its host plants is also being studied.

v) Introduced plants

Conservation managers have either neglected plants entirely, or have tended to concentrate on the 'obvious' problems, such as massive invasion by certain woody species. However, even in such cases, we often do not know what specific effects the invaders are having on indigenous species, and therefore how to choose priority targets for control. One study of *C. pubescens*, has shown that effects on indigenous communities are minimal except under the canopy of an individual *Cinchona* tree, until *Cinchona* density rises so high as to produce continuous cover by the species (Jäger, 1999). Most indigenous species do not grow under the canopy of *Cinchona* trees, and so most will eventually be displaced when *Cinchona* density rises high enough. The treatment of invasive plants in natural systems, as opposed to weeds in agriculture, has been neglected, relative to the amount of effort that has been devoted to the control of introduced mammals. Plant problems have been seen by land managers as less serious, and scarce resources have instead been directed at the more obvious problems caused by introduced herbivores and predators. In the long term, plants are almost certainly the greater threat in Galápagos. Control techniques for introduced vertebrates are comparatively well-developed, whereas we are still in the early stages of research into methods of control appropriate for the worst plant invaders in a sensitive island ecosystem. Current research into plant control focuses on identifying or developing cheap and effective manual and chemical techniques (Gardener *et al.*, 1999) that cause minimal adverse environmental effects, that are safe for use within a national park, and that do not require extensive training or complicated equipment (Tye, in press).

Where the conservation goal includes preservation of natural evolutionary processes, introductions of additional species, including biocontrol agents, should not be considered. However, in the short term, the damage done by some of the already-introduced species may outweigh concerns about the future contribution of biological control agents to the evolution of the Galápagos ecosystem. However, if further use of biological control is to be contemplated, the control agents will need to be very carefully tested for specificity. Biological control options for introduced plants may be limited, especially where the plant to be controlled forms a species pair with a Galápagos endemic. This applies to the invasives, *Lantana camara* and *Psidium guajava* for example, both of which have congeners endemic to Galápagos (*L. peduncularis* Andersson and *P. galapageium*).

2. What factors are causing the most damage to the plant diversity?

Determining what are the most damaging threat factors can be subjective, inasmuch as threats are extremely diverse, including direct human activities, effects of introduced animals, and effects of introduced plants. The risk assessment system being developed for Galápagos takes into account three main groups of characteristics, focusing mainly on the threat from alien taxa:

   i) The community type potentially affected;

   ii) The effect on the system (how significantly the alien taxa alters species composition or habitat structure, suppresses regeneration and persists over time);

   iii) The biological success (as assessed by maturation rate, seed production and viability, means of dispersal, establishment and growth rate) and importance of vegetative reproduction.

Other factors taken into account include:

- competitive ability
- fire risk
- resistance to control
- cost of control
- growth form
- year of naturalisation
- bioclimatic zones or other countries in which the species is a pest
- phylogenetic considerations (risk of hybridisation with indigenous species, difficulty of biological control) (Tye, in press)

A risk assessment system can also be used at the quarantine level to decide which species and products may be imported.

3. Species recovery and restoration ecology

New projects are examining the options for recovery of populations of some of the most endangered Galápagos plants. *Ex situ* cultivation of some of these is being initiated, as a preliminary stage prior to reintroduction as part of a broader recovery programme (see Appendix 4). Initial efforts will focus on *S. atractyloides* and *L. cratericola*, with efforts to establish both species in cultivation at CDRS. Plants of *S. atractyloides* held at Copenhagen Botanic Garden will eventually be used to supplement the genetic stock of this species in the wild.

Some plants may still be in decline even after the removal of introduced herbivores, as their populations seem to have dropped below the minimum required to achieve adequate recruitment to replace natural mortality (minimum viable population). Recovery efforts are then required to increase population size above that minimum. On Española island, the cactus *Opuntia megasperma* var. *orientalis* J.T. Howell may be in this position, as regeneration from seed seems to be currently prevented by a combination of seed predation by the Large Cactus Finch *Geospiza conirostris*, and a lack of giant tortoises to eat the fruit and cause an increased germination rate. The *Opuntia* and *Lecocarpus lecocarpoides* (B.L. Rob. & Greenm.) Cronquist & Stuessy (Asteraceae) were both severely depleted during the period when goats were present on the island. Both are currently in cultivation, at CDRS and Copenhagen respectively, and the aim for both species is to re-establish populations that are now extirpated, at sites within the recorded range distant from the remaining wild plants where natural re-establishment is unlikely to happen within the foreseeable future. The conservation aim in Galápagos is always to maintain viable populations of Galápagos plants in the wild. Cultivation elsewhere, or the use of tissue and seed banks, is regarded as only a temporary measure until wild populations can be adequately protected or re-established.

A project has recently examined control of the invasive elephant grass, *Pennisetum purpureum* Schumach. (Gramineae), combined with restoration of indigenous *S. pedunculata* forest. A variety of control techniques, with post-control treatments including seeding with wild species, have provided results on various methods for re-establishing indigenous vegetation in controlled sites.

Conservation Action

1. Short-term protection from the threats

The major short-term measure routinely used to protect critical sites and threatened species from introduced herbivores is fencing. This is considered short-term because it is envisaged as a temporary measure until permanent control of herbivores can be established. Many individual endemic species have been brought to the brink of extinction by the action of goats and other introduced herbivores. *S. atractyloides* and *L. cratericola* were both thought to have become extinct as a result of grazing by goats (and in the case of *L. cratericola*, invasion by *L. camara*) on Santiago and Floreana respectively. However, small populations of both species were recently rediscovered, and immediately protected by fences. *L. cratericola* is a small shrub, and its remaining site is a crater rim that was easily protected by a comparatively small fence. In contrast, the two populations of *S. atractyloides*, each representing a different variety of the species, occur in a crater and on a small hill and, in order to permit population growth, the crater and hill have each been surrounded by fencing.

Figure 17.3   Survey work being carried out on Santiago, highlighting the contrast between ungrazed and protected vegetation inside a herbivore closure plot and that of overgrazed vegetation outside the exclosure (Photo: Patricia Jaramillo).

As mentioned earlier, Santiago is the island that has been most seriously affected by introduced herbivores. The highlands of the island were formerly largely forested, but have been reduced to grassland or, in the dry season, bare earth (Figure 17.3). Most of the indigenous vegetation communities of the highlands have all but completely disappeared, and a series of fences has been constructed in an attempt to preserve small samples of the indigenous communities and their component species, until the menace of the goats can be tackled. At present, about 50% of the indigenous species of Santiago, and a higher proportion of the highland species, are represented within exclosures, and the fenced network will be expanded following identification of further key sites and the acquisition of further funding for this expensive form of conservation.

Similar action has been taken on Alcedo Volcano, where only two fences have been constructed, to protect the two last patches of the endemic tree-fern *Cyathea weatherbyana* C.V. Morton (Cyatheaceae). The remaining trees of other species are too sparsely distributed to protect by fencing, so individual trees of two species have been surrounded by tubes constructed out of chicken-wire. This very temporary measure is seen as adequate in the short-term, because a detailed plan for the complete eradication of introduced large mammals exists for the area.

Fencing in Galápagos is not a simple undertaking. For each of the two *S. atractyloides* sites, about 0.5 km of fence was required, for which materials had to be purchased on the mainland, transported to Galápagos and then to the uninhabited island of Santiago, unloaded in heavy surf and manually carried across difficult terrain to the remote construction site. These two fences were built along with other exclosures to protect a variety of community types. Altogether, 2.5 km of fencing were built during the operation, requiring a month's work by a team of over 40 people, at a total cost of about £30,000, of which the majority went towards the labour, transport and construction.

2. Long-term treatment or elimination of the threats

To date, total eradication of exotics has only been attempted in Galápagos for mammals, a few livestock diseases, some very recently introduced plants, and new colonies of insects on previously uninfected islands. Current research and management strategies are aimed at developing and implementing control and eradication techniques for a greater range of organisms and on larger scales. Initial efforts have been concentrated on the eradication of ungulates from northern Isabela, rats and fire ants *Wasmania auropunctata* from small islands, *C. pubescens* from Santa Cruz island, and invasive plants with small founder populations.

There have been many attempts at control or eradication of introduced mammals on various islands in Galápagos. Goats have been eradicated from several of the smaller islands, and more ambitious goat eradication attempts are now under consideration. Pigs have been almost eradicated from Santiago by hunting, and tracking with trained dogs. This is a tremendous achievement, to remove an established feral population of pigs from the largest island in the world (925 km^2). Small mammal eradication has been more problematic, while invertebrate eradication has only been attempted and achieved with fire ants in small founder populations using burning of the vegetation and, more recently, poisoned bait. In the medium term, a permanent ungulate control effort is probably the only way to protect threatened plants on the larger, inhabited islands, although little by little, eradication is becoming a more feasible option, both practically and politically.

i) Eradication and control of introduced plants

Invasive plant control should employ cheap and effective techniques that cause minimal adverse environmental effects and do not require extensive training or complicated equipment. In some cases the technique of choice is manual, such as uprooting saplings of *C. pubescens*. In other cases it is chemical, such as treatment of larger *Cinchona* trees by precisely targeted herbicide application by injection or basal bark application. In many cases, a combination of manual or chemical methods is required, such as felling a tree then painting the stump with herbicide, or cutting *Rubus* bushes or grasses and then spraying young regrowth with a rapidly degraded and minimally toxic herbicide (e.g. Glyphosate, Picloram and Triclopyr).

Plant eradication has so far only been attempted with recent introductions of species that were known to be invasive in other parts of the world. In 1996, seeds of tropical kudzu *Pueraria phaseoloides* (Roxb.) Benth. (Leguminosae) were brought by a farmer from mainland Ecuador, where it is commonly grown for forage and for ground cover. They were sown in one field, where they were noticed by CDRS staff. After enquiries to experts overseas revealed the enormous potential for damage by this known invasive, the decision was taken to eradicate it, with the agreement of the farmer. The plot was treated with glyphosate herbicide, which killed most of the plants. Survivors or late germinators were spot-treated, until the species was finally considered eradicated three years after the introduction, when no plants had been found for more than one year. Ironically, the field has now been sown with a newly-introduced, highly invasive grass, *Urochloa mutica* (Forssk.) T.Q. Nguyen (Gramineae). The relatively expensive effort required for this eradication of a small population, and the subsequent sowing with what may prove to be an equally problematic plant, reveal the need for prevention of introduction in the first place. This requires not only quarantine control but also education about the dangers associated with introducing new species to the islands (see Appendix 5).

Conservationists are now beginning to contemplate more ambitious plant eradication attempts, with plans being developed that may lead to an attempt to eradicate *C. pubescens*, using a combination of manual and chemical treatments. No plant eradication has been attempted on this scale before, in Galápagos or elsewhere, but several characteristics of *Cinchona* and Galápagos render it worth consideration. The plant is not regarded as useful by farmers; on the contrary, it is recognised as both a weed of agricultural land and the National Park. It is widely recognised as one of the most serious invasive species. It is conspicuous and therefore easy to find, and although it produces seed from about five years of age, it is recognisable before reaching that age. It has been introduced to only one island (Santa Cruz), and control attempts in the past, plus current trials, have produced a range of appropriate treatments. There are still many unknown factors, such as seed longevity in the soil, but if a long-term commitment and sufficient funding can be obtained, then eradication might be possible.

Most of the serious invaders in Galápagos were introduced deliberately as useful plants. As in the case of quinine, most of the species that are problematic in the National Park are also causing problems for farmers. To a large extent, therefore, there is little conflict regarding priorities for control or eradication of invasives. However, this is not true in every case, and potential conflict of values between farmers and conservationists needs to be taken into account. For instance, *P. purpureum* is a valued pasture grass and its simple removal from agricultural areas would be politically impossible. In such cases, research must then examine more comprehensive strategies such as replacement of the invasive with a more benign species or (especially for fruit crops) sterile cultivars.

Costs are far greater for management than for research, but budgets for both continue to be inadequate. As an example of relative research and management costs, the attempt to eradicate *C. pubescens* has been estimated at US$2 million over 15 years. Less than 10% of this would be required for research, and most would support a permanent team of 15–20 people working on control operations. After an initial 2–3 year intensive campaign to reduce the population to very low levels, the team would also be able to contribute to the control of other selected target species in the same areas during the remainder of the 15-year period. Similarly, the eradication of ungulates from northern Isabela island will be an enormous operation on the largest island of the archipelago, requiring teams of hunters, dogs, and helicopter and boat support, and lasting some three years. Costing some US$8 million, it will require a three-phase programme of ground-based hunting, then six months of intensive hunting from helicopters, followed by the use of radio-collared animals to help track down the remaining groups.

Despite such a high initial investment, in the long term, the most efficient means of dealing with a threat is to eradicate it entirely. If this is not done, permanent control programmes will be necessary, which, with the possible exception of biological control, will ultimately cost more. This cost-benefit balance between control and eradication is more crucial for the case of introduced species than for other types of threat (direct human activities), which are perhaps more easily controlled.

ii) Quarantine

An important part of the Galápagos conservation strategy is reducing the currently high rate of human-mediated introductions. Ecuador is in the process of establishing an international-standard quarantine system for Galápagos, which will tackle this problem. Almost all traffic to Galápagos comes from mainland Ecuador, where the first barrier must be established. The quarantine system will include inspection facilities at the mainland ports and airports from which planes and boats depart for Galápagos. There will be further inspection on arrival, as well as inspection of inter-island traffic. There will be comprehensive import controls by 'positive listing' of permitted species and products, assisted by importer licensing. The system should reduce both deliberate and accidental introductions of new species. Concurrently with the establishment of control, an education programme is raising awareness of the problems caused by introduced pests, so as to increase support for and voluntary compliance with the regulations. A risk assessment procedure will be employed for all proposed imports of new species and products, in order to reduce the possibility of new invasive species becoming established. The cost of establishing the Galápagos quarantine system has been estimated at US$6 million, to pay for infrastructure (inspection facilities and equipment), training and staff establishment.

iii) Restoration

Although little positive vegetation restoration has yet been done in Galápagos, this is likely to become much more important in the near future. As restoration research yields results, and as critical sites are protected or threats brought under control, attempts can be made to speed regeneration of the original plant populations and communities.

iv) Legal frameworks and management plans

Both the CDRS and GNPS contribute to broader planning at the local and regional level. Planning issues in Galápagos are co-ordinated by the Instituto Nacional Galápagos (INGALA), while other institutions involved include local and regional councils, chambers of commerce and tourism, etc. The Government of Ecuador recently passed the Special Law for Galápagos Province, which deals with tourism, agriculture, quarantine, immigration, invasive species and other issues that impinge on development and conservation in the archipelago. It requires Galápagos institutions and residents to take an active role in biodiversity conservation. Within this overall framework and in co-ordination with other local institutions and the Ministry of the Environment, GNPS and CDRS are contributing to the development of the regulations for the application of the law, especially those that deal with agriculture, quarantine, and tourism management. More inclusive participatory mechanisms have recently begun to involve the entire community in Galápagos in important planning issues. Examples of participatory planning that have proved extremely successful have been the production of the management plan for the Galápagos Marine Reserve, development of quarantine regulations, and the design of a comprehensive invasive species strategy.

v) Managing development and conservation in Galápagos

The problems faced by the Galápagos ecosystem stem ultimately from the presence of humans on the islands. In the long-term, the link between human presence and problems caused by introduced species has to be broken, and other damaging human activities curtailed. Galápagos has already come a long way towards controlling over-exploitation and habitat clearance, but this effort needs to be increased, to counter the rising pressure caused by a growing human population. The new quarantine system should help to slow down the rate of arrival of new species, although it can never entirely stop it. Environmental education at all levels, including school curriculum development, radio and TV programmes, community liaison activities and increased participatory planning, will help to nurture the conservation ethic among the resident population of Galápagos, as well as within Ecuador as a whole. These efforts should contribute to reducing the threats in the long-term.

The problems caused by the human presence in Galápagos have become rapidly more severe in recent years, along with the exponential growth of the resident population. The main source of income for the islands, and the reason for the rapid population growth, has been tourism. Although Galápagos is a model of well-controlled ecotourism, and the direct effects of tourism are minimal, the indirect effect of drawing people in to support the tourism industry is one of the most serious threats to the archipelago's biodiversity. However, tourism is both problem and solution. It is interest from outside the islands that maintains Galápagos in the spotlight of international concern and that strengthens the conservation argument, while tourism revenues pay for conservation. If the link between the presence of people and the problems of inappropriate development and invasive species can be broken, or at least weakened, then there is hope for the native plants and animals of Galápagos.

Lessons Learnt

Conservation in Galápagos has had some remarkable successes. Herbivores have been eradicated from many islands, uncontrolled land clearance has been halted, tourism is well-controlled, few species have become extinct. However, these successes represent small battles in a war that is currently being lost. Numbers of introduced species are still increasing, as is their distribution both within and between islands. Quarantine may slow introduction rates, but the hundreds of species already introduced represent a formidable problem.

More specifically, successes have mainly concerned conservation of threatened animal populations, including giant tortoises, land iguanas and nesting seabirds. Galápagos plants so far have survived by good luck, rather than by conservation action, with the obvious exception of the creation of the National Park itself, which has slowed the loss particularly of the humid vegetation zones. Key issues now to be tackled include preservation of the little humid vegetation that remains, including protecting it from introduced plants and animals, safeguarding remaining areas from encroachment, and restoring native vegetation where possible. Despite the fact that there have been few extinctions, large numbers of Galápagos endemics are now threatened. *In situ* protection, combined with emergency *ex situ* rescue for the most threatened, is the key activity. *In situ* protection requires determined action against introduced animals and plants.

In the long-term, perhaps the most cost-effective measure is to change attitudes. With fuller backing at the local and national and international levels, not only should funding increase, but also the size of the problems will be reduced, by Galápagos residents and visitors adopting lifestyles and behaviour patterns that favour rather than impede conservation.

Future Steps

Threatened plant management in an island system such as Galápagos requires much more than a narrow focus on the most threatened species. Effective management has ramifications throughout the ecosystem, including the human element. We need research on the native plant species and communities, to determine what is threatened and by what factors. We need research on the threats themselves, namely to determine how they function, what they threaten, and how to manage or eliminate them. We need 'emergency' alien species management to prevent the extinction of the most threatened species in the short-term. Long-term research and management is needed to deal with the root causes of the problems. Eradication and control will need to become much more comprehensive if the invasive species already introduced are to be prevented from causing irreparable damage to the Galápagos ecosystem. Ultimately, we need to manage human use of Galápagos better, if we are to tackle the ultimate source of the threats, which is our presence in the islands.

UPDATE – On a recent botanical survey carried out by the Charles Darwin Research Station and led by the author of this chapter, a previously unknown population of *Scalesia atractyloides* var. *darwinii* was discovered in a remote volcanic crater on the island of Santiago. 445 adult trees and 2,350 young plants were found (Randerson, 2000).

Acknowledgements

The work discussed in this chapter is being carried out by colleagues at CDRS and elsewhere, and the development of many of the ideas presented here has benefited from discussions with many of them, especially: Henning Adsersen, Robert Bensted-Smith, Karl Campbell, Charlotte Causton, Mark Gardener, Ole Hamann, Heinke Jäger, George Powell, Howard Snell, Pat Whelan and Sarah Wilkinson. Tom Allnutt, Rikke Frandsen and Sabine Kleikamp contributed to the map preparation and analysis. Clare Hankamer and Mike Maunder (RBG Kew) provided helpful comments on drafts. The background research was partly financed by the Darwin Initiative for the Survival of Species of the British Government, the European Union, Frankfurt Zoological Society, Fundación Natura, Keidanren Nature Conservation Fund and Monsanto Corporation. This article is contribution number 570 of the Charles Darwin Research Station.

References

Gardener, M.R., Tye, A. and Wilkinson, S.R. (1999). Control of introduced plants in the Galápagos islands, pp. 396–400. In: A.C. Bishop, M. Boersma and C.D. Barnes, (eds). *12th Australian Weeds Conference Papers and Proceedings*. Tasmanian Weed Society, Devonport, Tasmania.

Hamann, O. (1979). Regeneration of vegetation on Santa Fé and Pinta islands, Galápagos, after the eradication of goats. *Biological Conservation* **15**: 215–236.

Hamann, O. (1984). Changes and threats to the vegetation, pp. 115–131. In: R. Perry (ed.). *Key Environments. Galápagos*. Pergamon Press, Oxford, UK.

Hamann, O. (1993). On vegetation recovery, goats and giant tortoises on Pinta Island, Galápagos, Ecuador. *Biodiversity and Conservation* **2**: 138-151.

Hickman, J. (1985). *The Enchanted Islands: the Galápagos Discovered*. Nelson, Oswestry, UK.

Itow, S. and Mueller-Dombois, D. (1988). Population structure, stand level dieback and recovery of *Scalesia pedunculata* forest in Galápagos islands. *Ecological Research* **3**: 333–339.

Jäger, H. (1999). *Impact of the Introduced Tree* Cinchona pubescens Vahl. *on the Native Flora of the Highlands of Santa Cruz Island (Galápagos Islands)*. Diplomarbeit. thesis, University of Oldenburg, Germany.

Lawesson, J.E. (1988). Stand level dieback and regeneration of forests in the Galápagos Islands. *Vegetatio* **77**: 87–93.

Lawesson, J.E. (1990). Threatened plant species and priority conservation sites in the Galápagos islands, pp. 153–167. In: J.E. Lawesson, O. Hamann, G. Rogers, G. Reck, and H. Ochoa, (eds). *Botanical Research and Management in Galápagos. Monographs in Systematic Botany from the Missouri Botanical Garden* **32**.

Lawesson, J.E., Adersen, H. and Bentley, P. (1987). *An Updated and Annotated Check List of the Vascular Plants of the Galápagos Islands*. Rep. 16, Botanical Institute Univiversity of Aarhus, Risskov, Denmark.

Loope, L.L., Hamann, O. and Stone, C.P. (1988). Comparative conservation biology of oceanic archipelagoes. Hawaii and the Galápagos. *Bioscience* **38**: 272–282.

Mauchamp, A. (1997). Threats from alien plant species in the Galápagos Islands. *Conservation Biology* **11**: 260–263.

Mauchamp, A., Aldaz, I., Valdebenito, H. and Ortiz, E. (1998). Threatened species, a re-evaluation of the status of eight endemic plants of the Galápagos. *Biodiversity and Conservation* **7**: 97–107.

Porter, D. (1822). *Journal of a Cruise made to the Pacific Ocean. Volume 1*. Wiley and Halstead, New York, USA.

Randerson, J. (2000) Crater Trees. *BBC Wildlife* **8**: 38.

Simkin, T. (1984). Geology of Galápagos Islands, pp. 15–41. In: R. Perry (ed.). *Key Environments, Galápagos*. Pergamon Press, Oxford, UK.

Slevin, J.R. (1959). The Galápagos Islands. A history of their exploration. *Occasional Papers of the California Academy of Sciences* **25**: 1–150.

Tye, A. (1999). Revision of the threat status of the endemic flora of Galápagos: a preliminary analysis. *Galápagos Report* 3. WWF – Fundación Natura, Quito, Ecuador.

Tye, A. (2001). Invasive plant problems and requirements for weed risk assessment in the Galápagos islands, pp. 153–175. In: R.H. Groves, F.D. Panetta and J.G. Virtue (eds). *Weed Risk Assessment.* CSIRO Publishing, Colingwood, Australia.

Wiggins, I.L. and Porter, D.M. (1971). *Flora of the Galápagos Islands.* Stanford University Press, Stanford, USA

Chapter 18

An Integrated Approach to the Conservation of Cacti in Mexico

MEXICO

Héctor Hernández
Departamento de Botánica, Instituto de Biología,
National University of México
México

Carlos Gómez-Hinostrosa
Departamento de Botánica, Instituto de Biología,
National University of México
México

Introduction

Cacti are popular and recognisable components of the Mexican landscape. An increasing number of species of cacti are included in official listings of threatened plants (CITES, 1990; SEDESOL, 1994; Walter and Gillett, 1998). In order to effectively maintain wild populations of cacti, an accurate understanding of distribution patterns and population status is required. However, there has been no general agreement regarding the conservation status of individual species, due to the extremely scarce and fragmented knowledge on their biogeography, demography, and population genetics. The results of a long-term project with the aim of generating an effective information base needed for the conservation of cacti in the Chihuahuan Desert is outlined in this paper.

1. Diversity of cacti

The Cactaceae is a family of succulent plants comprising about 100 genera and 1,500 species (Barthlott and Hunt, 1993). With the exception of *Rhipsalis baccifera* (J.S. Muell.) Stearn, which is recorded from several tropical countries in the Old World, all species of the family are found in North and South America. In addition, economically important species, such as the prickly pears (*Opuntia* Mill. spp.), have been naturalised and proved invasive in different regions of the world (see Appendix 5), notably the Mediterranean Basin and Australia. The morphological diversity within the family is remarkable, including small globular or cylindrical; giant arborescent, candelabriform or columnar; and barrel-forms. The Cactaceae have adapted to live in a variety of areas with contrasting ecological and climatic conditions, ranging from deserts or semi-deserts to wet tropical forests. Nevertheless, most species occur in semi-deserts or in regions with a highly seasonal precipitation regime.

It has been hypothesised that the family originated in South America, as several taxa containing primitive morphological characters (e.g. *Pereskia* Mill. and *Maihuenia* Philippi ex K. Schum.) centre their distribution range in that region (Leuenberger, 1996, 1997). Currently, the family extends its natural distribution from the Alberta Province, Canada south to Patagonia, South America (see Table 18.1). However, there are great disparities in the distribution of species richness throughout this range. Mexico is considered the main centre of diversity for Cactaceae, with about 40% of the known species. In addition, northeastern Brazil (Andrade-Lima, 1981) and northern Argentina, together with Bolivia (Navarro, 1996), Peru and Chile have been determined as important secondary centres of diversity (Hernández and Godínez, 1994).

Table 18.1 Comparison of species richness of Cactaceae among different regions, states and countries in the American continent

Region/State/Country	Number of species	Source
Huizache	75	Hernández *et al.* (2000)
Mier y Noriega	56	Gómez-Hinostrosa and Hernández (2000)
Cuatro Ciénegas	48	Pinkava (1984)
La Paila	44	Villarreal (1994)
Mapimí	30	Cornet (1985); Ruiz de Esparza 1988)
Xichú	56	Bárcenas (1999)
Tehuacán-Cuicatlán Valley	76	Arias *et al.* (1997)
Arizona	78	Lehr (1978)
Texas	62	Correll and Johnston (1970)
Baja California	65	Rebman *et al.* (1999)
Baja California Sur	64	Rebman *et al.* (1999)
Guanajuato	92	Bárcenas (1999)
Central America	74	Bravo and Arias (1999)
Cuba	50	Hunt (1999)
Colombia	35	Hunt (1999)
Ecuador	44	Hunt (1999)
Venezuela	39	Hunt (1999)
Guyana	10	Hunt (1999)
Surinam	9	Hunt (1999)
Uruguay	52	Hunt (1999)

2. Patterns of distribution of Cactaceae in Mexico

There are 48 genera and about 570 species of cacti in Mexico. From the total number of genera occurring in this country, 15 (31%) are strictly endemic, and 20 additional ones are quasi-endemic, as their range extends beyond the Mexico-US borderline into southern Texas (Correll and Johnston, 1970; Hernández and Godínez, 1994). The level of species endemism at the country level (78%) is higher than that reported for the whole Mexican flora, 59% by Rzedowski (1991).

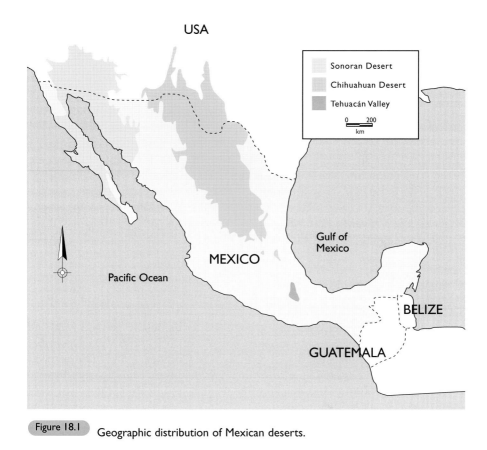

Figure 18.1   Geographic distribution of Mexican deserts.

The most important centres of species concentration of cacti are localised in the three major Mexican semi-desert areas: Chihuahuan Desert, Sonoran Desert, and in the Tehuacán-Cuicatlán Valley (Figure 18.1). However, a significant number of additional species, including many endemics, are found in the Balsas Basin and in the southern, driest portion of the Tehuantepec Isthmus. In these two dry, tropical regions, there is a considerably higher precipitation compared with the Mexican semi-deserts.

We have suggested (Gómez-Hinostrosa and Hernández, 2000; Hernández and Bárcenas, 1995, 1996; Hernández and Godínez, 1994; and Hernández *et al.*, 2000) that the Chihuahuan Desert Region has the highest number of species of cacti in the country, including the richest assemblages of rare and threatened species. Whilst this desert is recognised as having a relatively rich biodiversity, including a high proportion of endemic animal and plant species (Johnston, 1977; Morafka, 1977; Pinkava, 1984), it is still the least studied of the Mexican deserts. Most of the Cactaceae from this region are small, globular or cylindrical forms. The only exceptions to these patterns are two barrel species (*Echinocactus platyacanthus* Link & Otto and *Ferocactus pilosus* (Galeotti) Werderm.) and the arborescent *Myrtillocactus geometrizans* (Martius) Console and *Stenocereus griseus* (Haw.) Buxb. ex Bravo. However, the distribution of these four species is confined to the southern portions of this desert (Gómez-Hinostrosa and Hernández, 2000). It is likely that the low temperatures prevailing in the northern fragment of the Chihuahuan Desert during winter inhibit the existence of these and other species.

The predominance of small, usually cryptic species in the Chihuahuan Desert contrasts with the visual prevalence of large arborescent columnar or candelabriform species in the Sonoran Desert. This condition, however, reaches its maximum expression in the Tehuacán-Cuicatlán Valley (Arias *et al.*, 1997), where the landscape is dominated by large, arborescent species of cacti belonging to a diversity of genera (e.g. *Cephalocereus* Pfeiff., *Neobuxbaumia* Backeb., *Mitrocereus* (Backeb.) Backeb., *Pachycereus* (A. Berger) Britton & Rose, *Escontria* Rose and *Polaskia* Backeb.).

3. Conservation status

The Cactaceae are a uniquely diverse and characteristic element of the Mexican flora. However, they are recognised as being the most threatened family in the flora; it has been estimated that about 35% of the Mexican species are threatened with extinction (Hernández and Godínez, 1994). The primary threat to Mexican cacti is illegal collecting. For many years species of cacti have been the focus of attention of private collectors and dealers, often involving illegal export. There are a number of documented cases where natural populations have been totally or almost totally extirpated by irresponsible collectors. A recent example is the illegal harvesting and decline in the only known population of *Ariocarpus bravoanus* H.M. Hern. & E.F. Anderson. When originally described, the existence of only 230 individuals at the type locality was reported (Hernández and Anderson, 1992). A few years later, notices were received that this species was already present in private collections having been obtained through the European black market. More recently, members of the Sociedad Potosina de Cactología reported only 54 individuals surviving in the field, putting the species in the Critically Endangered category (*sensu* IUCN, 1994 see Appendix 2) (Manuel Sotomayor, pers. com.). Despite efforts by the Mexican environmental authorities in implementing a number of practical measures to prevent the illegal extraction of flora and fauna, illegal harvesting continues. For instance, between 1994 and 1996, more that 8,000 illegally collected Mexican cacti were confiscated at airports in Mexico, Germany, France, and the United States. Cacti populations are also being lost through the conversion of the original habitat. Large tracts of arid and semi-arid lands have been transformed by agriculture, goat raising, road construction, and mining.

Cacti usually have slow growth rates, long life cycles, and low recruitment rates. This makes them particularly vulnerable to collecting and habitat degradation. In addition, many species of cacti are particularly vulnerable because of restricted endemic localities or because they have extremely patchy, discontinuous distributions.

Unfortunately, within the Chihuahuan Desert the most significant protected areas (Cuatro Ciénegas and Mapimí Biosphere Reserve) do not correspond with the areas of high cactus species richness and endemism (Cornet, 1985; Gómez-Hinostrosa and Hernández, 2000; Hernández and Bárcenas, 1995, 1996; Hernández *et al.*, 2000; Ruiz de Esparza, 1988). On the other hand, as far as *ex situ* conservation is concerned, there have been important efforts in several Mexican institutional botanical gardens to include species of cacti in their collections. For instance, the Botanical Garden of the Institute of Biology, at the National University of México (UNAM), has maintained for several decades a cacti collection containing several hundred Mexican species, and an active program of propagation of endemic species is underway. These efforts, however, are insufficient to conserve the enormous genetical diversity of Mexican Cactaceae.

Conservation of Cacti in the Chihuahuan Desert

The Chihuahuan Desert Region as the main centre of diversity of cacti

The Chihuahuan Desert Region contains the richest assemblage of species of cacti in Mexico. Within this region there are a number of areas that are well known for having a high diversity of cacti, namely: Cuatro Ciénegas (Pinkava, 1984), Sierra de la Paila (Villarreal, 1994), and Big Bend (Everitt, 1993). However, the richest area is located in the southeastern portion of the Chihuahuan Desert, around the boundaries of the Nuevo León, Tamaulipas, and San Luis Potosí states (Hernández and Bárcenas, 1995).

Methods for the detection and analysis of areas of species concentration

i) Data sources
Conservation biologists frequently have to make decisions derived from poor quality data. For instance, most of the specialist cacti and succulent journals include a great deal of anecdotal information that cannot be verified. Accordingly, its practical value to science is limited. Also, valuable geographical information on cacti is often only transmitted orally within amateur circles and never reaches scientists or conservationists. Also, cacti are

very poorly represented in herbaria. This is due to the practical difficulties of collecting these plants and preparing them as herbarium specimens.

A prerequisite for biological data to be useful for conservation is that it is both taxonomically reliable and geographically accurate. Although there are numerous taxonomic problems among cacti still to be solved, there is an accepted and comprehensive list of cacti names (Hunt, 1999) and a monograph of Mexican Cactaceae (Bravo, 1978; Bravo and Sánchez-Mejorada, 1991 a and b). These base-line checklists have been verified and supplemented through considerable fieldwork and associated laboratory studies in order to improve the cacti collection at the National Herbarium of Mexico (MEXU). From these activities we have developed a Database of Cactus Collections from North and Central America. It stores the basic taxonomic, geographic, and descriptive information components usually recorded in herbarium labels, and currently includes over 20,000 records of specimens from 35 herbaria. This database has been the primary source of information for several studies of Mexican cacti (Bárcenas, 1999; Gómez-Hinostrosa and Hernández, 2000; Hernández and Bárcenas, 1995; etc.).

ii) Grid mapping

Mapping the distribution of species continues to be an important element of systematics, ecology, and conservation biology research. The equal-area grid approach is a useful tool for representing species presence/absence data and for analysing distribution patterns at both regional and global scales (Miller, 1994). This approach has been successfully used to study a number of plant and animal groups (e.g. Arita *et al.*, 1997; McAllister *et al.*, 1994; Myklestad and Birks, 1993). We used this method for analysing the distribution patterns of the species of cacti considered threatened under the Convention in International Trade in Endangered Species (CITES, 1990) and Hernández and Godínez (1994) in the Chihuahuan Desert Region (Hernández and Bárcenas, 1995, 1996). The region was divided into grids of 30 minutes latitude by 30 minutes longitude. Presence/absence data of 93 threatened species of cacti was used to determine species richness in each grid square. The study revealed that the threatened cacti are clearly concentrated toward the southeastern fragment of the Chihuahuan Desert (Hernández and Bárcenas, 1995). The majority of the species-rich grid squares are located along the southeastern fringes of this desert, in areas of moderate elevation (1100–1600 metres (m)).

iii) Estimation of species richness at a micro-regional scale

We determined the general pattern of distribution of threatened cacti within the Chihuahuan Desert (Hernández and Bárcenas, 1995). However, at the time of that study, basic inventories of the general diversity of the family were lacking, especially in the areas expected to be rich in species. More recently, we have concentrated our attention on two contiguous grid squares, the Huizache and Mier y Noriega grid squares (Figure 18.2), located in the south-eastern tip of the Chihuahuan Desert, in northern San Luis Potosí and southern Nuevo León and Tamaulipas (Gómez-Hinostrosa and Hernández, 2000; Hernández *et al.*, 2000). The Huizache was reported to be the grid square with the highest number of endangered species in the Chihuahuan Desert (14 species), whereas the Mier y Noriega grid square had ten species (Hernández and Bárcenas, 1995).

In order to assess the cacti diversity in each of these grid squares, a systematic sampling strategy was adopted (Gómez-Hinostrosa, 1998; Gómez-Hinostrosa and Hernández, 2000). Each grid square was divided into 25 sub-squares each measuring six minutes latitude by six minutes longitude. In each sub-square, an average of three sites were intensively sampled for species of cacti (Figure 18.2). All sites were geographically referenced by using a Global Positioning System (GPS), and samplings were taken in well-preserved, relatively undisturbed sites. Specimens of all species of cacti found in the samplings were incorporated into the National Herbarium of Mexico (MEXU) and the information was added to the Cactus Database. We judged that this method was effective for a plant group, such as the Cactaceae, which has a cryptic nature and a commonly discontinuous distributions. The data gathered using this method was used in the calculation of the indexes described below, as well as in the complementarity analysis.

The results revealed the existence of 75 species of cacti in the Huizache and 56 in the Mier y Noriega grid squares, excluding introduced species. These support our hypothesis that the Huizache grid square constitutes the most important concentration of species of cacti in Mexico and in the whole continent. This fact marks this region as an important focal point for the conservation of this plant family. Hernández *et al.* (2000) compared the diversity of Cactaceae in the Huizache with several other well-known regions in and outside Mexico (Table 18.1). The Huizache has a significantly higher number of species of cacti than several other areas of comparable size within the Chihuahuan Desert: Mier y Noriega, Cuatro Ciénegas, La Paila, Xichú, and Mapimí. Also, the Huizache, being a small area (2,855 km^2), concentrates a similar number of species of cacti as do the states of Arizona, Texas, Baja California, Baja California Sur, and Guanajuato (Lehr, 1978; Rebman *et al.*, 1999). Moreover, the Huizache has a similar diversity of species of cacti as all of the Central American countries combined, and a higher number of species than Cuba and most South American countries. The only area that competes with the Huizache is the Tehuacán-Cuicatlán Valley, located in the Mexican states of Puebla and Oaxaca. This valley, however, is more than three times larger (10,000 km^2) than that of the Huizache.

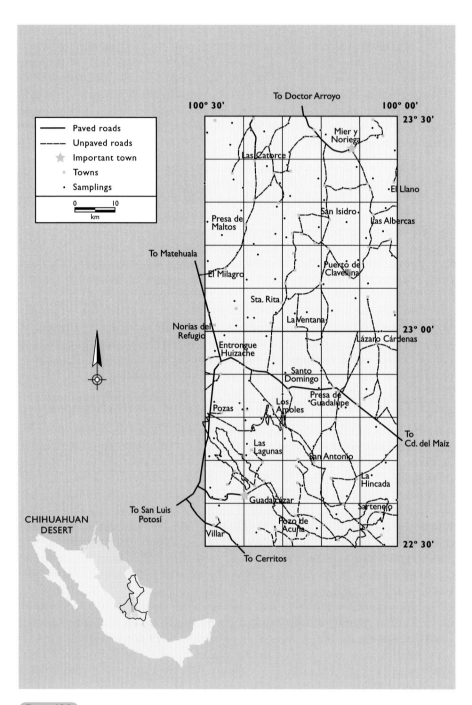

Figure 18.2   Localisation of the northern Mier y Noriega (23°30'–23°00') and southern Huizache (23°00'–22°30') grid squares. Solid dots indicate sampling sites.

iv) Rare versus common species

Rare species are commonly characterised as having small ranges, low local abundance, and/or a high degree of habitat specificity (Gaston, 1994; Rabinowitz, 1981). It has been usually assumed that rare species have an increased likelihood of extinction (Gaston, 1994). Accordingly, the concept of rarity has been inevitably linked to conservation considerations.

We have calculated the range size of most species of cacti occurring at the Huizache and Mier y Noriega grid squares (Gómez-Hinostrosa and Hernández, 2000; Hernández *et al.*, unpublished). A matrix was built recording presence/absence data in 30-minute latitude by 30-minute longitude grid squares, throughout the range of the species. A 30-minute cell in the Huizache latitude has an area of about 2,855 km^2, providing results with an acceptable level of resolution. Species relative range size was calculated by using the formula $IGE = Ss/Sm$, where IGE is the Index of Geographic Expansion, Ss the number of grid squares from which the species has been recorded, and Sm the number of grid squares occupied by the species with the most extensive range. The following examples taken from Gómez-Hinostrosa and Hernández (2000) illustrate the use of this index. *Mammillaria heyderi* Muehlenpf. was used as a reference for the calculation of the index, as it turned out to be the species with the largest geographic range ($Sm = 40$) of all the analysed species. The narrowly endemic *Turbinicarpus subterraneus* (Backeb.) A. Zimmerman has only been recorded in two grid squares in the whole Chihuahuan Desert; thus it has an $IGE = 2/40$ ($IGE = 0.05$). Conversely, *E. platyacanthus*, which is relatively widespread in the southern half of the Chihuahuan Desert Region and in some areas south of the Trans-volcanic Belt, has been recorded in 31 grid squares and has an $IGE = 0.78$. It is important to emphasise that in the calculation of this index the whole distribution range of the species in Mexico was taken into account.

When the range size of a set of species from a given region is analysed using this method, their IGE values vary in a continuous fashion. For instance, in the case of the species of cacti from the Mier y Noriega grid square, values varied continuously from 0.05 to one. We have used the median of the IGE values as a practical way to set a boundary between rare and common species. Thus, in Mier y Noriega we found that 51.3 % of the species of cacti are rare, as their IGE values are below the median. We found a similar set of values for the species occurring in the Huizache grid square (Hernández *et al.*, unpublished), confirming the fact that the Cactaceae have restricted distribution ranges. The unusually high incidence of narrow endemisms among Mexican Cactaceae is probably comparable to that of the Aizoaceae and other succulent plant families in South Africa.

v) Regional frequency

An additional parameter providing further evidence about whether a given species is rare or common is its frequency at the regional or local scale. This can be calculated as the frequency with which a species is found in a determined number of samplings. In this case we used the formula $IRF = sl/nl$, where IRF is the Index of Regional Frequency, sl the number of localities in which a given species was found, and nl the total number of localities sampled. As in the IGE, we used the median of the IRF values to set a limit between frequent and infrequent species.

We divided the Huizache and Mier y Noriega grid squares into smaller subsquares, and an average of three localities in each of these sub-squares were sampled. The set of data derived from these samplings were used for the calculation of the IRF for each of the cactus species occurring in these areas. For instance, in the Mier y Noriega grid square the stenoendemic *T. subterraneus* was only found in one of the 80 sampled localities. Consequently, it has a low IRF ($1/80 = 0.01$). This value contrasts with that of *E. platyacanthus*, a species found in 73 out of the total number localities sampled in this grid square ($IRF = 0.9$).

vi) Complementarity analysis

Figure 18.3 shows the distribution patterns of the species of cacti occurring in the Huizache and Mier y Noriega grid squares. In both grid squares, the areas with the highest species richness are located in the lowland valleys, where the lowest precipitation is recorded. Diversity decreases toward the top of the mountain ranges, where a greater precipitation and lower mean temperatures prevent the development of xeric vegetation types.

There is no doubt that these two grid squares deserve special attention due their outstanding richness of species of cacti. However, they cover an extremely large area (approx. 5,700 km^2) to make any conservation action practicable. In order to optimise the conservation of biodiversity in these areas, a complementarity analysis was conducted (Humphries *et al.*, 1991; Pressey *et al.*, 1993). The complementarity analysis, as applied in the studied areas, was delineated by Gómez-Hinostrosa and Hernández (2000). The principle assumes the designation of a first priority area, defined as the sub-square with the highest number of species. Then other areas are defined in order of priority according to their contribution of additional species not found in the areas of higher priority. Thus, by means of this procedure all 50 sub-squares in the Huizache and Mier y Noriega grid squares were ranked according to their importance in terms of their particular species numbers and composition. In this particular aspect of our study we calculated the complementarity for all the species of cacti in the area, and separately for a subset corresponding to the threatened species.

The results of the complementarity analysis of endangered species are shown in Figure 18.4. Here, the total number of the species that we regarded rare or endangered were considered (31 spp.). According to this analysis all species are comprised in a minimum of ten sub-squares (shaded areas in Figure 18.4). The

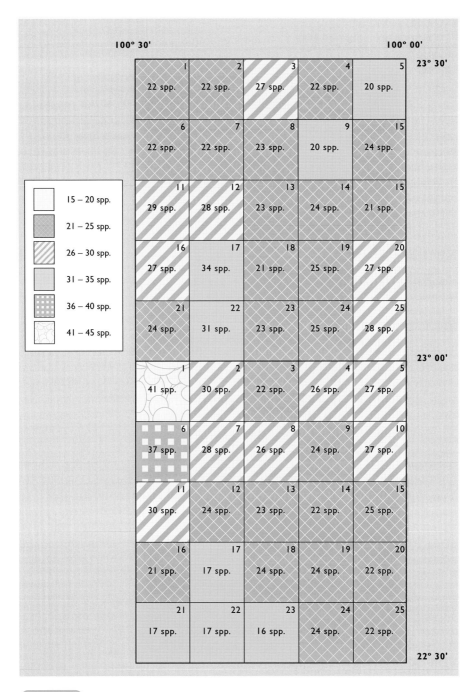

Figure 18.3   Patterns of geographic distribution of Cactaceae in the Mier y Noriega (23°30'–23°00') and Huizache (23°00'–22°30') grid squares. The figures indicate numbers of species by sub-square.

ten sub-squares, which correspond only to one fifth of the total area, were ranked in order of importance. Thus, for example, sub-square 1 of the Huizache grid square is first priority (P1), as this holds the highest number of endangered species (15 spp.). Sub-square 13 is second priority (P2), because it has five additional species not present in P1. With three additional species not present in P1 and P2 combined, sub-square 2 is third priority. Each of the remaining seven sub-squares have a moderate contribution to the total endangered diversity of species of cacti of these areas, and they are ranked as P4 or P5 because they provide either two or one additional species (Figure 18.4). It is particularly interesting to notice that sub-squares 1, 13, and 2 of the Huizache grid square contain 74% of the threatened species in the area; thus, conservation actions have to be concentrated in these areas in particular.

Conclusions

The Real de Guadalcázar Protected Area

The results of our research revealed that the studied areas, which correspond mostly to the Municipality of Guadalcázar, San Luis Potosí, are critical for the conservation of Mexican Cactaceae. Most of its areas are reasonably well preserved, the diversity of other plant and animal taxa is comparatively greater than other parts of the Chihuahuan Desert, and there is a variety of vegetation types, resulting in a high degree of environmental heterogeneity. In 1996 we made a proposal to the Government of San Luis Potosí for the conservation of this region. The proposal was focused toward the conservation of the richest, most intact, and less populated areas within the Municipality of Guadalcázar. The configuration of the natural area was planned to guarantee the conservation of all species of cacti (Figure 18.5). In addition, using the results of the complementarity analysis and the state of conservation of the different areas, two separate core areas were proposed in order to maximise the conservation of the areas with high species richness, especially of those that are rare and threatened (Figures 18.4–18.5).

In September 1997, the State Government declared the creation of the *Real de Guadalcázar Protected Natural Area* (Anonymous, 1997). The reserve has a surface area of 188,758 ha, including the two proposed core areas. Among its objectives, the conservation of the Cactaceae was explicitly stated. The official recognition of this natural area is an important event for the conservation of the Cactaceae, from both national and international perspectives. However, it currently has the status of a state reserve, and it is still necessary to recognise it at the national and international level, perhaps as a biosphere reserve. In addition, a great deal of action will have to be implemented by the Government of San Luis Potosí and by local organisations before this reserve becomes a reality.

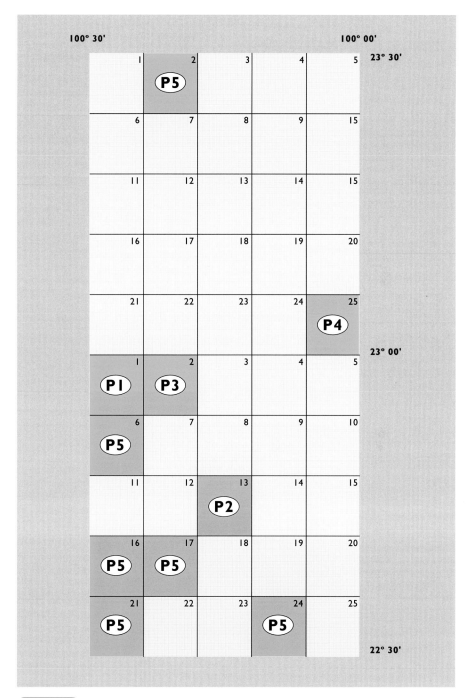

Figure 18.4   Priority areas for the conservation of endangered cacti in the Mier y Noriega (23°30′–23°00′) and Huizache (23°00′–22°30′) grid squares based upon complementarity analysis, where P1 = highest priority.

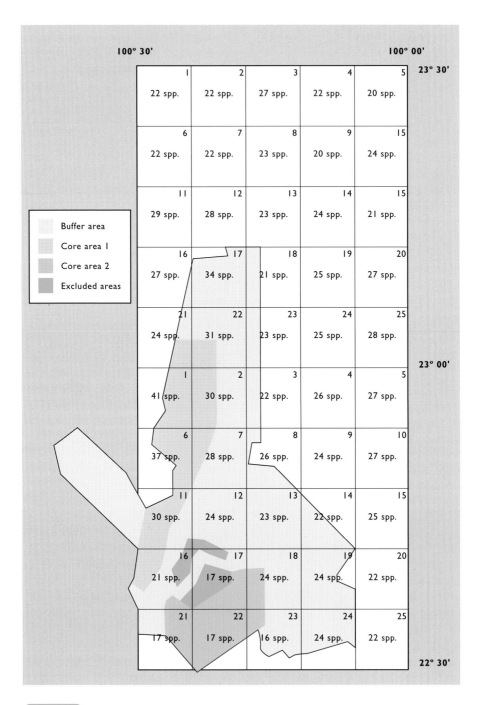

Figure 18.5 Configuration of the Real de Guadalcázar Protected Natural Area.

Acknowledgements

We wish to thank Rolando Bárcenas for continuous support along all stages of the Chihuahuan Desert Project, and Fernando Chiang for reviewing the manuscript.

References

Andrade-Lima, D. (1981). The caatinga dominium. *Revista Brasileira de Botânica* **4**: 149–153.

Anonymous (1997). Declaratoria del Área Natural Protegida bajo la modalidad de Reserva Estatal con características de Reserva de la Biósfera, la región históricamente denominada "Real de Guadalcázar", ubicada en el Municipio del mismo nombre. *Periódico Oficial del Gobierno de San Luis Potosí* **80**: 1–12.

Arias, S., Gama, S. and Guzmán, U. (1997). *Cactaceae A. L. Juss. Flora del Valle de Tehuacán-Cuicatlán.* 14. Mexico City: Instituto de Biología, Universidad Nacional Autónoma de México, Mexico.

Arita, H. Figueroa, F., Frisch, A., Rodríguez, P. and Santos-Del-Prado, K. (1997). Geographical range size and conservation of Mexican mammals. *Conservation Biology* **11**: 92–100.

Bárcenas, R.T. (1999). *Patrones de distribución de cactáceas en el estado de Guanajuato.* Tesis de Licenciatura, Facultad de Ciencias, Universidad Nacional Autónoma de México, Mexico City, Mexico.

Barthlott, W. and Hunt, D. (1993). Cactaceae, pp. 161–97. In: K. Kubitzki, J. Rohwer and V. Bittrich (eds). *The families and genera of vascular plants. II. Dicotyledons.* Springer-Verlag, Berlin, Germany.

Bravo, H. (1978). *Las Cactáceas de México. Vol. l.* Universidad Nacional Autónoma de México, Mexico City, Mexico.

Bravo, H. and Arias, S. (1999). Sinópsis de la familia Cactaceae en Mesoamérica. *Cactáceas y Suculentas Mexicanas* **44**: 4–19.

Bravo, H. and Sánchez-Mejorada, H. (1991*a*). *Las Cactáceas de México, Vol. ll.* Universidad Nacional Autónoma de México, Mexico City, Mexico.

Bravo, H. and Sánchez-Mejorada, H. (1991*b*). *Las Cactáceas de México, Vol. lll.* Universidad Nacional Autónoma de México, Mexico City, Mexico.

CITES (1990). *Appendices I, II and III to the Convention.* U.S. Fish and Wildlife Service, Washington, DC, USA.

Cornet, A. (1985). *Las cactáceas de la Reserva de la Biosfera de Mapimí.* Instituto de Ecología A.C., Mexico City, Mexico.

Correll, D. and Johnston, M. (1970). *Manual of the vascular plants of Texas.* University of Texas, Austin, USA.

Everitt, J.H. (1993). *Trees, shrubs, and cacti of south Texas*. Texas Tech University, Lubbock, Texas, USA.

Gaston, K.J. (1994). *Rarity*. Chapman and Hall, London, UK.

Gómez-Hinostrosa, C. (1998). *Diversidad, distribución y abundancia de cactáceas en la región de Mier y Noriega, México*. Tesis de Licenciatura, Facultad de Ciencias, Universidad Nacional Autónoma de México, Mexico City, Mexico.

Gómez-Hinostrosa, C. and Hernández, H.M. (2000). Diversity, geographical distribution, and conservation of Cactaceae in the Mier y Noriega region, Mexico. *Biodiversity and Conservation* **9**: 403–418.

Hernández, H.M. and Anderson, E.F. (1992). A new species of *Ariocarpus* (Cactaceae). *Bradleya* **10**: 1–4.

Hernández, H.M. and Bárcenas, R.T. (1995). Endangered cacti in the Chihuahuan Desert: I. Distribution patterns. *Conservation Biology* **9**: 1176–1188.

Hernández, H.M. and Bárcenas, R.T. (1996). Endangered cacti in the Chihuahuan Desert: II. Biogeography and conservation. *Conservation Biology* **10**: 1200–1209.

Hernández, H.M. and Godínez, A.H. (1994). Contribución al conocimiento de las cactáceas mexicanas amenazadas. *Acta Botánica Mexicana* **26**: 33–52.

Hernández, H.M., Gómez-Hinostrosa, C. and Bárcenas, R.T. (2001). Diversity, spatial arrangement, and endemism of Cactaceae in the Huizache area, a hot-spot in the Chihuahuan Desert. *Biodiversity and Conservation* **10**(7): 1097–1112.

Humphries, C., Vane-Wright, D. and Williams, P. (1991). Biodiversity reserves: setting new priorities for the conservation of wildlife. *Parks* **2**(2): 34–38.

Hunt, D. (1999). *CITES Cactaceae Checklist*. Royal Botanic Gardens, Kew, UK.

Johnston, M.C. (1977). Brief resume of botanical, including vegetational, features of the Chihuahuan Desert Region with special emphasis on their uniqueness, pp. 335–59. In: R.H. Wauer and D.H. Riskind (eds). *Transactions of the symposium on the biological resources of the Chihuahuan Desert Region, United States and Mexico*. National Park Service, Washington, DC, USA.

Lehr, J.H. (1978). *A catalogue of the flora of Arizona*. Desert Botanical Garden, Phoenix, Arizona, USA.

Leuenberger, B. (1986). *Pereskia* (Cactaceae). *Memoirs New York Botanic Garden* **41**: 1–141.

Leuenberger, B. (1997). *Maihuenia*. Monograph of a Patagonian genus of Cactaceae. *Bot. Jahrb. Syst.* **119**: 1–92.

McAllister, D.E., Schueler, F.W., Callum, M.R. and Hawkins, J.P. (1994). Mapping and GIS analysis of the global distribution of coral reef fishes on an equal-area grid, pp. 155–75. In: R.I. Miller (ed.). *Mapping the Diversity of Nature*. Chapman and Hall, London, UK.

Miller, R.I. (1994). Possibilities for the future, pp. 199–205. In: R.I. Miller (ed.). *Mapping the Diversity of Nature*. Chapman and Hall, London, UK.

Morafka, D.J. (1977). A biogeographical analysis of the Chihuahuan Desert through its herpetofauna. *Biogeographica* **9**: 1–313

Myklestad, A. and Birks, H. (1993). A numerical analysis of the distribution patterns of *Salix* L. species in Europe. *Journal of Biogeography* **20**: 1–32.

Navarro, G. (1996). Catálogo ecológico preliminar de las cactáceas de Bolivia. *Lazaroa* **17**: 33–84.

Pinkava, D. J. (1984). Vegetation and flora of the Bolsón of Cuatro Ciénegas region, Coahuila, Mexico: summary, endemism and corrected catalogue. *Journal of the Arizona-Nevada Academy of Science* **19**: 23–47.

Pressey, R., Humphries, C., Margules, C., Van-Wright, R. and Williams, P. (1993). Beyond opportunism: key principles for systematic reserve selection. *Trends in Ecology and Evolution* **8**: 124–128.

Rabinowitz, D. (1981). Seven forms of rarity, pp. 205–17. In: H. Synge (ed.). *The Biological Aspects of Rare Plant Conservation.* Wiley, New York, USA.

Rebman, J., Resendiz M.E. and Delgadillo, J. (1999). Diversity and documentation for the Cactaceae of Lower California, Mexico. *Cactáceas y Suculentas Mexicanas* **44**: 20–26.

Ruiz de Esparza, R. (1988). Lista de especies vasculares, pp. 225–39. In: C. Montaña, (ed.). *Estudio integrado de los recursos vegetación, suelo y agua en la Reserva de la Biosfera de Mapimí.* Instituto de Ecología, A.C., Mexico City, Mexico.

Rzedowski, J. (1991). Diversidad y orígenes de la flora fanerogámica de México. *Acta Botánica Mexicana* **14**: 3–21.

SEDESOL (1994). Norma Oficial Mexicana NOM – 059 – ECOL – 1994, que determina las especies de flora y fauna silvestre, terrestres y acuáticas en peligro de extinción, amenazadas, raras y las sujetas a protección especial y que establece especificaciones para su protección. *Diario Oficial de la Federación* **438**: 2–60.

Villarreal, J.A. (1994). Flora vascular de la Sierra de la Paila, Coahuila, México. *Sida* **16**(1) 109–138.

Walter, K.S. and Gillett, H.J. (eds) (1998). *1997 IUCN Red List of Threatened Plants.* Compiled by the World Conservation Monitoring Centre. IUCN – The World Conservation Union, Gland, Switzerland and Cambridge, UK.

Chapter 19

The Conservation Status of the Coco de Mer, *Lodoicea maldivica* (Gmelin) Persoon:

a flagship species

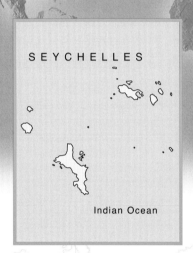

SEYCHELLES

Indian Ocean

Frauke Fleisher-Dogley
Botanic Garden, Mahe, Republic of the Seychelles

Tony Kendle
Plant Sciences Laboratory, University of Reading, UK

Introduction

The Seychelles archipelago is an island group of international importance for biodiversity conservation located in the Western Indian Ocean (Figure 19.1). There are approximately 115 islands, although the precise figure depends on the definition of what constitutes an island. Some of these are coralline in origin, but the group also includes a truly unique cluster of ancient granitic islands that are derived from remnants of Gondwana that have been separated for 130 million years. They therefore support relict flora that do not always share the characteristics of plants normally found on islands, for example plant species adapted for animal and wind dispersal. The granitic islands have been identified as a Centre of Plant Diversity (WWF and IUCN, 1994–1997) and are a biodiversity hotspot (Myers *et al.*, 2000).

There are approximately 250 indigenous flowering plants on the islands, of which about 35% are endemic. There are about 1,100 endemic organisms in total. Since human settlement 200 years ago, only eight indigenous species are believed to have become extinct, but about 20% of the flora is highly

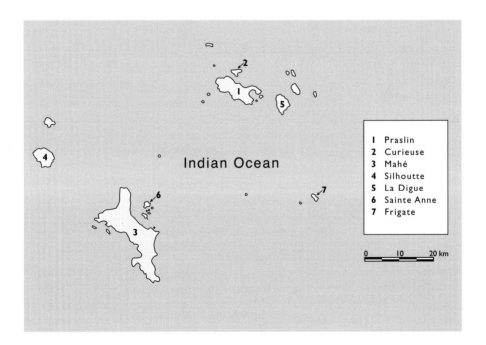

1	Praslin
2	Curieuse
3	Mahé
4	Silhoutte
5	La Digue
6	Sainte Anne
7	Frigate

Figure 19.1 Location map of the Seychelles archipelago.

endangered. The major threats are habitat destruction, introduced invasive plants and direct exploitation. Some plants are naturally rare because they are restricted to high altitudes above 500 m with limited possible ranges.

The main industries on the islands today are tourism and fisheries, both dependent on environmental quality. It is not surprising therefore that the Government of the Republic of the Seychelles has ratified key conventions such as the Nairobi Convention (East African Action Plan), the Convention on Trade in Endangered Species (CITES), the Montreal Protocol, the Law of the Sea, the Convention on Climate Change and the Convention on Biological Diversity (CBD) (see Appendix 1). Implementation of the National Biodiversity Action Plan (NBAP) under the CBD has already started and more than 47% of the land area has been designated as protected areas. Two world heritage sites are situated in the Seychelles: Aldebra and the Vallée de Mai. The latter is one of the main population centres for the coco de mer, *Lodoicea maldivica* (Gmelin) Persoon (Palmae). The need for sustainable management of this species has been identified in the NBAP.

The coco de mer is one of the best examples of a relict species, and this has become an internationally recognised 'flagship' species for the Seychelles. It is a spectacular endemic palm that was for many years known only through the discovery of gigantic seeds washed ashore in other countries (these seeds were not viable as a healthy nut is too heavy to float). Since no one knew where the seeds had come from, they were called 'coco de mer' (coconut of the sea) and assigned a mythical origin. Their true origin was not discovered until some years after the initial exploitation of the Seychelles (Beaver and Chong Seng, 1992).

Although the locations of the plants are now known, the almost mythological status is maintained by the extraordinary ecology and growth characteristics of this palm. It belongs in a monotypic genus, and grows up to 30 m tall with leaves up to 10 m in diameter. The seed is the largest known in the plant kingdom, weighing up to 17 kg, and it can take years to mature on the plant. It is gravity dispersed and germination tubes are sometimes remarkably long (up to 4 m). Despite stories of individuals reaching 800 years or more in age, evidence is lacking and *L. maldivica* probably never exceeds 400 and rarely 300 years (Savage and Ashton, 1983). The resemblance of the female nut and the male inflorescence to human sexual organs, rarely goes unnoticed.

The endemic fauna associated with *L. maldivica* habitat includes birds such as the black parrot (*Coracopsis nigra barklyi*) and the seychelles bulbul (*Hypsipetes crassirostris*). Two gecko species (*Phelsuma* sp., and *Ailuronyx sechellensis*) and the tiger chameleon *Chamaeleo tigris* are found in these areas. Among the Seychelles snails, only one, *Pachnodus praslinus*, is found in *L. maldivica* forest (Beaver, 1992). There may be complex interactions between the palm and this fauna that are yet to be discovered.

The palm grows on ridges, slopes and valleys, but has been most studied in the more accessible valleys. In such locations, it probably once formed the dominant canopy species in almost pure stands. Undergrowth in these areas of forest is limited by the lack of light and the thick leaf litter. Epiphytes, such as lichens, are found in tiny crevices in the bark. Ferns grow around the inflorescence in the crown of the trees. Higher on the slopes, *L. maldivica* is found at a lower density and before settlement it probably grew in mixed stands with other endemic species, notably: *Deckenia nobilis* H. Wendl. ex Seemann (Palmae), *Verschaffeltia splendida* H. Wendl. (Palmae) and *Pandanus hornei* Balf.f. (Pandanaceae). In these areas, indigenous understorey species would include *Nephrosperma vanhoutteanum* Balf.f. (Palmae), *Phoenicophorium borsigianum* Stuntz and *Pandanus sechellarum* Balf.f. Today, although these species are still found, the slopes and ridges also contain many naturalised introduced species, such as *Cinnamomum verum* J. Presl. (Lauraceae) and *Chrysobalanus icaco* L. (Chrysobalanaceae).

Conservation Issues

The characteristics of the coco de mer help to explain why the plant deserves the status of a flagship species. It is not the most endangered endemic plant of the islands, but it is one of the most striking plants that exist anywhere and it is actively exploited for the tourist market, making it the first, and often the only, endemic plant that visitors can name. Images of the nuts and flowers are used widely in tourist marketing. Visits to one of the best preserved remnants of the natural forest at the Vallée de Mai, Praslin island, are features of many tourist trips, and the massive nuts and nut products are sold as high priced souvenirs.

Whole nuts fetch high prices as souvenirs and are extensively traded in the island tourist markets. The price ranges from £90 (sterling) for unpolished nuts to £350 (sterling) for large, well-shaped, polished nuts. The annual income from the sale of the nuts is estimated to be £200,000 (sterling).

There are many traditional uses of the palm. It was harvested for building timbers, the leaves were used for thatch and the down from young leaves for stuffing pillows. Fishermen used half nuts as bailers in their boats and whole nuts as water containers. The empty shells have even been used as vessels and spoons in shops and households. Small quantities of the kernel were part of the ingredients in the preparation of aphrodisiacs. The 'jelly' of the immature nut was eaten as a special dessert. However, the society of the Seychelles is relatively young, and even such cultural uses are arguably recent in origin. They do not play a large role in the local economy compared with the sale of the nuts.

Currently, the Government of the Republic of Seychelles, through the Division of Environment, regulates the marketing of the nuts. The registration of the shape of the coco de mer nut as a international trademark is also proposed. Almost immediately following independence in 1978, the Coco de Mer Management Decree was developed as a management tool and the nut was declared as a licensable product. Under this Decree it is an offence to deal in immature nuts. The amendment of the Breadfruit and other Trees Act in 1994 required permission from the Division of Environment before a coco de mer could be felled. In 1994, the parliament approved an amendment to the Coco de Mer Decree to strengthen the control of poaching and illegal trade, with one officer recruited to implement the Decree. In line with the law, inventories have been conducted from 1995 to 1998 and the populations are now continuously monitored. Data concerning the reproduction of the palm, which includes the harvest and sale of the nut, are collected and analysed at the end of each year. The possession of a coco de mer nut has to be supported by a 'Coco de Mer Permit' and a tag (Figure 19.2). Police inspections are carried out on a quarterly basis at sales outlets to control illegal trade.

Figure 19.2 *Lodoicea maldivica* (Gmelin) Persoon (Palmae) nut showing the authorised permit for sale attached.

Despite these efforts, a black market is known to exist and the over-collection of nuts may threaten the viability of the wild populations. There is a worrying possibility that an international market is developing for aphrodisiac use that could lead to excessive harvesting of even immature nuts. The illegal sale of just four nuts at a discounted price of £50 (sterling) will provide the equivalent of a forestry worker's monthly salary.

Population statistics (Table 19.1) show that natural stands are heavily dominated by juveniles even though the period to maturity should represent a small part of the lifespan of the plant. Seventy percent of the total population of trees is immature. This is a sign that the palm went through a population collapse before state control of the habitats was introduced (Figure 19.3). The current growth of so many juveniles is superficially a positive sign, although it is far from clear whether they will all reach maturity. However, more detailed inspection suggests that there are almost no germinating nuts or very young seedlings in the wild. This reflects the growth in trade of the nuts. Currently, 100% of the retrievable nuts are removed in the state forest on the island of Praslin (outside of the Vallée de Mai). Only a small number of nuts, fallen in inaccessible sites, are left behind to germinate. This over-exploitation may threaten the species unless sustainable harvesting methods are used.

At a political level, the need for protection of the palm was never questioned. However, the majority of the Seychellois regard the coco de mer as abundant. Indeed, with a total population of 24,448 trees the species is not under immediate threat of extinction. However, only 14% of the total population are mature nut-bearing female trees, and regeneration will be effectively impossible if all the viable seeds are being removed.

Such problems are compounded by the naturally restricted distribution that makes the species vulnerable to stochastic events. A major fire in 1989 partly destroyed the largest wild population on Praslin, and another fire occurred in 1999. The proximity of human populations to these sites, as well as access by tourists, greatly increases the risk from such disasters. To date, disease problems have not been an issue, but vulnerability to introduced diseases or pests is another concern.

As well as species-focused legislation, attention is also given to designating protected areas. In 1979, the first National Parks were created, but the responsibilities for management of the habitats where coco de mer are found are complex. The World Conservation Union (IUCN) Category of Threat (see Appendix 2) for *L. maldivica* is Vulnerable (Oldfield *et al.*, 1998). Although locally abundant, it is found naturally only on Praslin and Curieuse islands (Table 19.1). On Praslin, there are two concentrations in Fond Ferdinand and Vallée de Mai. Only the latter is currently located within Praslin National Park.

Table 19.1 The location, size and demography of populations of *Lodoicea maldivica* (Gmelin) Persoon (Palmae) as of June 1997

Location	Inventory results (1995–1996)				Proportion of trees		
	Female	Male	Immature	Total	Female	Male	Immature
Praslin: National Park							
Fond Ferdinand	986	769	6,721	8,476	12%	9%	79%
Private land	898	1,030	1,770	3,698	24%	28%	48%
Vallée de Mai	518	644	4,920	6,082	9%	11%	80%
Subtotal Praslin	**2,402**	**2,443**	**13,411**	**18,256**	**13%**	**13%**	**74%**
Curieuse	892	1,401	3,718	6,011	15%	23%	62%
Subtotal Praslin/Curieuse	**3,294**	**3,844**	**17,129**	**24,267**	**14%**	**16%**	**70%**
Other Islands:							
Mahé	70	51	8	129	54%	40%	6%
Silhouette	7	2	12	21	33%	10%	57%
La Digue	7	1	0	8	88%	12%	0%
Sainte Anne	8	3	1	12	67%	25%	8%
Ile Moyenne	0	0	9	9	0%	0%	100%
Frigate	0	2	0	2	0%	100%	0%
Total population	**3,386**	**3,903**	**17,159**	**24,448**	**14%**	**16%**	**70%**

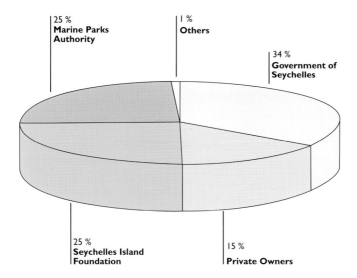

25 %
**Marine Parks
Authority**

1 %
Others

34 %
**Government of
Seychelles**

25 %
**Seychelles Island
Foundation**

15 %
Private Owners

Figure 19.3   Chart showing division of land ownership of *Lodoicea maldivica* (Gmelin) Persoon (Palmae) sites.

The management of Praslin National Park is under the responsibility of the Division of Environment. Although both sites are government property, Fond Ferdinand is managed by a non-governmental organisation, the Praslin Development Foundation (PDF). The Vallée de Mai, a World Heritage Site, is managed by the Seychelles Island Foundation. Other coco de mer trees are scattered throughout Praslin on private properties. Under current legislation, private landowners have constraints on felling and harvesting of nuts, but not on indirect habitat modification.

The population on the island of Curieuse is scattered and reduced following deforestation and subsequent soil erosion. Curieuse is designated as a National Park under the management of the Marine Parks Authority, with an historical focus more on coastal rather than terrestrial protection.

Action Taken

Conservation management of the coco de mer is moving from the establishment of a baseline of legislative protection towards a more strategic approach. In 1998, the Government of the Republic of the Seychelles, as part of the NBAP, recognised the need for a sustainable management plan for the species. The protection of the natural habitat of the species is only a beginning. Whenever the exploitation of a natural product is involved, active conservation measures, including monitoring, are needed. The Coco de Mer Management Plan is intended to provide a focus for this work.

A continuous monitoring programme with adequate documentation is important. The data also need to be analysed regularly to allow for adjustments to the management plan. Even though resources are scarce, the process of implementing these initial stages has raised the profile of this species and initiated further discussions on conservation work. Starting work, even at a small scale, and maintaining momentum can catalyse the creation of new projects and assist in identifying new sources of funding. The Management Plan is not yet complete, but there are several issues that have been addressed:

- A review was undertaken to determine whether the designation of Praslin National Park provided adequate coverage of the distribution of the natural habitat of the species. It was decided to expand the extent of the protected area by a further 133 ha to encompass a second major valley population at Fond Ferdinand. Much of this additional area is already state-owned forestry land, but stakeholder involvement played a major role in the implementation of this protected area expansion. The early involvement of the PDF in proposals to extend the Praslin National Park speeded up the processing of the matter in other Ministries and helped to make funds available to purchase 9 ha of private property. Legal ratification of this boundary change is underway.

- Within the National Park, a fire observation tower has been erected and fire fighting equipment purchased. A system of fire breaks has been established and its effectiveness is being monitored.

- The viability of the population on Curieuse Island was assessed. It was found to be too small and subject to degradation because of lack of vegetation cover. A reforestation project involving the planting of 500 coco de mer was funded by the Dutch Trust Fund and supervised by the World Bank and this began in October 1996. In December 1998, the planting of 536 nuts was completed and so far 41% of the nuts have germinated, although 29 nuts have been illegally removed.

Practical Issues, Costs and Lessons Learnt

The coco de mer is an exceptional plant that requires exceptional treatment and practical costs are not directly applicable to other circumstances. Planting a single seed involves digging a hole 1 m × 1 m in size in steep terrain and the nut itself carries a 'cost' equivalent to a week's salary. In fact, all conservation work on the Seychelles involves logistical challenges that distort costs. The islands carry a global responsibility for biodiversity protection, but in population terms, the Seychelles is one of the smallest countries in the world, and staff numbers and resources are very limited. Like many small island developing states, it is faced with huge logistical problems, such as extremes of topography and isolation of sites. However, there are some insights that have emerged that can provide lessons for others.

Much has been written about *L. maldivica*, but many of the often-repeated facts about its growth and ecology remain unconfirmed by scientific data. Little is known about the germination, growth and longevity of this species apart from work by Savage and Ashton (1983). The most obvious constraint in research is the relatively long duration between the different stages in the lifecycle of this species. The use of archive information, aerial photographs and software in population modelling may help to minimise the research period required, but it will still need one to two human generations just to collect the full data and during this period environmental conditions, such as climate, are likely to change dramatically. We may never have a classic 'model' of the demography of this species based on stable and complete data, and dealing with this uncertainty, and therefore the inability to predict the effect of conservation action on the viability of these populations, is one of the challenges facing the management team.

A good example of where priorities need to be resolved by judgement, rather than proof, relates to the recovery programme on Curieuse. All available nuts from the island were dedicated to this project, but even with this security it was clear that the population would remain highly vulnerable and fragmented for decades. A decision was therefore made to supplement the supply of nuts with seed from Praslin Island. There was a risk that some genetic diversity would be obscured in the process, but conversely there was also a risk that the Curieuse population would be lost through genetic decline or stochastic events if segregation was maintained. Studies to determine the degree to which the two populations are really isolated in terms of pollen flow are still awaited, but some conservation measures have been taken in the meantime.

This example raises some interesting issues related to conservation philosophy since the line between *in situ* and *ex situ* conservation is somewhat blurred. Does assisted regeneration within a palm's native area constitute a positive or negative action (Friedmann, in Johnson, 1996)? To what extent

Box 19.1 **Palms: their conservation and sustained utilisation**

The Palm Specialist Group of the Species Survival Commission (IUCN) has produced a review of global conservation needs for palms, *Palms: Their Conservation and Sustained Utilization* (Johnson, 1996). The study identified a total of 224 threatened palm species, about 10% of the world's 2,200 species. The largest proportions of these threatened palms are found on tropical islands, with, for instance, 69 species from Madagascar and 19 from Borneo. The two main trends driving this decline are habitat loss and over-harvesting. A number of areas rich in endemic palms are suffering massive levels of habitat destruction; these include the Peruvian Amazon, Madagascar, Indonesia and Malaysia. Major harvests of edible palm hearts and rattan are derived from wild populations in South America and Asia respectively (see Clayton *et al.*, Chapter 23).

The study made the following recommendations:

Research

- Study palms of poorly known forests with high palm diversity e.g. Atlantic Forests, Brazil; Yunga Forests, Bolivia; Irian Jaya, Indonesia; central Africa.
- Study threatened non-economic palms, e.g. *Roystonea* O.F. Cook and *Bactris* Jacq. ex Scop. in the Caribbean and South America; *Raphia* P. Beauv. in Africa.
- Conduct taxonomic and ecological research, e.g. national floristic accounts required for Venezuela and Malaysia; rattan inventories required for South East Asian countries.
- Study threatened and under-developed economic palms, including domestication of potentially useful species.

***In situ* conservation**

- Establish or strengthen protected areas – priorities include: Madagascar, Papua New Guinea and Peninsula Malaysia.
- Introduce legal incentives and regulations for the management of utilised wild populations.

***Ex situ* conservation**

- Establish living collections, with an emphasis on the establishment of breeding populations – priorities include: Colombia, Cameroon and Madagascar.
- Support appropriate re-introductions (see Appendix 4).

Education

- Promote value and conservation of palms through locally appropriate media.

(Editors, 2001)

should the coco de mer be domesticated to ensure a supply of a harvestable crop, or does its value reside in its wild status? Are we concerned just with preservation of the genetic stock, or is the associated habitat equally important? For global recommendations for palm conservation made by the Palm Specialist Group, see Box 19.1.

Future Steps

The formulation of the Management Plan is proceeding, but this document needs to be one that recognises uncertainty and makes provisions accordingly. There needs to be room for incorporation of ongoing scientific and demographic studies, and flexibility to respond to these findings, but a momentum needs to be maintained with regard to practical conservation work. Ideally, the Management Plan will serve as a tool that will help to focus and direct research when it happens. Some key points are outlined below:

- Regeneration of coco de mer on Curieuse Island is good, but growth rates are slow and it will be many years before the canopy closes or litter accumulation and soil protection will begin. Attention is turning to the use of other faster growing indigenous plants as nurse species. A common fault of species-focused conservation work is that the process may overlook ways in which the different components of the ecosystem may be mutually supporting.

- In the past, most attention has been focused on the World Heritage Site of the Vallée de Mai where the palm grows in dense, almost monospecific, stands. The stand on Curieuse is in drier conditions and the plant community is probably more typical of that seen on the slopes and ridges of Praslin where *L. maldivica* is less vigorous and forms part of a mixed forest with more complex ecological interactions.

- Census data on nut production are now accumulating, but future research needs to look in more detail at questions about ecology and growth rate so that the likely severity of threats and environmental changes can be assessed. For future studies, it is planned to analyse data based on three population zones, notably valley, slope and ridge. As well as reproductive behaviour, sample plots will be established to determine parameters such as plant density and cover and the presence and nature of competing species.

- Another issue to be addressed is the need to review policies on *ex situ* collections. Coco de mer is now found on many more islands than previously thought but collections have built up in an ad hoc way. A few palm trees have been introduced on the others islands, including a respectable *ex situ* population in the botanic gardens of Victoria on Mahe. Scattered trees in cultivation, planted as both ornamentals and trees for nut harvesting, can also be found on Silhouette, La Digue, St. Anne, Fregate and Moyenne islands. *Ex situ* populations also exist in other countries such as Sri Lanka. Establishing new populations may be a very sensible way of dealing with the problem of stochastic events, but if this decision is made, a proper germplasm collection should be established linked to a full environmental impact assessment procedure.

References

Beaver, K. and Chong Seng, L. (1992). Vallée de Mai. Space Publishing Division, Mont Fleuri, Seychelles.

Johnson, D. (ed.) and the IUCN/SSC Palm Specialist Group. (1996). *Palms: Their Conservation and Sustained Utilization. Status Survey and Conservation Action Plan.* IUCN, Gland, Switzerland.

Oldfield, S., Lusty, C. and MacKinven, A. (1998). *The World List of Threatened Trees.* World Conservation Press, Cambridge, UK.

Savage, A.J.P. and Ashton, P.S. (1983). The population structure of the double coconut and some other Seychelles palms. *Biotropica* **15**(1): 15–25.

WWF and IUCN (1994–1997). Centres of plant diversity. *A guide and strategy for their conservation.* 3 volumes. IUCN Publications Unit, Cambridge, UK.

Chapter 20

Propagation Techniques for the Conservation of an Endangered Medicinal Tree in Africa:

Prunus africana *(Hook.f.) Kalkman*

Hannah Jaenicke
International Centre for Research in Agroforestry (ICRAF), Nairobi, Kenya

Moses Munjuga
ICRAF, Nairobi, Kenya

James Were
ICRAF, Nairobi, Kenya

Zacharie Tchoundjeu
ICRAF-Cameroon, Yaounde, Cameroon

Ian Dawson
ICRAF, Nairobi, Kenya

Introduction

Prunus africana (Hook. f.) Kalkman (Rosaceae) is a geographically widespread tree, restricted to highland forests (primarily Afro-montane forest 'islands') in mainland Africa (Angola, Cameroon, Democratic Republic of Congo, Ethiopia, Kenya, Malawi, Nigeria, Somalia, South Africa, Sudan, Swaziland, Tanzania, Uganda, Zimbabwe) and outlying islands (Bioko, Grand Comore, Madagascar, Sao Tomé) (Kalkman, 1965). *P. africana* is long-lived and may grow to a height of more than 40 m and diameter of greater than 1 m. Populations show unusual size class distributions, suggesting that natural regeneration is episodic as a result of forest disturbance, and tree density is generally low (Ewusi *et al.*, 1997). The species is valued for its pale red heartwood that matures to a warm dark red and is used for plywood production, furniture and tool making. Most importantly, however, *P. africana* is highly sought after for its medicinal properties. Alkaloid compounds in its bark are extracted and processed into drugs that are used in the treatment of prostate disorders that are common in older men. The annual over-the-counter trade for the final pharmaceutical product is estimated at over US$220 million, primarily in Europe, Japan and North America (Cunningham *et al.*, 1997). To supply this market, over 3,000 tonnes of bark is collected annually from natural stands in Cameroon, Kenya and Madagascar.

Bark is usually collected by villagers – often in the dry seasons when labour requirements on farms are low. They sell to middlemen who then sell to the companies involved in processing and exporting bark. Sometimes, collectors may be employed directly by the exporting/processing company itself.

In Cameroon, when commercial bark harvesting began in the early 1970s, a single company was awarded a monopoly for extraction. Efforts were made to sustainably manage trees by harvesting two opposite quarters of bark, in a five-year rotation. However, trade liberalisation in 1985 resulted in harvesting concessions being awarded to 50 entrepreneurs, leaving little opportunity for sustainable harvesting. This led to destructive harvesting by felling of trees and stripping all bark, and a rapid depletion of the natural resource base of the species. In Madagascar and Kenya, extraction is similarly unsustainable, although until recently in Kenya the primary purpose for cutting has been for timber or agricultural land clearance. In 1995, *P. africana* was listed on Appendix II of the Convention on International Trade in Endangered Species of Wild Fauna and Flora (CITES), which means that bark and other products from the tree are regulated at export and import in an effort to ensure sustainable harvesting. Nevertheless, statistics show an increase in unsustainable bark harvesting from natural forest stands in all three countries, even in areas such as national parks or forest reserves, which are protected ecosystems under national legislation. In Cameroon, moves have been taken to issue a much-reduced quota for bark collection from the Mount Cameroon area (see Sunderland *et al.*, Chapter 21), but it is not yet clear whether this will reduce overall tree cutting or simply shift exploitation to other regions or countries. In Madagascar, natural populations have been decimated except in remote inaccessible areas (Nouhou Ndam, Mount Cameroon Project, pers. com.).

The isolated 'island' distribution of *P. africana* in highland forests implies large genetic differences between populations. This has been confirmed by DNA-based analyses of populations from Madagascar, Cameroon, Ethiopia, Kenya, Uganda and South Africa. Studies indicated large differences between countries and significant differences between populations from within countries where multiple populations were tested (Cameroon and Madagascar) (Barker *et al.*, 1994; Dawson and Powell, 1999). Furthermore, chemical analyses of bark extracts show significant differences between populations from Cameroon, Democratic Republic of Congo, Kenya and Madagascar (Martinelli *et al.*, 1986). Therefore, continued over-exploitation from natural populations, leading to local extinctions, will lead to the loss of important genetic variation in the species.

Agroforestry and the Environment

According to Cunningham (1994), the emphasis for the conservation of Afromontane forest should be on providing alternative sources of supply of useful products outside core protected areas through cultivation by small-scale farmers. The production of trees with a marketable value on farms benefits farmers through the provision of an income. Further, an increased diversity of products on-farm can buffer environmental and economic changes. Taking into consideration the increasing threat to natural forest ecosystems through deforestation, the transfer of germplasm from the wild into on-farm niches can help preserve valuable genetic resources, particularly if attention is paid to the origin and genetic diversity of cultivated material. In addition, the cultivation of threatened forest species takes pressure off their natural resource base, thereby promoting the conservation of natural stands and other associated species (in the case of *P. africana*, this includes birds and mammals that feed on the fruits). Research indicates that, in some areas, the number of trees planted on smallholder farms has increased together with human population density; natural forest has been cleared and more trees planted to partly compensate this loss (Arnold and Dewees, 1995). Therefore, agroforestry may be a particularly appropriate method for conservation in highland areas of Africa where high population density and the pressure on natural forest are especially high.

In Cameroon, farmers have shown considerable interest in growing *P. africana*, with more than 3,000 farmers planting the species over the past ten years (Cunningham *et al.*, 1997). Seedlings were obtained from government, community or non-governmental organisation nurseries, but were often of unknown origin and genetic base. Furthermore, knowledge on the handling of seed or alternative methods for propagation remains limited, both in Cameroon and elsewhere. These factors are important because flowering and seed set in natural stands is episodic, with good seed production occurring only every two to three years, and the timing of seed production within years is unpredictable. Furthermore, unlike orthodox species, the seed of *P. africana* cannot be dried and stored for a prolonged period of time. Obtaining sufficient quantities of high-quality germplasm for planting is therefore a limiting factor in most areas where a potential for cultivation exists.

At the International Council for Research in Agroforestry (ICRAF), Nairobi, a co-ordinated strategy to promote the sustainable on-farm cultivation of *P. africana* has been adopted, which includes research into the distribution of genetic variation in the species and methods for its on-farm management. Here, we report on two aspects of work relating to the provision of appropriate planting material to farmers. The first aspect relates to on-going studies to increase the storage life of *P. africana* seed, while the second concerns the development of alternative propagation methods for the species. The implications of these results for the conservation and sustainable utilisation of the species are discussed.

1. Seed storage studies

When ripe, *P. africana* fruits are purple, ellipsoid and measure approximately 8 mm × 12 mm. Fruits are normally collected when they start turning from green to purple. If immediately depulped and sown in tree nursery beds, germination levels may exceed 90%. However, extremely low levels of germination result if the seed is treated in an orthodox manner, with drying and storage before sowing (Sunderland and Nkefor, 1997).

Our research has been undertaken to determine the effect of different levels of fruit maturity and drying on seed germination over time. At collection, fruits were sorted into three distinct colours: green, purple-green and purple, corresponding to increasing levels of fruit maturity. Fruit were depulped or left whole and dried to different moisture contents. Initial germination rate was highest in purple depulped seed (Table 20.1).

After storage at 5°C for two months, seed generally showed low percentage germination, with no definite pattern emerging with respect to moisture content. As previously, purple depulped seed exhibited the highest level of germination. Similarly, after one year, purple depulped seed showed the best germination rate (Table 20.2, other data not shown). Some treatments showed higher germination after two months storage than initial tests, which may be attributed to after-ripening effects, such as the breakdown of germination inhibiting substances.

Table 20.1 **Initial germination percentages for *P. africana* (Hook.f.) Kalkman seed of different levels of maturity after desiccation to different moisture levels (m.c. = moisture content, n = 50 seeds for each treatment, 4 replicates)**

Fruit colour/condition	Germination (%)		
	8 % m.c.	15 % m.c.	25 % m.c.
Green, with pulp	16	4	5
Green, depulped	<1	0	1
Purple-green, with pulp	7	2	8
Purple-green, depulped	4	2	2
Purple, with pulp	0	3	2
Purple, depulped	67	72	49

Table 20.2 **Germination percentages of purple depulped seed of *P. africana* (Hook.f.) Kalkman after storage at 5 °C (n = 50 seeds for each treatment, 4 replicates)**

Moisture content (%)	8	15	25
Germination % after 2 months storage	72	48	52
Germination % after 1 year storage	37	38	12

This study indicates that only mature seed – collected when their colour has changed to purple – germinate at relatively high rates and can be kept viable in storage for a considerable length of time. Collection of green or purple-green fruits should be avoided. High germination rates can only be obtained with depulped seed. This is likely to be due to (so far unidentified) physical or chemical germination inhibitors present in the pulp.

To maintain viability for as long as possible, depulped seed should be stored in a cool environment, with approximately 15% moisture content. Care should be taken when preparing *P. africana* seed for storage because they are very susceptible to fungal attack. An appropriate broad-spectrum fungicide should be applied to the seeds before storage. Experiments are currently ongoing to determine germination rates for seeds that have been in storage for longer than one year.

2. Vegetative propagation studies

Although experiments to extend the storage life of *P. africana* seed go some way toward meeting the demand for germplasm, vegetative propagation methods, such as cuttings, air-layering or grafting, are also useful approaches for the convenient provision of material. Furthermore, vegetative methods have an important application for the conservation of specific genetic information present in endangered localities, where trees are reproductively isolated or physiologically immature and cannot set seed. Through cloning of individuals and production of trees on farms nearby, this genetic information can be conserved locally. Testing the steps described below or similar approaches are a start in determining suitable propagation methods for other threatened tropical tree species. The reader is referred to the very informative book by Hartmann *et al.* (1997) for general information about plant propagation.

i) Rooted stem cuttings

In order to develop the optimal procedure for rooting stem cuttings, a sequence of experiments was undertaken with juvenile plant material in Yaounde, Cameroon.

Single-node cuttings were harvested from the main stem of one-year old stock plants. In each sub-experiment one factor was tested:

* substrate
* leaf area
* the application of the plant hormone indole-3-butyric-acid (IBA).

The standard rooting substrate was a 1:1 sand-sawdust mixture, the standard leaf lamina area was 20 cm^2, and the standard IBA concentration was 50 microgrammes (µg) in a 10 microlitres (µl) droplet of industrial methylated spirit, which was applied to the base of the cuttings with a micropipette. This method ensured that all cuttings regardless of size or 'wettability'received the same amount of IBA. Before inserting the cuttings into a closed frame ('non-mist propagator', Leakey *et al.*, 1990), the methylated spirit was evaporated in a stream of cold air. Thirty cuttings were used for each treatment, replicated three times. The experiment was set under approximately 60% shade provided by black polyester shade netting.

a) Effects of different rooting substrates

Three rooting substrates were tested:

* washed river sand
* rotted sawdust
* and a sand-sawdust mixture.

Three weeks after setting, root development started in the sand and in the sawdust substrates. Cuttings in the sand-sawdust mixture started rooting in week four. At the end of ten weeks, the sand treatment was clearly inferior to the other two treatments (Table 20.3).

Table 20.3 **Effect of substrates on rooting success of *P. africana* (Hook.f.) Kalkman ten weeks after setting. Standard deviation (SD) = 12.69**

Substrate	Sand	Sawdust	Sand : sawdust (1:1)
Rooting (%)	68	84	78

b) Effects of leaf area

Templates cut from graph paper were used to determine leaf areas of 0, 5, 10, 20 and 25 cm². Leaf area was found to be crucial in rooting success. No leafless cuttings rooted during the six-week period of the experiment, indicating clearly that at least some leaf area is necessary for the survival of *P. africana* juvenile cuttings. At the end of the experiment, cuttings with leaf areas of 20 and 25 cm² had significantly higher mean number of roots than cuttings with smaller leaf areas. This may relate to the leaf's role in the production of carbohydrates and auxins.

Data are consistent with findings in Kenya, where cuttings with leaf areas of 40–50 cm² increased rooting success by a further 30%. Rooting dropped to almost zero when cuttings had leaf areas of above 51 cm² (Figure 20.1, taken from Ndeti, 1999; Table 20.4). An average single leaf has an area of 35–45 cm².

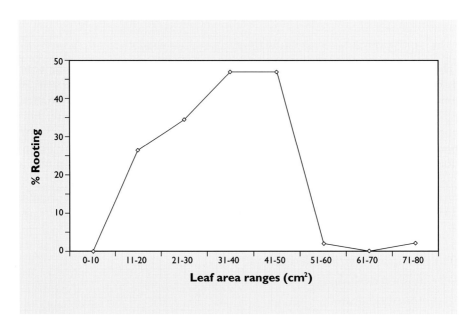

Figure 20.1   Effect of leaf area on rooting percentage of *P. africana* cuttings. Summary of cuttings treated with 0–80 µg indole-3-butyric-acid (IBA) (Ndeti, 1999).

Table 20.4 Effect of leaf area on average root numbers of *P. africana* (Hook.f.) Kalkman six weeks after setting. Standard deviation (SD) = 2.96

Leaf area (in cm²)	5	10	20	25
Root numbers	5	7	11	14

c) Effects of IBA concentrations

Six concentrations of IBA were chosen for this experiment: 0, 50, 100, 150, 200 and 300 µg per cutting.

The stimulatory effects of IBA treatments on rooting were already apparent at the first assessment in the third week of the experiment. Rooting was significantly faster with the application of IBA concentrations of 100–150 µg. By the tenth week, 88% of cuttings treated with 100 µg IBA had rooted, while the value was 82% for those treated with 150 µg IBA, and only 74% for cuttings treated with the highest concentration of 300 µg IBA. Only 39% of cuttings treated with no auxin had rooted (Table 20.5). Application of 50 µg IBA improved rooting compared with the control by a factor of two. In experiments in Kenya (Ndeti, 1999), rooting was already doubled by applying as little as 10 µg IBA to the cuttings, with a plateau at higher concentrations.

Table 20.5 **Effects of indole-3-butyric-acid (IBA) on rooting success in *P. africana* (Hook.f.) Kalkman after ten weeks. Standard deviation (SD) = 7.2**

IBA (µg)	0	50	100	150	200	300
Rooting (%)	39	82	88	82	87	74

ii) Air-layering

During air-layering, branches, usually 1–2 cm in diameter, are ringbarked, the wounded area covered with moist moss or sawdust, wrapped in plastic to prevent desiccation, and left to root for several weeks. This method is particularly relevant for the propagation of mature trees, which do not easily root from stem cuttings.

Initial experiments with air-layering of *P. africana* in different substrates have shown root development of 80% of layers within five weeks in a peat-based substrate with the application of approximately 300 µg IBA in a slurry of Seradix3® (0.8% IBA in talc). This treatment also led to the development of stronger roots and best survival rates of layers after potting (>70%).

Practical Issues and Costs

In order to carry out successful propagation studies, a well-organised nursery with a few non-mist polypropagators (Leakey *et al.*, 1990; Longman and Wilson, 1993) is necessary. Each 1 × 3 m polypropagator is made from locally available materials (timber and polyethylene sheets), costing about US$100. A well-trained full-time nursery technician is essential. If a reliable power supply is available, a refrigerator should be used for the short-term storage of seed.

Lessons Learnt

Although *P. africana* seed germination is initially high, long-term seed storage results in a significant loss of viability. However, using appropriate collection and storage practices, approximately 35% germination can be obtained after one year in storage, which is high enough to make the efforts involved in such storage worthwhile. Studies on vegetative propagation from leafy cuttings of juvenile material by leafy cuttings indicate that high rooting success can be achieved by using a rooting substrate of rotted sawdust or sand mixed with rotted sawdust, leaving one leaf on the cutting, and applying a moderate amount of rooting hormone. Air-layering is also a promising technique that requires further development before it may be applied in order to multiply planting material from older trees. In the west of Cameroon and in other parts of Africa where farmers have suffered from a lack of seed, vegetative propagation techniques offer a solution for cultivation. Both seed storage and vegetative propagation studies are still in their initial stages and ongoing research will allow methods to be further refined.

Future Steps

Future on-farm experimentation will show the feasibility of the propagation methods developed here. Additional experiments will focus on the proper management of rooted layers during air-layering, and on grafting. The latter has shown promising results in pilot trials. Reports of pest and disease problems, especially in the nursery stage, have also to be evaluated.

A monograph on the current status of knowledge on the ecology, production, management and economics of *P. africana* is currently being prepared by the University of Bangor (UK) in collaboration with the Mount Cameroon Project, ICRAF and others.

Acknowledgements

The authors would like to acknowledge that this paper resulted from research funded through grants from several agencies, including the UK Department for International Development (DFID) and the Danish International Development Agency (DANIDA).

References

Arnold, J.E.M. and Dewees, P.A. (1995). *Tree Management in Farmer Strategies: responses to agricultural intensification.* Oxford Science Publications, Oxford University Press, Oxford, UK.

Barker, N.P., Cunningham, A.B., Morrow, C. and Harley, E.H. (1994). A preliminary investigation into the use of RAPD to assess the genetic diversity of a threatened African tree species: *Prunus africana*, 221–230. In: B. Huntley (ed.). *Botanical Diversity in Southern Africa.* National Botanical Institute, Pretoria, Republic South Africa.

Cunningham, A.B. (1994). Integrating local plant resources and habitat management. *Biodiversity and Conservation* **3**: 104–115.

Cunningham, M., Cunningham, A.B. and Schippmann, U. (1997). *Trade in* Prunus africana *and the Implementation of CITES.* German Federal Agency for Nature Conservation, Bonn, Germany.

Dawson, I.K. and Powell, W. (1999). Genetic variation in the Afromontane tree *Prunus africana*, an endangered medicinal species. *Molecular Ecology* **8**: 151–156.

Ewusi, B.N., Tako, C.T., Nyambi, J. and Acworth, J. (1997). Bark extraction: the current situation of the sustainable cropping of *Prunus africana* on Mount Cameroon, pp. 39–54. In: G. Davies (ed.). *A Strategy for the Conservation of Prunus africana on Mount Cameroon.* Technical papers and workshop proceedings. Limbe Botanic Garden, Limbe, Cameroon, 21–22 February 1996.

Hartmann, H.T., Kester, D.E., Davies, F.T. and Geneve, R.L. (1997). *Plant Propagation: Principles and Practices.* 6th edition. Prentice Hall International, Upper Saddle River, New Jersey, USA.

Kalkman, C. (1965). The old world species of *Prunus* sub-genus *Laurocerasus. Blumea* **13**(1): 33–35.

Leakey, R.R.B.*et al.* (1990). Low-technology techniques for the vegetative propagation of tropical trees. *Commonwealth Forestry Review* **69**(3): 247–257.

Longman, K.A. and Wilson, R.H.F. (1993). *Rooting cuttings of tropical trees.* Tropical Trees: Propagation and Planting Manuals, Volume 1. Commonwealth Science Council, Commonwealth Secretariat Publications, London, UK.

Martinelli, E.M., Seraglis, R. and Pifferi, G. (1986). Characterization of *Pygeum africanum* bark extracts by HRGC with computer assistance. *Journal of High Resolution Chromatography and Chromatography Communications* **9**: 106–110.

Ndeti, J.N. (1999). *The Status of* Prunus africana *in Kakamega Forest and the Prospects for its Vegetative Propagation.* MPhil Thesis. Moi University, Eldoret, Kenya.

Sunderland, T. and Nkefor, J. (1997). Conservation through cultivation. A case study: the propagation of pygeum – Prunus africana. *Tropical Agriculture Association Newsletter December 1997*: 5–12.

Chapter 21

Conservation through Cultivation:
the work of the Limbe Botanic Garden, Cameroon

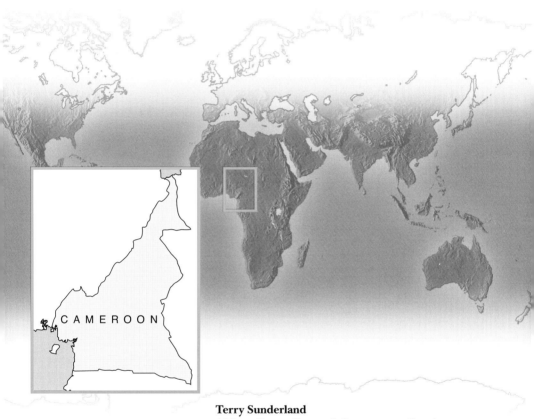

CAMEROON

Terry Sunderland
African Rattan Research Programme, Royal Botanic Gardens, Kew, UK

Paul Blackmore
Limbe Botanic Garden, Limbe, Cameroon

Nouhou Ndam
Limbe Botanic Garden, Limbe, Cameroon

Joseph Nkefor
Limbe Botanic Garden, Limbe, Cameroon

Introduction

The Limbe (formerly Victoria) Botanic Garden, Cameroon, was established in 1892 by the German colonial administration with the direct objective of introducing exotic tropical plants of economic and medicinal potential, to support the economy and population of the new colony of 'Kamerun'. These initial objectives of the Limbe Botanic Garden reflected the predominant role of colonial botanic gardens at that time. Indeed, through the networks emanating from these botanic gardens, the economic development of the tropics was significantly changed, precipitating global exchange and movement in plant material on a scale that had never been previously witnessed, and that subsequently changed the economic nature of the world (Hobhouse, 1999).

Today, the activities of botanic gardens have changed greatly. The recognition of more holistic concepts in which plants, animals and people play an integrated role has become the driving force behind many botanic garden activities. Indeed, the increasing destruction of tropical forests and other natural habitats, coupled with the loss of culture and displacement of indigenous peoples, has resulted in massive reductions of many plant, and animal species, as well as the loss of the intrinsic indigenous knowledge relating to these species. In addition, the fact that areas of high biodiversity also correspond to gene pools and centres of diversity for important crops and medicinal plants has focused the energies of many botanic gardens towards biodiversity monitoring and conservation with particular emphasis on highlighting the intrinsic relationship between plants and people.

Complementary to these activities is the role that botanic gardens are playing in the development of policies controlling the exploitation of plant resources as well as through coherent conservation strategies. The Limbe Botanic Garden is attempting to address these very issues, particularly in the Mount Cameroon region.

The historical role of the Limbe Botanic Garden

The Limbe Botanic Garden is situated between the foothills of Mount Cameroon, an active volcano, and the Atlantic coastline. The prevalent tropical climate and highly fertile volcanic soils offer excellent growing conditions for most tropical crops. The German settlers recognised the agricultural potential of the coastal region and converted large areas (up to 1,000 km^2) of the lowland forest around Mount Cameroon to large-scale commercial plantations. The establishment of the Botanic Garden was essential to this activity, and being situated close to the coast, the garden acted as a port of entry and centre for the introduction, propagation, acclimatisation and distribution of a wide range of crops. Lowland cash crops, such as cocoa (*Thedoroma cacao* L. (Sterculiaceae)), rubber (*Hevea brasiliensis* (A. Juss.) Muell.-Arg. (Euphorbiaceae)), sugar cane (*Saccharum officinarum* L. (Gramineae)), pepper (*Piper nigrum* L. (Piperaceae)) and banana (*Musa* L. spp. (Musaceae)), became vital to the development of the colony and still play a major role in the economy of Cameroon today. In addition to exotic crops, further selection and breeding work was carried out on a number of important indigenous species, most notably the African oil palm, *Elaeis guineensis* Jacq. (Palmae), and wild rubber species such as *Funtumia elastica* (Preuss) Stapf and *Landolphia* Pal. spp. (both Apocynaceae) – resources that were essential to European industrial development.

Whilst contributing to this early economic development, the Limbe Botanic Garden was also the base for the scientific study of the flora of the area and an herbarium was established in the Garden. This provided the basis for many later taxonomic studies as well as contributing to the understanding of the use and economic value of the local flora in particular. The Garden also established the first agricultural school in West Africa where Cameroonians were able to train in agriculture, forestry and horticulture.

The biological importance of Mount Cameroon

The main massif and foothills of Mount Cameroon cover an area of 1,500 km². Mount Cameroon is the highest mountain in West Africa (4,095 m) forming part of an isolated chain of volcanic uplands (Figure 21.1). The annual rainfall of the area ranges from 12,000 mm to the south-west of the main massif to 2,000 mm to the north-east. Mount Cameroon provides a biological mosaic that contributes to extremely high speciation and biological diversity that is centred in the Guineo-Congolian regional area of endemism, one of the main Pleistocene refugia postulated for Africa (Gartlan, 1989; White, 1983). The massif and surrounding foothills contain around 3,000 higher plant species with 49 of these endemic to Mount Cameroon itself (Cable and Cheek, 1998). This is the last area in Africa where natural vegetation remains unbroken from lowland forest at sea level to the sub-alpine grassland at the summit with all the vegetational gradients in between. Since 1800AD, at least ten volcanic eruptions have created a vegetation succession mosaic of various ages and over a range of altitudes. The wildlife of the region is also extremely rich with at least two endemic birds, three endangered primates, one of which, Preuss' guenon (*Cercopithicus preussii*) is endemic, as well as a small population of forest elephant *Oxodonta africana* (Stuart, 1986).

Threats to the forest

With rapid population growth and continued migration to the area around Mount Cameroon, clearance of natural vegetation to provide land for subsistence farming has become the biggest threat to the forest. With the

Figure 21.1   Mount Cameroon.

majority of the lowland areas already converted to commercial plantations, the pressure on the remaining areas of fragmented forest for farmland has increased as most families maintain a 'chop farm' mainly producing annual and semi-perennial crops for subsistence use.

Both timber and non-timber forest products (NTFPs) are indiscriminately exploited at levels that are unsustainable. As a result, the availability of many economically important plants has decreased significantly in the past decades. The collection and sale of products, such as honey, medicinal plants, vegetables and herbs and spices, from farm-fallows, forest, and the montane grassland, contributes greatly to the household incomes of the area. In the upper villages on Mount Cameroon itself, at least half of the households surveyed by the project in 1991, are involved in selling forest products. Although the marketing of many of these products often follows a complex chain of traders and transporters, the resulting comparatively small cash income, is important. Hunting for bush meat is also a major activity around farmland and in the forest. Although some bush meat is consumed at the subsistence level, it is also an important source of cash, supplying a lucrative market in Douala, over 80 km away, and other urban centres.

Conservation Initiatives on Mount Cameroon

The Man and the Biosphere Conference in Sweden in 1972 identified Mount Cameroon as one of the top 20 conservation priorities in the world. Later, in 1982, the United Nations Food and Agriculture Organisation (FAO) produced a highly favourable report on the potential of the Limbe Botanic Garden, which at that time was in a state of considerable neglect, to address local conservation and development issues. However, it was not until 1988 that the British Overseas Development Administration (ODA) (now the Department for International Development (DFID)) advised by the Royal Botanic Gardens, Kew (RBG Kew), signed a bilateral aid agreement with the Government of Cameroon (GoC) to form a project with the objective of rehabilitating the Limbe Botanic Garden whilst contributing to conservation efforts on the mountain. Thus the Limbe Botanic Garden and Rainforest Genetic Conservation Project was created.

Until its completion in 1993, this project concentrated on the restoration and development of the Limbe Botanic Garden and Herbarium, and acted as a focal point for conservation efforts for two important identified areas of biodiversity: Mount Etinde and the lowland forest of Mabeta-Moliwe to the east of Limbe. Experience gathered during this initial project suggested that the need for local participation in the conservation and development of the Mount Cameroon area was essential. This led to creation of the greater Mount Cameroon Project, which became operational in 1994; a multilateral initiative funded by the ODA, GoC, the German Technical Co-operation (GTZ) and the World Bank's Global Environmental Facility (GEF). The main aim of this project is to conserve the unique biodiversity of the mountain with the support of the local people. The Limbe Botanic Garden is a component of this greater Mount Cameroon Project and, as such, has an invaluable role to play in the conservation and development of the area.

The current role of the Limbe Botanic Garden

Since 1988, the role of the Limbe Botanic Garden has been to establish itself as a centre of science, education, conservation and recreation and since then, great physical, structural and institutional improvements have been achieved. As an integral component of the Mount Cameroon Project, the Limbe Botanic Garden has made a significant contribution to the Project by providing a strong resource base and technical expertise in the areas of biodiversity surveys, monitoring, environmental education, database management, training and plant domestication. Through these activities, Limbe Botanic Garden has helped to support the activities of the Mount Cameroon Project in the development of systems for the sustainable management of the rich biodiversity on Mount Cameroon and the surrounding region (Ndam, 1994).

> **Box 21.1　The work of the Limbe Botanic Garden in the context of national and global legislation**
>
> As a Technical Unit within the Ministry of Environment and Forestry, the Limbe Botanic Garden has a responsibility to operate within the context of national and global legislative requirements. The following examples illustrate this:
>
> - The cultivation and development of rattans is recommended by Project 59 of the National Forestry Action Programme of Cameroon (1995).
> - Replanting (through forest enrichment) of *Prunus africana* (Hook.f.) Kalkman (Rosaceae) is stipulated by the Cameroon Forestry Law of 1994: Article (88) 2.
> - The Convention on Biological Diversity (CBD), which Cameroon ratified in November 1993, is also adhered to by Limbe Botanic Garden (see Appendix 1). With the stated objective of the CBD being, *'the conservation of biological diversity, the sustainable use of its components, and the equitable distribution of benefits deriving from this use.'*
>
>   Limbe Botanic Garden also complies with the following specific articles:
>
>   ...*'inventories and monitoring of biodiversity'* (Article 7);
>
>   ...*'the implementation of ex situ conservation'* (Article 9), and expanded research;
>
>   ...*'training and education in relevant areas'* (Articles 12 and 13).
>
> - The Cameroon National Forestry Law (no. 94/01), which was promulgated in October 1994, also broadly encompasses these desired objectives and Articles 13 and 56 relate directly to the role of Limbe Botanic Garden with regard to *'sustainable exploitation of forest products'* and *'ex situ conservation'* respectively.
> - In 1993, *P. africana* was included in the Convention on International Trade in Endangered species (CITES) Appendix II, meaning that, in theory, the export of *Prunus* bark is controlled at both the source and destination points. Currently, there is no Scientific Authority for Plants in Cameroon, meaning that current export licenses are issued without reference to actual estimates of sustainability. Limbe Botanic Garden has applied to become the Scientific Authority for Plants in Cameroon and, if successful, would be more effective in influencing the currently unsustainable nature of exploitation.

Since 1993, there has been a growing awareness of the long-term potential of the Limbe Botanic Garden as a centre for research, training, conservation and biodiversity monitoring for the national, regional (Central and West Africa) and international scientific communities (Box 21.1). However, with the understanding that donor funding is not infinite, and the recognition of the need to develop a robust financial base, recent attempts at fulfilling these objectives have led to the development of a Biodiversity Conservation Centre at Limbe Botanic Garden.

The Biodiversity Conservation Centre will focus the ability of the institution to attract funding and wider institutional support through direct commercial activities, including the provision of technical services, consultancies and the undertaking of mutually beneficial collaborative research. To provide the optimum environment for the development of the institution, there is a need for Limbe Botanic Garden to have further institutional autonomy from the GoC in terms of the management of both financial and human resources. To this end, a draft decree has recently been prepared and is currently under discussion within the GoC.

i) The role of scientific research

Both scientific and sociological research has provided a greater insight into the natural resources of Mount Cameroon, their utilisation, and management, and the dissemination of research findings remains a priority for the institution. However, it is essential that research activities continue to contribute to the base-line information necessary for the achievement of the management objectives of the wider project. Also, a more formal approach to the issue of research, in terms of establishing protocols for collaboration with outside agencies, is currently being established.

ii) Living collections

The living collections inherited in 1988 were, in the main, undocumented and accordingly of little scientific value. Many of the dominant relic tree species were either of common taxa that were planted for shade, such as the ubiquitous *Terminalia catappa* L. (Combretaceae), or exotic weed species that were able to establish themselves during extended periods of the garden's decline, such as *Cedrela odorata* L. (Meliaceae) and *Leucaena leucocephala* L. (Leguminosae). Development of these inherited collections has entailed establishing a number of planting themes with significant emphasis placed on the local use of plants by people. Since 1988, over 5,000 new accessions have been planted according to these themes. Examples include a field of useful products, local and exotic medicinal plants, fruit trees (Asian and South American), and collections of important agricultural crops, such as land races and species of *Musa*. Such collections provide an invaluable educational resource as well as an aesthetically pleasing backdrop to their scientific function. Today, the living collections are well documented, with the plant records being held on a database, and most new accessions are of wild stock.

A number of ecological plantings have also been established at Limbe (or been allowed to develop naturally) reflecting the attempts of many other botanic gardens to recreate natural landscapes and ecosystems. Within the living collections at Limbe is an area of mangrove colonisation (*Avicennia germinans* L. (Verbenaceae)), which is being allowed to develop naturally, an aquatic plant collection, an example of lava flow colonisation and an area of semi-natural forest. Such emphasis on ecological displays by botanic gardens, examples of which are becoming difficult to find in their natural state, reflects the growing concern about rapid man-made environmental changes that have accelerated significantly in the past fifty years. Such areas also provide an interesting and interactive educational resource.

Figure 21.2 The nursery at Limbe Botanic Garden.

iii) *Ex situ* conservation

The Limbe Botanic Garden is focusing its *ex situ* programme not only on threatened taxa, such as West African orchids, ferns, *Cola* Schott & Endl. spp. (Sterculiaceae), and Madagascan palms. Selection of target taxa for the *ex situ* programme is based more broadly on their biological diversity, level of threat and suitability for cultivation (Figure 21.2). The selection focus is also on major economic species of the region. The exploitation of many of these species is contributing to species-level resource depletion, as well as to overall forest degradation. For example, Africa's largest collection of pan-African and Mascarene palms, with particular emphasis on Madagascan taxa (which are under severe threat in their natural range) is housed in Limbe. Limbe is an ideal repository for the conservation of Madagascan palms. The national botanic garden of Madagascar, Tsimbazaza, is situated at an altitude of over 1,000 m above sea level and therefore does not provide ideal growing conditions for many of the lowland forest palms of Madagascar, many of which are endangered.

The Mount Cameroon region is the centre of diversity for both *Diospyros* L. (Ebenaceae) and *Cola* spp. *Cola*, in particular, is an extremely important genus from both an economic and cultural perspective and 50% of the known species are represented within the living collections. The collections of African orchids and ferns are also comprehensive and represent the high diversity found in the Mount Cameroon region.

More recent developments include a rattan arboretum, containing over 60% of the rattans of Africa, and plantings of a number of species of *Raphia* Pal (Palmae). These collections are not only important scientifically, but through interpretation have also proved extremely popular with visitors and students alike.

The Conservation through Cultivation Programme

The Conservation through Cultivation Programme is perhaps the most important contribution of the Limbe Botanic Garden to the conservation and development of the Mount Cameroon region, and provides a useful model for other such initiatives in the tropics. As a result of uncontrolled exploitation, the degree of threat to a NTFPs is acute, yet little research has been undertaken on determining appropriate methods for the sustainable extraction or production of such products. The primary aim of the Conservation through Cultivation Programme is to conserve wild populations of NTFPs by developing community-based cultivation systems, reducing the harvesting pressure on those resources, whilst ensuring that the benefits of exploitation accrue to local communities. In order to achieve this, efforts so far have concentrated on identifying the optimum method of bulk propagation for each target species for the purposes of replanting and reintroduction (see Appendix 4) for direct management by local communities. The type of cultivated system, used for these species depends on the autecology of the species concerned. For example, the majority of tree species are introduced to farm-fallows and multi-tiered agroforestry systems, whilst others, such as the delicate climbers, *Gnetum* L. spp. (Box 21.2), are cultivated in intensive 'market gardens'.

The objectives of the Conservation through Cultivation Programme are achieved through the following process:

- the development and promotion of cost-effective and scientifically sound methods of germplasm collection;
- the development of simple and effective methods of bulk propagation;
- the development and promotion of cost-effective and transferable methods of cultivation, especially within existing land-use patterns, such as farm-fallows;
- establishing accessible means of extension to allow local communities to benefit directly from the research;
- establishing a regular monitoring and follow-up programme with extension officers.

Box 21.2 Conservation through Cultivation Programme case study: *Gnetum africanum* Welw. & *Gnetum buchholzianum* Engl. (Gnetaceae)

Gnetum in Africa is represented by two, almost identical, species: *G. africanum* and *G. bouchholzianum*. These are slender climbers, found in both secondary and closed-canopy forest, the leaves of which provide the source of the important forest vegetable known locally as 'eru'. They are harvested destructively by cutting the base and then pulling the vine from the tree canopy. Once harvested, they are wrapped into bundles and transported throughout Southern Cameroon, Nigeria, and as far as the United Kingdom and North America.

Eru was originally eaten by the Bayangi people from the south-west province of Cameroon. The development of roads and the displacement and migration of people during and after the colonial period has resulted in an increased demand. In order to satisfy local, national and now international demand, the harvesting pressure on eru has increased to unsustainable levels and in many areas both species are locally extinct (Nkefor *et al.*, 1999).

Currently all the eru that is traded on a commercial scale, is being harvested directly from wild stocks, with the main source coming from the Yaounde region, 400 km from Limbe (Ngatoum and Bokwe, 1994). The harvesting, transportation and processing of eru supports a considerable economy providing employment and income to thousands of people. Although the demand for eru is high in Cameroon, over half of the eru harvested is transported directly to Nigeria, via Idenau, where it is either consumed by the domestic population or further exported to the west for the expatriate West African population.

Work schedule for Limbe Botanic Garden:

- Early in 1994, through funding from the Earth Love Fund (Oxford), germplasm of *Gnetum* was collected and experimental propagation trials were started. Problems with propagation were solved by mid-1995, with vegetative propagation being the most efficient means of multiplication, and large scale production started.

- By mid-1995, one intensive, commercial trial farm, and one agroforestry style planting was established within the Garden.

- In 1996, the Garden obtained funding from North Carolina Zoological Park for the collection of a wide range of natural source material to be used in the establishment of the first *Gnetum* gene bank in Africa. This is an important tool for the conservation of a wide genetic base, and for selection of superior provenances.

- October 1998 – first harvest and tasting trials. This provided essential data on annual yield and quality of the cultivated product when compared with wild harvested material.

- December 1998 – funding was obtained from CARPE (Central African Regional Programme for the Environment) for the training of 40 farmers and establishment of four trial farms in Mount Cameroon region.

- January 1999 – funding was obtained from CARPE for the production of a 'State of Knowledge' document for the species.

- March 1999 – production of a basic eru propagation and cultivation manual and extension leaflets for local farmers.

1. Selection of target taxa

The selection of species involved is based on existing local demand and the level of threat, rather than to fulfil the recommended development of new markets for forest products. As such, the programme has strict principles for the selection of target species. These are based on the following social, institutional and biological criteria:

i) Demand from the local population

In the past, many attempts at encouraging local farmers to improve upon existing agricultural practices in the area were from a 'top down' perspective. Certainly, in the Mount Cameroon region, this approach was limited in its success. As the Botanic Garden began to have a higher local profile in the early 1990's, local farmers groups began to lobby the Garden for access to its stock material of high-value NTFPs, as these were over-exploited in the wild and not traditionally cultivated. In general, both these farmers groups and the wider rural communities, in which they operated, lacked both the technical capacity and the local organisation to undertake the cultivation of these high-value NTFPs without support. Many saw the application of a focused scientific approach to addressing plant-related problems as highly relevant in this respect. The species originally targeted for this programme were those most vigorously requested by local people; hence the approach was, and continues to be, distinctly 'bottom up'.

ii) Wider policy recommendations

Over-exploitation of forest products, particularly those of enough value to contribute to the formal forest economy, eventually draws the attention of policy-makers. A number of these products, such as *P. africana* (Box 21.3; Figure 21.3) and rattans (Box 21.4) have been identified as targets for cultivation activities and, being a government institution, the Limbe Botanic Garden has a responsibility to adhere to policy recommendation at both the national and global levels. Despite the claims of many, such policy is not always in contrast to the needs of local communities. For example, the cultivation of rattans is recommended by Project 59 of the National Forestry Action Programme of Cameroon (1996) and the replanting (through forest enrichment) of *P. africana* is stipulated by the Cameroon Forestry Law Article (88)2 of 1994 (see Box 21.1). Both these resources were identified by many local communities in the Mount Cameroon region as high priorities for cultivation.

iii) Requests from collaborating partners

In respect to the above criteria, other institutions have also developed an interest in the cultivation of high-value NTFPs. For example, the International Council for Research in Agroforestry (ICRAF) focuses on the introduction of tree crops to farming systems and the Limbe Botanic Garden has provided planting stock for the establishment of a field gene bank of *P. africana* (see Jaenicke et al., Chapter 20). In addition, the Cameroon Development Corporation (CDC), responsible for the commercial agriculture around Mount Cameroon, is seeking to increase its resource base.

Box 21.3 Conservation through Cultivation case study: *Prunus africana* (Hook.f.) Kalkman (Rosaceae)

The bark of the montane trees species, *Prunus africana*, is used for the treatment of benign prostate hyperplasia, and has a current market value of around US$220 million per annum (Cunningham *et al.*, 1997). To date, the exploitation of *Prunus* bark is undertaken solely from the wild and is concentrated primarily in Cameroon, Kenya, Madagascar, Tanzania and, to a lesser extent, the Democratic Republic of Congo. More recently significant amounts of raw and macerated bark have been exported from the island of Bioko, Equatorial Guinea. *P. africana* is a common component of the montane forests of Mount Cameroon.

Plantecam, a subsidiary of a French pharmaceutical company, exploits around 2,000 tonnes of bark from Mount Cameroon, supplied by their own workers as well as independent contractors. Previous guidelines that were said to ensure sustainable harvesting have broken down and there has been a significant increase in uncontrolled and illegal exploitation of *P. africana* in recent years, and such exploitation has proved extremely destructive to wild populations. In many cases, to remove the maximum quantity of bark, trees are stripped entirely, or even felled (Figure 21.7). Many of the local farmers on Mount Cameroon will retain individuals of *Prunus* during clearance operations yet do not actually cultivate the species, as there were considerable biological constraints to doing so. These include unpredictable reproductive events (masting) and the presence of germination inhibitors in the pericarp. In the early 1990s, many of these farmers expressed real interest in planting the species if provided with suitable stock. Initial work on the optimum means of bulk propagation concentrated on seed production and was extremely successful, leading to the establishment of a number of cultivated trials and incorporation into agroforestry systems:

- In collaboration with the International Council for Research in Agroforestry (ICRAF), a 5 ha living gene bank of *P. africana* was established at Tole. This gene bank is maintained and monitored by Limbe Botanic Garden and ICRAF (see Jaenicke *et al.*, Chapter 20).

- An experimental 8.8 ha plantation of *P. africana* has been established in Moliwe by the Cameroon Development Corporation. This plantation is monitored by Limbe Botanic Garden.

- A further 1.5 ha plantation of *P. africana* has been established by a 'Women in Development' co-operative in the North-West Province of Cameroon. In addition, 200 assorted timber tree species were also supplied for an agroforestry programme with the same co-operative (see Burnley, 1999).

- One thousand *P. africana* seedlings were supplied to a 'Women in Development' co-operative in Fako Division, South-West Province, Cameroon. This material was distributed by the co-operative to many women for small-scale planting on farm-fallow areas and crop association programmes.

- Further *P. africana* material was supplied to the Mbalmayo Forest Research Division to undertake further trials in vegetative propagation.

Through the provision of planting stock by the Limbe Botanic Garden, plantations of *P. africana* and a rattan, *Laccosperma acutiflorum* (Becc.) J. Dransfield (Palmae), have been established. Progress of these plantations is monitored and evaluated by Limbe Botanic Garden.

Figure 21.3 Nouhou Ndam (Limbe Botanic Garden) inspecting a *Prunus africana* (Hook.f.) Kalkman (Rosaceae) trunk showing bark removal.

iv) The taxa must be an indigenous species of economic or cultural importance

The programme focuses on species native to West and Central Africa and does not work with the introduction of exotic species to community agricultural systems.

v) The taxa must be harvested directly from the wild

Species that have been domesticated have already had the preliminary research on their ecological and cultural requirements undertaken, either implicitly by local people, or explicitly by another research centre and, as such, are not appropriate for inclusion into the Conservation Through Cultivation Programme. It is important to stress that duplication of effort (and often competition for funds) is a particular problem between scientific and conservation-orientated institutions in Cameroon and this programme is aimed at original research that will not needlessly be duplicated elsewhere, nor duplicate the existing efforts of others.

vi) The taxa must be considered threatened by extreme harvesting pressure

There must be evidence that harvesting is leading to biological threat as well as affecting future marketable supplies. Initial priority is given to those taxa that are locally threatened, but globally threatened taxa, such as *P. africana*, are also included in the programme.

Box 21.4 Conservation through Cultivation case study: African rattans

Rattans are climbing palms, the flexible stems of which are used for furniture manufacture and basketry. The global trade in rattan is valued at some US$6.5 billion/annum (ITTO, 1997) with the majority of this trade being concentrated in South East Asia. In the lowland forest areas of Africa, a proportion of the 20 or so known species of rattan contribute greatly to indigenous subsistence strategies as well as to a thriving cottage industry. However, this exploitation, based on an 'open access' approach, has led to significant local scarcity, particularly around urban centres of production. As one of the highest value non-timber forest products in the region, rattan has long been requested by local communities for inclusion into agroforestry systems (Ndoye, 1994; Sunderland, 1999a, b) and was one of the priority species requested by many communities in the hinterlands of Mount Cameroon. In collaboration with the African Rattan Research Programme, initial efforts aimed at bulk propagation have concentrated on seed production and addressing the complicated dormancy exhibited by many of the species (Sunderland and Nkefor, 2000). This work has led to the following:

- The establishment of an *ex situ* rattan collection in Limbe Botanic Garden. Over 60% of the known species are represented and this display provides an invaluable educational and scientific resource.

- In collaboration with the Cameroon Development Corporation and the African Rattan Research Programme, a one hectare silvicultural trial of *Laccosperma acutiflorum* (Becc.) J. Dransfield (Palmae) has been planted in association with obsolete rubber. This trial was planted in 1998 and initial growth rates have been encouraging.

Future cultivation efforts with a number of species of rattan will concentrate on enrichment planting of logged forest and the incorporation of rattan into agroforestry systems. As a climber requiring arboreal support, rattans are well suited to this multi-strata approach to their cultivation.

The species selection process is augmented by market studies indicating the local demand and the rate of consumption of certain NTFPs. In addition, discussions with local harvesters via consultations with Garden staff and Mount Cameroon Project geographical officers, who are in continued close contact with villages and farmers groups in the region, also provide considerable insight into the market dynamics of the NTFP trade.

2. Conservation through Cultivation methodology

The implementation of the Conservation through Cultivation Programme is undertaken through following a structured research strategy for each target species. The primary research methods are discussed below.

3. Preliminary research

The preliminary research entails an extensive literature search to obtain as much available information as possible on the target taxa and provides the necessary background for the programme to implement a coherent research strategy for each species selected. Particular attention is paid to taxonomic data, abundance and distribution, ecological and socio-economic aspects of the species. This preliminary research is undertaken at Limbe Botanic Garden, which houses an excellent natural resource library, and in close collaboration with partner institutions, such as the RBG Kew.

An eco-geographic survey is also undertaken to determine distribution and abundance of each species. This involves preliminary study of the Botanical Research and Herbarium Management System (BRAHMS) database (see Pearce and Bytebier, Chapter 4) held in Limbe, that holds some 9,600 botanical records from Mount Cameroon, as well as wider studies of herbarium records held in Yaounde and other herbaria. This is augmented by further and more detailed inventory data held at the project, particularly of the Mount Cameroon region.

This preliminary study of the target taxa aims to produce both a clear taxonomic description of the species concerned and an understanding of the geographical distribution and ecological variation. The studies also include an assessment of the phenology of the taxa to ensure that seed collecting trips can be planned as accurately as possible.

4. Field collection of germplasm and sampling method

At these early stages in the cultivation process, the emphasis is placed on obtaining as wide a representation for conservation, of the species concerned, rather than beginning the process of selection for desired traits, although preliminary selection of superior genotypes is undertaken during the monitoring stage. For each species, the initially aim is to obtain the largest sample of the gene pool possible. The important principle here is that random samples are taken from the entire geographical range of the population, taking care to target as many ecotypes as possible. For example, in the case of *P. africana*, consideration is given to its distribution over distance and altitude as within these two geographical parameters are different ecological conditions giving rise to a wide variety of ecotypes.

Great care is taken to maintain the collections separately during the domestication process. Detailed collection data (provenance-related) is obtained during germplasm collection; the information collected must consist of geographical data (altitude, longitude, latitude, soil type, ground conditions, aspect and vegetation type) as well as data concerning the collectors, i.e. collection date, collectors' names, etc. This information is vital to both the conservation and the scientific value of the material.

5. Germination tests and propagation trials

Propagation and multiplication trials are carried out to identify the most cost-effective methods of mass production. Seed dormancy is one of the greatest problems to the programme with, for instance, many rattan species often taking up to nine months to germinate. It is possible to overcome seed dormancy by using a range of pre-sowing treatments, however the experimental process takes a considerable amount of time. Often, as in the case of *Gnetum* spp., vegetative propagation through stem cuttings is the most appropriate means of

Figure 21.4   Propagation trials in the nursery at Limbe Botanic Garden.

multiplication as the seed biology of these species has proven to be somewhat complicated. The converse was the case with *P. africana*, with seed propagation giving 80–90% germination rates, whilst stem cuttings resulted in a <10% rooting rate. Experimentation with the various types of propagation and husbandry regimes (e.g. the influence of shade, variations in growing media) (see Jaenicke *et al.*, Chapter 20) entails a substantial amount of structured trial and error before the most appropriate and transferable methods of propagation are determined. Each trial is monitored on a weekly basis. The records are maintained in a spreadsheet format so that treatments per species under each husbandry regime can be easily compared (Figure 21.4)

6. Distribution and monitoring

The success of the Conservation through Cultivation Programme depends on effective plant distribution and post-planting monitoring. The majority of the plants produced are made available to local communities at a minimum cost. From experience, it has been found that plants given away for free are often not maintained or valued in the same way as when they are purchased directly. No matter how small the investment, when a plant is purchased it is highly valued and is often well maintained. The sale of plant material also helps Limbe Botanic Garden to recover a small proportion of the investment costs of the propagation and cultivation trials.

In many cases, distribution of plant material is undertaken through the Mount Cameroon Project village network as well as through the existing extension services, farmers' co-operatives and, more recently, non governmental organisations (NGOs). A condition of plant distribution through these agents is that Botanic Garden staff monitor the growth and development of the plants distributed.

7. Characterisation and evaluation of germplasm

Characterisation is the observation of characters that are highly heritable and that are maintained in a range of environments. Such characters are important as they determine the qualities of the germplasm, both on a genetic level and on an exploitable level (i.e. high quantities of alkaloid activity for medicinal species; good nutritional value for edible species, high growth rates for rattan). Evaluation data is the observation and measurement of how the germplasm interacts with the new environment. Such characters are important as they indicate the survival or loss of certain genotypes and their characteristics in different conditions (Ford, 1986) and this rationale forms the basis of the monitoring programme for target taxa by Limbe Botanic Garden.

Once wide representation of each species has been obtained through broad sampling, farm trials are established and the preliminary selection for desired characteristics is undertaken. This has worked well with short-lived crops, such as *Gnetum* spp., and the material now being cultivated by communities is both high-yielding and fast growing; the result of considerable experimentation and selection at the Botanic Garden nursery. The provision of a wide range of provenance material of *P. africana* by Limbe Botanic Garden has also led to the establishment of the field gene bank by ICRAF (Figure 21.5). This trial is undergoing continued monitoring for the selection of superior provenances that are both relatively fast growing and for high levels of active ingredient present in the bark (there is a vast amount of variation in the active ingredient in the bark of *P. africana*, particularly between populations (Barker *et al.*, 1994)).

Figure 21.5 Joseph Nkefor, of Limbe Botanic Garden, with seedlings of *Prunus africana* (Hook.f.) Kalkman (Rosaceae).

8. Publication, dissemination and extension

Dissemination of the research findings of the Conservation through Cultivation Programme has taken place through formal scientific meetings (Nkefor *et al.*, 1999; Sunderland and Nkefor, 1997). Informal presentations of the work at the Botanic Garden to farmers' groups and extension agents have also proved to be a good means of dissemination and interaction with our target audience.

Once appropriate and transferable methods of bulk propagation and cultivation have been determined for each target species, it is necessary to impart this newly acquired knowledge to those able to implement it. One of the most important contributions of the programme has been the training of trainers and extension workers in propagation methods. This training has taken place both on a formal and an informal basis and has entailed the training of university/professional students in horticulture and related subjects, including eight students from the Regional College of Agriculture, Bambili, the Forestry School, Mbalmayo, and the University of Buea. Field staff from the Ministry of Environment and Forestry (MINEF) have also been trained. Recently, the need for a full-time extension officer based in the garden has been identified. An important principle for the employment of this officer is that while being, she/he will maintain their link with the National Extension Program at the Ministry of Agriculture, although he/she will be based at the garden, and operating within the Mount Cameroon region. This linkage will provide a route for the transfer of products and technologies to a wider, indeed potentially national, audience.

Social and Environmental Benefits

1. Benefits for local people

In summary, this programme has led to the following benefits for local people:

- the supply of plant material for which there is a guaranteed income from well-established and stable markets;

- the economic empowerment of local people by providing the means by which they can enter the lucrative markets of products, such as *Gnetum* spp. and *P. africana*, through supply. This has traditionally been denied them through the uncontrolled exploitation of such material from the wild, predominantly by outside parties (Figures 21.6 and 21.7);

- an increased diversification of products grown by both individuals and groups leading to greater crop diversification and hence economic security;

- in many cases, a guaranteed supply of raw material ensures that price fluctuations are minimised. Hence, those crops bought in markets by the majority of urban people (*Gnetum, Cola* spp., etc.), remain affordable and within the price range of the majority of families;

- traditionally in South West Province, it has been forbidden for women to own land directly. However, with the formation of women's co-operatives, they have been able to purchase or obtain land for long-term cultivation (Burnley, 1999). In Cameroonian common law, planting an economic tree on a piece of land implies tenure over that land. The Conservation through Cultivation Programme has enabled the empowerment of women, through the supply of economic species;

- the use of Limbe Botanic Garden as an educational resource has been highlighted; benefiting both visitors and recipients of current literature.

2. Environmental benefits

- The primary long-term environmental benefit is that harvesting pressure on selected wild populations of high-value NTFPs is reduced with alternative supplies hopefully coming from cultivated sources. The benefits are not only confined to individual taxa but also through the maintenance of ecological integrity. For example, the continued felling of *P. africana* in the wild to exploit the maximum bark yield causes large and discontinuous gaps in the forest, affecting dynamic processes.

Figure 21.6   Bundles of *P. africana* (Hook.f.) Kalkman (Rosaceae) bark being carried out of the forest.

- Individual taxa are protected from the serious levels of endangerment by over-harvesting through *ex situ* conservation efforts.

- Planting trees on otherwise fallow areas has had a direct impact on the amount of reforestation in South West Province, both at a commercial and subsistence level. Greater potential yields per hectare are anticipated through the low input/high output systems encouraged.

- Mixed cropping with tree species and herbaceous agricultural crops with a multi-storey canopy is far closer to natural ecological processes. This has a direct positive impact on soil erosion, with a corresponding increase in species/ha thus encouraging a more balanced, albeit, impoverished, ecosystem.

- Planting of such species within the designated themes of Limbe Botanic Garden highlights the institutional policy of presenting the inextricable link between plants and people.

Lessons Learnt

Initial problems have centred on the technical question of determining the optimum method of bulk propagation for each target taxa. To solve these problems, variations in approach are developed, i.e. if seed propagation is a problem, clonal propagation is adopted and *vice versa*. A rigorous approach to experimental design and recording of results is encouraged.

The distribution of plant material has been hampered by the fact that many local farmers groups, NGOs and even extension agencies have no access to transport and are unable to move plant material from the garden nursery to the planting site. However, the DFID component of the Limbe Botanic Garden has provided significant support with this.

Threats to the success of the Conservation through Cultivation Programme

There are two main threats to the long-term success of the programme. Firstly, although during the preliminary study stage, current and future markets are evaluated as far is practicable, the demand for current target taxa due to changes in market requirements cannot necessarily be anticipated when encouraging the development of tree-based cultivation systems that may take perhaps 15–25 years before the economic benefits are realised. Of particular concern is the case of *P. africana* where it is feasible that alternative remedies for the treatment of benign prostate hyperplasia might render the currently marketed *Prunus* – based remedies obsolete.

The second main threat is concerned with the future institutional aspects of Limbe Botanic Garden itself. At some point in the future, DFID will relinquish its current financial responsibility for the institution and the programmes that are implemented by it. It is essential that if the full benefits of the Conservation through Cultivation Programme are to be felt long-term, then equally long-term alternative sources of funding, managed within a well-resourced institution, must be identified. Current attempts at establishing Limbe Botanic Garden as a regional Biodiversity Conservation Centre will hopefully address this concern.

Additional Comments and Proposed Next Steps

The next step of the programme is to continue to identify suitable species to strengthen and augment the work undertaken so far and to use the expertise gathered for the benefit of other threatened species. Equally important, is the monitoring of the existing plantings, especially in plantation situations, to determine the viability of such an approach. A monitoring programme has been established close to the Botanic Garden and is at the implementation phase. For many of the target taxa a cost-benefit analysis should be undertaken to determine the economic viability of the cultivation of such products.

It is important that this work is viewed within the context of research, education and development, rather than as conservation *per se*. The entire *ex situ* programme of the Botanic Garden, including the Conservation through Cultivation Programme, has been identified to meet human needs and to provide solutions to developmental issues related to plants. This consideration has provided the immediate criteria behind species selection and subsequent development of a conservation programme, as opposed to the consideration of purely scientific criteria. In this respect, the methods and activities of the Conservation through Cultivation Programme have potential applications for similar institutions in the tropics whose objectives are focused on the intrinsic link between people and plants and providing plant-based solutions to real development problems.

Acknowledgements

The Earth Love Fund (Oxford) provided initial funding for the launch of this programme in 1994. The Overseas Development Administration (now DFID) of the British government has funded the programme since then. The North Carolina Zoo and the Central African Regional Programme for the Environment (CARPE) should also be acknowledged for financial assistance particularly with the work on *Gnetum* spp. Heartfelt thanks are also extended to the Friends of the Limbe Botanic Garden for their active participation in establishing the trials of *Gnetum* and to the Cameroon Development Corporation (CDC) and the International Council for Research in Agroforestry (ICRAF) for long-term collaboration on trials of *P. africana*. Finally, the nursery staff of the Limbe Botanic Garden should also be thanked for their relentless efforts to ensure this programme is a continuing success.

Figure 21.7   Bundle of *Prunus africana* (Hook.f.) Kalkman (Rosaceae) bark.

References

Barker, N.P., Cunningham, A.B., Morrow, C. and Harley, E.H. (1994). A preliminary investigation into the use of RAPD to assess the genetic diversity of a threatened African tree species: *Prunus africana*, 221–230. In: B. Huntley (ed.). *Botanical Diversity in Southern Africa*. National Botanical Institute, Pretoria, Republic of South Africa.

Burnley, G. (1999). The role of women in the promotion of forest products, pp. 139–142. In: T.C.H. Sunderland, L. Clark and P. Vantomme (eds). *The Non-Wood Forest Products of Central Africa: current Research Issues and Prospects for Conservation and Development*. Food and Agriculture Organisation (FAO), Rome, Italy.

Cable, S. and Cheek, M. (1998). *The Plants of Mount Cameroon: a conservation checklist*. Royal Botanic Gardens, Kew, UK.

Cunningham, H., Cunningham, A.B. and Schippman, U. (1997). *Trade in* Prunus africana *and the implementation of CITES*. German Federal Agency for Nature Conservation, Bonn, Germany.

Ford, B.L. (1986). *Plant Genetic Resources: an introduction to their conservation and use*. Edward Arnold Publishers Ltd., London, UK.

Gartlan, S. (1989). *La Conservation des Ecosystèmes forestières du Cameroon.* IUCN Programme pour les Forêts Tropicales. IUCN, Gland, Switzerland.

Hobhouse, H. (1999). *Seeds of change: six plants that changed mankind.* Macmillan Publishers, London, UK.

Ndam, N. (1994). The rehabilitation and development of the Limbe Botanic Garden, Cameroon. In: Huntley, B. (ed.). Botanical Diversity in Southern Africa. *Strelitzia* 1: 305–30.

Ndoye, O. (1994). *New employment opportunities for farmers in the humid forest zone of Cameroon: the case of palm wine and rattan.* Paper prepared for the Rockefeller Fellow Meeting, Ethiopia, Addis-Ababa. November 14–18.

Ngatoum, D. and Bokwe, A. (1994). *Rapport de la Mission: Effectuée autour du Mont Cameroon Relatif au Recensement de Certains Especes des Produits Forestiére Secondaires en Voie de Disparation.* Unpubl. report for MINEF. Government of Cameroon.

Nkefor, N.P., Ndam, N., Blackmore, P.C. and Sunderland, T.C.H. (1999). The conservation through cultivation programme at the Limbe Botanic Garden: achievements and benefits, pp. 79–86. In: T.C.H. Sunderland, L. Clark and P. Vantomme (eds). *The Non-Wood Forest Products of Central Africa: Current Research Issues and Prospects for Conservation and Development.* Food & Agriculture Organisation (FAO), Rome, Italy.

Stuart, S.N. (ed.) (1986). *Vegetation in the montane forests of Cameroon.* International Council for Bird Preservation, Cambridge, UK.

Sunderland, T.C.H. and Nkefor, J.P. (2000). *Technology transfer between Asia and Africa: rattan cultivation and processing.* Central African Regional Programme for the Environment (CARPE), Washington, DC, USA.

Sunderland, T.C.H. (1999a). The rattans of Africa, pp. 227–236. In: R. Bacilieri and S. Appanah (eds). *Rattan cultivation: achievements, problems and prospects.* CIRAD-Forêt & FRIM, Kuala Lumpur, Malaysia.

Sunderland, T.C.H. (1999b). New research into African rattans: an important NWFP from the forests of Central Africa, pp. 87–98. In: T.C.H. Sunderland, L. Clark and P. Vantomme (eds). *The Non-Wood Forest Products of Central Africa: Current research issues and prospects for conservation and development.* Food & Agriculture Organisation (FAO), Italy, Rome.

Sunderland, T.C.H. and Nkefor, J.P. (1997). Conservation through cultivation. A case study: the propagation of pygeum – *Prunus africana. Tropical Agriculture Association Newsletter. December 1997*: 5–12.

White, F. (1983). *The Vegetation of Africa.* UNESCO. Paris.

Chapter 22

Peasant Nurseries:
a Concept for an Integrated Conservation Strategy for Cycads in Mexico

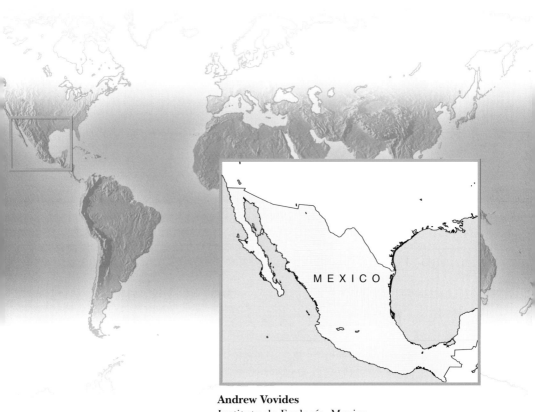

Andrew Vovides
Instituto de Ecología, Mexico

Carlos Iglesias
Instituto de Historia Natural de Chiapas, Mexico

Miguel Pérez-Farrera
Instituto de Investigaciones Biológicas de la
Universidad Veracruzana, Mexico

Marivo Vázquez Torres
Instituto de Investigaciones Biológicas de la
Universidad Veracruzana, Mexico

Uwe Schippmann
Bundesamt für Naturschutz, Bonn, Germany

Introduction: Status and Trade of Cycads

There are approximately 200 species and 11 genera of cycads distributed amongst three families (Stevenson *et al.*, 1995; Norstog and Nicholls, 1997). They are found within tropical and subtropical areas, with Mexico a global centre of diversity holding about 40 species in three genera: *Ceratozamia* Brongn., *Dioon* Lindley and *Zamia* L. (all Zamiaceae). Many of these species are threatened by over-collection and habitat destruction and are in need of *in situ* conservation.

Over 80% of the 40 species of Mexican cycads are endemic, and 12 are classified as Endangered (E), 12 as Vulnerable (V) and the remaining as Rare (R), or Indeterminate (I) using the World Conservation Union (IUCN) 1981 categories (Vovides *et al.*, 1997). They are threatened largely by habitat destruction for agricultural expansion and illegal extraction from the wild for trade. Cycads, with their palm-like habit, are popular ornamental and landscaping plants and the international trade worldwide generates several million dollars annually. Although cultivated material for a few species is available in large quantities, for instance the Japanese species, *Cycas revoluta* Thunb. (Cycadaceae), the demand for wild-collected material exists for other species, for instance the Mexican species of *Zamia*. Several thousand cycads were exported out of Mexico during the 1970s and 1980s (Gilbert, 1984; Vovides, 1986), with over 40 tons of adult *Zamia furfuracea* Ait. per month being exported to the USA to satisfy the landscaping industry. Other Mexican species have been re-exported by the USA to South Africa and to a lesser extent to Europe (Gilbert, 1984). Other examples of unsustainable trade include the leaf crowns of large plants of *Dioon edule* (Lindley) Mex. which are removed from adult plants and sold in the major cities of Mexico (Figure 22.1), and the removal from Chiapas in 1985 of over 2,000 plants of *Ceratozamia norstogii* Brongn., that were then exported to the USA only one year after this species was described (Vovides, 1989).

Project History and Development

The cycad 'cottage industry' as a solution to unsustainable harvesting

Cycads, as trade items from the agricultural landscape, require conservation management within a socio-economic context. Issues such as land tenure, are important to project development and long-term success. For example, the Ejido system, put in place in Mexico in the 1920s during the agrarian reform following the revolution, assigns lands to local communities as common property.

Figure 22.1   Plant sellers with decapitated crowns of *Dioon edule* Lindley (Mex.) (Zamiaceae) for sale on the streets. Photo: Glafiro Alanis.

Mexico ratified the Convention on International Trade in Endangered Species (CITES) in March 1992 and created the Comisión Nacional para el Conocimiento y Uso de la Biodiversidad (CONABIO) which works very closely with the Convention on Biological Diversity (CBD) (Glowka *et al.*, 1994; Appendix 1). In May 1994, CONABIO published the first official list of threatened and endangered species (Anon., 1994), this being based, for plants, largely on published sources (Hernández and Godínez, 1994; Vovides 1981, 1988) and consultation with experts. Cycads are officially protected by law and at the time of writing, a cycad sub-committee has been created by the Mexican authorities which consists of a panel of experts that liase with the authorities on matters concerning cycad protection and conservation. Cycad biodiversity studies are still sporadic, but some inroads on sustainable management and population studies have been made (Pavón, 1999; Pérez-Farrera *et al.*, in press; Vovides, 1990; Vovides and Iglesias, 1994).

The need to establish cycad nurseries for the benefit of both the cycad populations and local villagers was in response to the constant poaching of threatened cycad species. The trade resulted in little financial gain for villagers who were paid small sums per plant by the poachers, and, ultimately, the trade was having a detrimental impact on the small and vulnerable cycad populations.

Project history

This project was started in the early 1990s by independent researchers of the Jardín Botánico Francisco Javier Clavijero of the Instituto de Ecología, A.C. (IE), and the Instituto de Investigaciones Biológicas of the Universidad Veracruzana (UV). It is hoped that this project will become a model for the 30 or so botanic gardens in Mexico. The villagers were invited to take part in local projects in which wild-collected seeds, harvested from community-owned Ejido lands, were propagated in low-investment and low technology nurseries. The project is an attempt to create awareness and an ownership value of the cycad habitat amongst the farmers. The prospect of being able to sell propagated plants has created an incentive for the farmers to protect the wild populations. These 'cottage industry' activities are designed to complement the farmers' traditional agricultural practices and are not intended to replace them. Thus, an incentive amongst the farmers to conserve has been initiated and reduction of collecting pressure in the wild has already been achieved (Vovides, 1997; Vovides and Iglesias, 1994). Local commitment has been demonstrated with, for instance, the cessation of illegal collecting in the Monte Oscuro area, after reports to the authorities by the farmers led to arrests and plant confiscation on two occasions.

These nurseries have great potential in the future as *in situ* Wildlife Management Areas (UMAS) under Mexican federal environmental law. These are permitted and registered with the environmental authorities (Secretaría de Manejo de Recursos Naturales y Pesca – Instituto Nacional de Ecología (SEMARNAP-INE)) providing that the following conditions are met:

- conservation of the natural habitat as a seed source
- reintroduction of nursery produced plants to compensate for seed removal from the habitat.

A number of institutions, such as IE, Instituto de Historia Natural de Chiapas (IHN) and UV, have assisted the development of the nurseries through supportive research and providing horticultural expertise on cycads. In the case of IE and UV, these institutions have over 20 years of research expertise on the biology, especially the demography and reproductive systems, of *D. edule*, *Ceratozamia mexicana* Brongn. and *Z. furfuracea*. For example, the Mexican National Cycad Reference Collection (registered with the Mexican Association of botanic gardens) is located in the botanic garden of the IE, established during 1980, and that now holds about 40 native Mexican cycad taxa.

The role of these nurseries are in accord with Article 9 of the CBD on *ex situ* conservation which states that '*ex situ* conservation should preferably be in the countries of origin' as this facilitates research, rescue and propagation of threatened species; regulates and names *ex situ* collections that do not threaten ecosystems and *in situ* populations (in our case the national cycad collection at the botanic garden); and the provision of finance for *ex situ* conservation especially in developing countries. The CBD also refers to public education and awareness (Article 13) as well as technical and scientific cooperation (Article 18).

Funding and project partnerships

By the end of 1995 a proposal to the Deutsche Gesellschaft für Technische Zusammenarbeit (GTZ) for a two-year project was approved for funding, and was eventually extended to three years. Before this, the project leaders raised funds from their individual institutions or other private and public sources. The Mexican partners are the IE, the Instituto de Investigaciones Biológicas (UV), both in Xalapa, and the IHN in Chiapas. The GTZ project included two nurseries in Veracruz assessed by Mario Vázquez of the UV and two nurseries in Chiapas assessed by Miguel Angel Pérez-Farrera of the IHN. Further funding for these nurseries in Veracruz and Chiapas has been obtained from various national and international sources including CONACYT for Cienega del Sur; local authorities and CONABIO for Monte Oscuro; Mexican Fund for Nature Conservation (FMCN), MAB-US, CONABIO; and Fauna & Flora International (FFI), UK, for the Chiapas nurseries.

Project Case Studies

1. Veracruz: Monte Oscuro Nursery

This nursery was established during 1990 with farmers whose lands contained *D. edule* habitats (Figures. 22.2 and 22.3). The project was based on field studies of cycad population ecology (Vovides, 1990) and propagation and cultivation experience in Mexican botanic gardens. Difficulties arose in securing initial funding for the project, with sporadic support being provided by the Jardín Botánico Franciseo Javier Clavijero and a private donation from Tim Gregory and Associates, a group of independent Californian cycad enthusiasts interested in conservation This nursery began with 24 members, although by 1992 only five members remained. Only the family heads (males) ran and tended the nursery on a weekly rota system, whereby one member attended the nursery once every five weeks. Co-ordination of more intensive work was supervised by project co-ordinators approximately twice a month. The nursery has produced approximately 15,000 plants of *D. edule*, varying in age from one to nine years, during the period 1990–1998. The project also involves the conservation of an 80 ha cycad habitat in tropical deciduous forest, and an experimental reintroduction of nursery-produced *D. edule* (Figure 22.4) (see Appendix 4).

Figures 22.2 and 22.3   The nursery at Monte Oscuro, Veracruz, for production of *Dioon edule* Lindley (Mex.) (Zamiaceae). Photo: Andrew Vovides.

Figure 22.4 Reintroduction of artificially propagated *Dioon edule* Lindley (Mex.) (Zamiaceae) seedlings from Monte Oscuro nursery into its natural habitat by Carlos Iglesias. Photo: Andrew Vovides.

Figure 22.5 The nursery at Cienega del Sur, coastal Veracruz, for production of *Zamia furfuracea* Ait. (Zamiaceae). Photo: Andrew Vovoides.

Monitoring following reintroduction is necessary, with the first year being the most critical. However, parallel studies on naturally dispersed seed of *D. edule* showed that during the first year there is a percentage of pre-germination mortality through predation and seed loss and, to a lesser extent, post-germination mortality through dehydration during the dry season (Pavón, 1999).

2. Veracruz: Cienega del Sur

Mario Vázquez Torres (UV) set up the *Z. furfuracea* nursery at Cienega del Sur (Fig 22.5) in 1991, based upon previous ecological studies by university students. This locality, on the Veracruz coast, was subjected to systematic extraction of cycads at a rate of 40 tons per month during the 1980s. Twenty-five members of different families developed the nursery, although by 1994 only nine members were involved, with the majority belonging to one family. At this nursery, men, women and children of the family are involved.

Approximately 28,000 plants have been produced at this nursery from wild-collected seed, with 2,500 first generation plants produced between 1998–1999. *Z. furfuracea* has a shorter life cycle than most other cycads and plants may take only three to four years to reach reproductive age. Approximately 2,000 five-year old *Z. furfuracea* were reintroduced into the coastal dune habitat adjacent to the nursery in 1997. The establishment of the Cienega del Sur Nursery has had a direct effect in the adjacent coastal region of Los Tuxtla during the last 2–4 years. Farmers have independently initiated similar nursery projects to propagate *Z. furfuracea*. Several of these new nurseries have linked up with the team from UV for assistance whilst others are still working independently. As a result, there has been unco-ordinated seed collection amongst the farmers, and to date an estimated 600,000 *Z. furfuracea* plants have been produced in the unsupervised nurseries in Los Tuxtla (M. Vázquez, pers. com.).

3. Veracruz: Tlachinola

During 1995, two nurseries were created at Tlachinola (Figure 22.6), a small peasant community in cloud-forest vegetation near the city of Xalapa, for the propagation of *C. mexicana*, based upon previous ecological studies (Sánchez-Tinoco, 1988; Sánchez-Rotonda, 1993). Approximately 4,700 cycads have been propagated from seed while the adjacent intact cloud-forest habitat has approximately 1,700. The population is distributed in the remaining patches of cloud-forest that survive on very steep slopes. The cycad population scattered amongst the remaining cloud-forest is estimated to cover at least 10 ha. The cloud-forest is under threat from firewood extraction and coffee production; the latter has led to the destruction of the understorey vegetation. It is hoped that the *in situ* conservation of cycads will prevent any further destruction.

Figure 22.6　　The nursery at Tlachinola, Veracruz, for production of *Ceratozamia mexicana* Brongn. (Zamiaceae). Photo: Andrew Vovides.

Members rotate working at the nursery on a weekly basis unless heavy jobs, such as plant bagging and maintenance, necessitate all members being present. Assessment is carried out by the team from the UV twice a month to ensure horticultural protocols and general nursery maintenance is being carried out.

4. Chiapas: La Sombra, El Camanario, San Andrés Quintana Roo, Tres Picos and Mount Ovando

During 1994, Miguel Angel Pérez of the IHN set up nurseries in Chiapas at La Sombra (Figures 22.7 and 22.8), El Camanario, San Andrés Quintana Roo and Tres Picos for the propagation of *C.* aff. *norstogii* D.W. Stev. and *Dioon merolae* De Luca, Sabato & Vázq. Torres. These nurseries are part of the Biosphere Reserve La Sepultura and were based upon ecological studies of *D. merolae* (Pérez, 1994). A further nursery on Mount Ovando, part of the Biosphere Reserve El Triunfo, was created for the propagation of *Ceratozamia matudae* Lundell and *Zamia soconuscensis* Schutzman, Vovides & Dehgan.

Originally, La Sombra was started with 36 members from several family nurseries. At present, there are 11 members operating these nurseries. At El Campanario, seven members owning seven properties have agreed to integrate into one nursery within the next three years. A total of 25,000 *C.* aff. *norstogii* and 1,000 *D. merolae* have been produced by La Sombra and 1,000 *D. merolae* by El Campanario.

Figure 22.7 Dr Knut Norstog, visiting scientist, examining *Dioon merolae* De Luca, Sabato & Vázq. Torres (Zamiaceae) seedlings at La Sombra. Photo: Andrew Vovides.

The nursery at La Sombra in the Sepultura Biosphere Reserve, Chiapas for the production of *Ceratozamia mirandai* (ined.) (Zamiaceae). Photo: Andrew Vodides.

5. Chiapas: Las Golondrinas

The Las Golondrinas nursery, El Triunfo Reserve, is run by farmers already receptive to alternative methods of agriculture, integrating subsistence farming with growing organic coffee for health food stores in Mexico, the USA and Europe (Figure 22.9). This activity is assessed by a team led by a Catholic missionary who gives courses on soil conservation, organic fertiliser and pesticide production. Twenty-six farmers began the nursery, and there are now 28 members. The nursery has produced 600 plants of *Z. soconuscensis,* 70 plants of *C. matudae,* both of which are endemic to this region of Chiapas. In addition, it has propagated indigenous understorey palms, 115 plants of *Chamaedorea oblongata* Mart. (Palmae), 225 plants of *Chamaedorea quezalteca* Standl. & Steyerm., and 390 plants of *Chamaedorea graminifolia* H. Wendl. One of the main members (Don Lorenzo) of the nursery was a field guide to the first author and Bart Schutzman during the discovery of *Z. soconuscensis* in 1985.

Figure 22.9   Collaborators and project partners for the nursery at Las Golondrinas on the El Triunfo Biosphere Reserve, Chiapas. *Ceratozamia matudae* Lundell and *Zamia soconuscensis* Schutzman, Vovides & Dehgan (both Zamiaceae) are produced here. From left to right: two nursery collaborators, Rigoberto Hernandez, Dr Knut Norstog and Jesus de la Cruz. Photo: Andrew Vovides.

Training and Marketing

Training

The training programme has focused on horticultural and monitoring skills. These have enabled participating farmers to provide monitoring data to guide harvesting levels, including: monitoring of cone and seed ripeness of the cycads (*Dioon* cones can take over two years to mature), seed collection, selection, and collection of germination and cultivation data (Pérez, 1994; Pérez and Vovides, 1997). In addition, basic practical horticultural training was given through regular talks and workshops. This commitment to sustained resource management is fundamental to the success of the project because only through sustainability of utilisation will the project succeed.

This is especially important in view of the threatened population status of the cycad species involved and is fundamental for the granting of CITES permits. In addition, talks have been given to the farmers and, in some cases, a guided tour of the botanic garden at Xalapa with an emphasis on the importance of cycads as a source of sustainable income. Formal courses in basic horticultural practice were provided for the farmers from the Cienega del Sur, Tlachinola and Monte Oscuro nurseries at the Jardín Botánico Clavijero in 1996. The UV held a similar course at the Cienega del Sur nursery in 1997. Both training courses covered aspects of practical horticulture, such as: soils, composting, seed collecting, selection and pre-treatment of seeds, containers, and use of artificial shade. The courses also provided opportunities for the peasant farmers to meet, exchange ideas and problem-solve. Nursery and collecting permits were discussed at a meeting organised during February 1996 by the Mexican permitting authorities (SEMARNAP, PROFEPA and INE) at Cienega del Sur and attended by members of the Monte Oscuro and Tlachinola nurseries.

Members have met to discuss the proliferation of unsupervised *Z. furfuracea* nurseries that have been established in the adjacent area of Los Tuxtla. Producers and relevant authorities have met to assign seed collecting areas and quotas for *Z. furfuracea* to each nursery, and to create an implementation committee for native species protection. Ten meetings with the farmers have taken place since 1995, including two visits from GTZ, GTZ-ProTrade and an independent Mexican marketing agency, Multicomercio Internacional. Exchange visits between the Chiapas and Veracruz nurseries and the production of a video to demonstrate nursery activities are planned.

Project Marketing

Further improvements in marketing and exports in order to increase the benefits to the farmers and habitat conservation are planned. Although, the intermittent production has been encouraging to the farmers, a full breakdown of each nursery's success is not possible until each nursery is independent and establishes regular sales. The oldest nursery, Monte Oscuro, has sold about 1,100 *D. edule* plants in the last four years, worth 15,000 pesos (US$1,930). These sales were mainly from the botanic garden shop of the Jardín Botánico Clavijero, with others sold at conferences, symposia and town fairs in Mexico. Two hundred plants were exported from this nursery to the USA during 1995, 500 to Germany in 1998, and a further 500 were sent to Germany in 2000. The nursery at Cienega del Sur sold about 40 plants at the town fair of Xalapa during 1996, 70 plants to Germany in 1998, 1,000 plants to the town authority of Coatzacoalcos to be used in urban plantings in June 1999, and a further 800 to Germany during January 2000. The more recently established Chiapas nurseries sold plants to the USA at a price of US$7 for a three year old plant. Chiapas has the advantage of maintaining a number of rarer and more sought after species. The 500 eight-year old *D. edule* exported to Germany represented the project at the International Trade Fair for Plants (IPM) at Essen, Germany during January 1999 and February 2000.

Marketing solutions are proposed through GTZ's subsidiary office, Protrade, that has approved a marketing assessment project with a horticultural and marketing expert (Reiner Schrage), an assigned Mexican based agent (Juan Carlos Andrade) and a private buyer under a Private Partnership Project (PPP). Their duties include contacting buyers, co-ordinating shipments with the GTZ-funded nurseries in Veracruz and Chiapas, obtaining national, CITES and phytosanitary permits, and co-ordinating the customs and freighting requirements. It is hoped that this agency and the nurseries will be self-sufficient as a sustainable management unit in the near future.

The PPP has inspired interesting potential buyers for *D. edule*, not just out of interest for the plant, but also for the conservation message carried i.e. production by farmers in order to conserve the cycad forest habitats in the tropics. Whilst sales have occurred, it is too soon to compare the income from cycad sales with income from other agricultural activities of the peasant communities. However, an example can be mentioned here: lemons, the main agricultural activity of the Monte Oscuro farmers, have slumped in price in recent years. A sale of around 100 cycads, at US$4 each, brought an income of almost 4,000 pesos for 1995, whereas no income came from the lemon crop during that period.

Discussion

The role of botanic gardens

Botanic gardens are playing an ever increasing role in the conservation, monitoring and collaboration with authorities regarding CITES (Akeroyd *et al.*, 1994). Botanic gardens should play a vital role in helping to ensure the survival of the world's remaining plant life by serving and reaching out to the wider community with a clear commitment to positive conservation action (Hopper, 1997). Botanic gardens can give technical advice to nurseries and to establish a significant link between the nurseries and reforestation or conservation areas. This is essential for plant propagation on a larger scale for reforestation or reintroduction. In spite of high financial aid given to reforestation projects in Mexico, the important role of botanic gardens has been underestimated with respect to this problem (Gómez-Pompa *et al.*, 1997).

The botanic garden has played an important role in giving training and technical assistance at all levels during this project:

- it has provided practical horticulture courses and ongoing assessment to farmers;
- it has facilitated the interchange of ideas and experiences between nurseries;
- Horticulture students from the Royal Botanic Gardens, Kew (UK) have already visited the Monte Oscuro nursery and students from the Orto Botánico of Naples have visited the Cienega del Sur nursery. This has contributed towards training of horticulture specialists in botanic gardens.

Conservation through sustainable utilisation

This concept has been highlighted by many conservationists and plays a major role in the framework of the CBD (see Article 10, Appendix 1). It has been postulated (and questioned by others) that putting an economic value to populations can create an incentive for their conservation. Despite being an elegant concept, there has been a lack of positive examples, especially in the field of over-utilised plant resources (however, for further discussion see Fleisher-Dogley and Kendle, Chapter 19; Sunderland *et al.*, Chapter 21; Clayton *et al.*, Chapter 23; Kamatenesi-Mugisha and Bukenya-Ziraba, Chapter 24). The cycad nursery projects in Veracruz and Chiapas can be regarded as an encouraging case study in which 'conservation through utilisation' has taken place. The prospect of being able to sell propagated plants has created an incentive for the farmers to protect the wild populations that they had previously been heavily over-collecting. The fact that non-official Ejido reserve habitats are being protected by the farmers themselves as well as strengthening conservation activities through propagation in the Biosphere Reserves of Chiapas is very encouraging. However, the project can only be regarded as a full success if the following conditions can be met:

- Further scientific monitoring and applied demographic research on the impact of seed removal on the wild populations and the establishment of reintroduced plants have to be continued to ensure the viability of wild populations and therefore the sustainability of utilisation; and
- Sales of propagated plants and financial revenues have to take place on a regular basis in order to create a circular process with regular new sowing, planting, selling and reintroduction.

Importance of previous studies to the project

To attain any level of sustainable management of cycads, or any other plant group, ecological studies must have been carried out prior to the start of the project to give an understanding of the demography and phenology of the species (Dehgan, 1983; Norstog, 1992; Pérez, 1994; Vovides, 1990, 1993). Studies on population structure, regeneration, phenology and germination trials have allowed the establishment of a workable germination protocol for use by farmers. In Costa Rica, a similar sustainable management project concerning the propagation of *Z. skinneri* Warsz. ex A. Dietr. was frustrated through insufficient knowledge of the species' lifecycle. It appears that under deep rainforest shade these plants only produce seed cones once every six years. However, it was later found that appropriate management in secondary forest with higher light levels in the understorey resulted in more frequent coning (Ocampo, pers. com.).

Importance of reintroduction studies and monitoring

Reintroduction trials were initiated during autumn 1997 in the Monte Oscuro nursery. Three hundred *D. edule* plants of different age classes (two to seven years) were reintroduced into three microhabitat plots (rocky substrate exposed to full sun, under tree shade in deeper soil and an intermediate area that was partly rocky with shade). This enabled us to determine the optimum plant age for reintroduction. Older plants represent a drain on nursery resources while younger plants may be more vulnerable when re-introduced. At the time of writing only 12 plants have perished (after two very exceptional droughts) and it appears that two year old *D. edule* plants can be safely reintroduced with a survival rate of over 80% after two years. Since the plants re-introduced are from seed of the same cycad population there is little chance of mixing the different populations. However, population genetic studies are contemplated in a separate research proposal. We also need to know the genetic variation of plants managed for each cycad population.

Project achievements

As a result of the project, the farmers have put aside about 80 ha of tropical thorn-forest at Monte Oscuro for conservation and have discouraged crown-lopping and removal of cycads. Conservation of some of the coastal habitat of cycads in the Cienega del Sur area has occurred. Nursery produced plants have already been introduced into habitat at both Cienega del Sur and Monte Oscuro. The nurseries in Chiapas, now running for three years, are situated within established Biosphere Reserve buffer zones.

One of the major problems has been the small number of sales in the propagated cycads in the early years of the project. This could have been compensated for if the registration necessary for commercialisation been set in place before the marketing evaluation had already been carried out. Registration was only carried out recently (1997–98).

The dependency of each farmer on the habitat must be consolidated to achieve meaningful conservation. The establishment of mother plants off-site will diminish the incentive for habitat protection. It is interesting to note that despite few sales of propagated plants having taken place, new nurseries outside of, but influenced by the GTZ project, have been established by other communities. Their establishment carries the risk that illegal harvesting of seeds and propagation of plants will be carried out without any scientific assistance or knowledge of sustainability thresholds obtained through close monitoring of the mother stock. These new projects have clearly benefited from the project experiences; they have improved their propagation techniques (e.g. more space for plants) and have assigned a sales manager at an early stage of establishment. This may have a detrimental effect in the long run through 'market flooding'.

Whilst cycads such as *Z. furfuracea* are apparently common in the international market and are propagated in countries outside Mexico, such as the USA and Africa (R. Schrage, pers. com.), the home market for cycads is virtually unexplored. It was observed that the well established, older nurseries, especially Cienega del Sur and to a lesser extent Monte Oscuro, have a tendency to expand on the range of species propagated by introducing and propagating cycad species not present in their respective regions. Gifts of seed from other nurseries in Mexico have helped fuel this practice. Occasionally, the introduced species grow better in the nurseries than in their native habitat (as is the case of *D. edule* in Cienega del Sur). This practice obviously contradicts the principle goal of the project to propagate, conserve and protect threatened plants growing close to the communities. In addition, it has been observed that a number of nurseries use species, such as *Z. furfuracea*, that quickly become reproductive and produce potential mother stock within 3–4 years. First generation *Zamia* seedlings were observed in Cienega del Sur that arose from propagated plants. This could break the intended link between the farmers and the wild populations and reduce the interest in maintaining each population and its habitat. Nevertheless, species like this may have to be incorporated into the nurseries to diversify production in order to generate a steady income over the years to offset the slow-growing species.

A more efficient marketing system is being worked out through the PPP and the link between the farmer, nursery and marketing is being solved; the newly created agency in Mexico will co-ordinate the GTZ financed nurseries for exportation of plants. The current legislation pertaining to nursery registration and permitting needs to be revised in order to prevent competition for limited seed resources from wild populations. Since the establishment of the Cienega del Sur nursery, there has been a proliferation of nurseries. Some are co-ordinated into a production unit 'Unidad de Productores de Zamia' (UPZ) by a local marketing agent, others are working alone. Permits are apparently being issued to these nurseries without any apparent previous ecological study or supervision and some of these nurseries may not have the plants growing naturally on their lands (we know of at least two). This may create competition (by collecting plants and seed off-site) with the farmers who are managing natural populations on their properties.

If successful, this cottage industry nursery model will need to be diversified to incorporate other products (orchids, bromeliads, palms, cacti, etc.) in order to reduce duplication with similar nurseries that may cause competition between farmers over a limited resource.

Opinion of project participants

Both the opinions of the farmers and the project leaders have been sought in order to review and assess the success and shortcomings of this project. The farmers have had a positive attitude towards establishing this project, and this has largely been achieved through a mutual exchange of ideas and communication. Perhaps the largest concern has been the low quantities of plants being sold. This is one of the reasons for nursery members leaving the scheme during the early months of nursery establishment. The remaining farmers have realised that although cycads are slow growing they are comparatively low maintenance, and do not perish as easily as agricultural crops, but actually gain value with age. In the El Triunfo Reserve where farmers were sensitised to forest conservation over ten years ago they are now cultivating organic coffee, and they embraced the idea of a nursery to produce local cycads and palms. Sporadic sales at Cienega del Sur and Monte Oscuro has encouraged the farmers to continue, and the visit of the marketing expert (Reiner Schrage) and the first exports related to the IPM fair has raised their expectations. The need for reintroduction of plants by the farmers was initially met with many questions, such as, 'Why cultivate potentially saleable plants just to put back into the wild?', but has since been resolved after discussions with the project leaders. Concern at Cienega del Sur about the proliferation of nurseries in the adjacent Los Tuxtla region has arisen, especially relating to competition for wild seed collecting. This problem still needs to be solved. A more efficient law enforcing system is needed as well as a modification to the terms and conditions by which nursery permits are issued.

It is felt that national markets as well as international markets should be investigated for potential sales. Cycads should be promoted for use in Mexican parks and gardens to contribute to a greater public knowledge and appreciation of this spectacular group of plants. More information also needs to be made available to schools and libraries in Mexico, such as the interactive CD-ROM on the cycads of Mexico (Gómez-Pompa *et al.*, 1994, 1999).

Recommendations for establishing similar projects

The GTZ-funded cycad nursery project is an ideal model for the concept of 'conservation through utilisation', and any such template will have to take into consideration the socio-economic idiosyncrasies of the participating communities. For example, at the Cienega del Sur nursery the whole farming family is involved with the day-to-day running of the nursery, especially the women and children. At Monte Oscuro the situation is reversed, with only the (male) head of each family involved with no participation by spouses or children.

Scientific monitoring and assessment of the impact of seed harvesting has to be carried out on a constant basis. Any utilisation without parallel scientific evaluation cannot be regarded as sustainable. This is also emphasised by the fact that all cycads are protected under CITES. In the case of exports of propagated plants from Mexico to other countries, the provisions of CITES have to be observed. For example, in the case of Appendix I taxa (*Ceratozamia*), the CITES authorities of both importing and exporting countries will only grant permits if they are convinced that the trade is 'not detrimental to the survival of the species' (Art. IV.2.a, CITES).

The following points are meant as guidance for establishing similar projects:

- To ensure the link between the product and the retention of natural habitat the project must ensure that the product can be utilised and that it cannot be easily propagated on off-site nurseries. It is, therefore, often better to select species that are slow growing (like *Dioon* species) that do not set seed frequently.

- The biology of the species, especially its reproductive system, demography and population status, has to be clarified by research well before utilisation begins.

- Site selection is important. Wild populations should be geographically close to the nursery as this lessens the need to maintain mother plants in the nurseries since seed collection from wild stock is an easier option.

- Land property rights and the understanding of socio-economic structures, such as the Ejido lands, are very important factors in conservation problems and their possible solutions.

- Propagation and horticultural protocols should be established beforehand and simplified for local management. Given their often long-term commitment to local communities, botanic gardens are an ideal agency to provide this expertise.

- The monitoring of the long-term effects of utilisation needs to be carried out on a constant basis as an integral part of the project.

- It is advisable to continuously assess the needs of project participants in terms of horticultural and conservation training and nursery administration to reduce frustration during the project. Contact should be established and maintained preferably by one person who is committed to the project and community, and who can talk on the same level as the farmers.

- The establishment of a marketing strategy early on in the project is essential, and the collaboration of specialists in horticultural marketing is necessary. This also leads to a need for further training and improved management skills in the nursery in order to produce high quality plants that meet the buyers' specifications. Again, botanic gardens are ideally placed to transfer necessary horticultural practices to the nursery.

- Extensive educational and conservation programmes need to be created for peasant communities that have similar resources in order to diversify the sustainable management concept to include cacti, orchids, palms, medicinal plants and other useful species found in natural forests. This avoids over-production of one single species leading to the flooding of markets and competition for wild seeds. Incorporating ecotourism activities into projects may be a possibility, especially when working within Biosphere Reserves.

- Continuous financing, especially in the early stages of nursery establishment, is essential for long-term project survival. The Global Environment Facility (GEF), administered by the World Bank, suggests at least 10–15 years investment into such projects (GEF, 1998).

Conclusions

The goal of this German-Mexican co-operative project has been to demonstrate how species conservation can be linked with the economic interests of local people. The establishment of the nurseries in village communities in Veracruz and Chiapas where the nurseries are dependent on long-term habitat conservation to ensure regular seed harvests from *in situ* mother stock is a tangible achievement. Parallel to this is the continued assessment of the effects of seed harvesting on populations and the reintroduction of nursery-grown plants into their natural habitat to compensate for this removal. The project has led to a decline in the poaching of plants and seeds in most of the nursery management areas, with project farmers reporting poachers to the police as they see the benefits of managing and protecting their mother stock. The farmers value the cycads as a renewable resource and have committed themselves to the long-term development and management of their nurseries. The scientific monitoring and technical assistance through the IE, UV and IHN staff has ensured that the utilisation of a highly threatened plant resource like Mexican cycads can be carried out in a cautious way that has (so far) not damaged the populations.

In rounding off, the experience gained over the last ten years in working closely with farmers, training them in new horticultural and management techniques and conservation concepts has been challenging. We learnt many aspects of the farmers' outlook and idiosyncrasies that were respected. What stood out most prominently was the mutual trust and belief in the project from the project leaders, collaborators and the farmers themselves. The project was not financed continuously over the ten years described and there have been lean years. However, the fact that the communities were visited occasionally during the thin years to continue assessment and to inspire encouragement gave credibility to the project leaders and collaborators.

Acknowledgements

The authors wish to thank various persons and institutions that have collaborated with this project over the years through finance to the institutions directly involved as well as to the peasant farmers themselves:

- GTZ-Germany project No. PN 93.2208.2-06.201 for the five nurseries in Veracruz and Chiapas;
- FFI project No. 96/64/15, FMCN No. B2-134;
- CONABIO project No. FB177A/C120/94;
- US Fish and Wildlife Service project No. BL99-004, MAB-US s/n, to IHN for the Chiapas nurseries;
- CONACYT projects No. D1308-N9201 and 29379N;
- SEP project No. 91-30-001-830;
- CONABIO project No. FB567/Q039/98 for the coastal nurseries of Veracruz and Tlachinola;
- Tim Gregory and Associates (private donation), CONABIO Project. No. B-140, SEDESOL-Solidaridad s/n for the Monte Oscuro nursery;
- Instituto de Ecología, A.C. Botanic Garden, Instituto de Historia Natural de Chiapas, Universidad Veracruzana for use of vehicles and installations.
- Fairchild Tropical Garden for the purchase of *Dioon edule* plants and their promotion amongst Garden Members as a conservation effort.
- Montgomery Botanical Center for encouragement and hosting the Cycad Specialist Group (IUCN/SSC).

Thanks are also given to Reiner Schrage for reading and commenting on the manuscript.

A special recognition is due from the authors to the farmers themselves for without their collaboration and enthusiasm there would not have been a project.

References

Akeroyd, J., McGough, N. and Wyse Jackson, P. (1994). *A CITES Manual for Botanic Gardens.* Botanic Gardens Conservation International, London, UK.

Anon. (1994). *Norma Oficial Mexicana NOM-CRN-001. ECOL/1994 que determina las especies y subespecies de flora y fauna silvestres, terrestres y acuáticas en peligro de extinción, amenazadas, raras y las sujetas a protección especial y que establece especificaciones para su protección.* Diario Oficial 16 de Mayo 1–25 México, D.F., Mexico.

Dehgan, B. (1983). Seed morphology in relation to dispersal, evolution, and propagation of *Cycas* L. *Botanical Gazette* **144**: 412–418.

GEF (1998). *Global Environment Facility: GEF evaluation of experience with conservation trust funds.* GEF/C.12/Inf.6. GEF Council, Washington, DC, USA.

Gilbert, S. (1984). *Cycads: status, trade, exploitation, and protection 1997–1982.* Traffic (USA), WWF, Washington, USA.

Gómez-Pompa, A., Heywood, V., Pattison, G. and Rzedowski, J. (1997). Sínopsis de la mesa 1: Formación de colecciones nacionales y áreas críticas para nuevos jardines botánicos. In: Los Jardines Botánicos y las Colecciones Nacionales. *Amaranto* **10**(3): 72–73.

Gómez-Pompa, A., Vovides, A.P., Ogata, N. and Gonzalez, J. (1994). *Las Cycadas de México.* (Interactive CD-ROM) Gestión de Ecosistemas, A.C. Mexico City, Mexico.

Gómez-Pompa, A. *et al.* (1999). *Cycads: the endangered living fossils.* (Interactive CD-ROM) Gestión de Ecosistemas, A.C. Mexico City, Mexico.

Hernández, H.M. and Godínez, A.H. (1994). Contribución al conocimiento de las cactáceas Mexicanas amenazadas. *Acta Botánica Mexicana* **26**: 33–52.

Hopper, S.D. (1997). Future plant conservation challenges, pp. 11–20. In: D.H. Tochell and K.W. Dixon (eds). *Conservation into the 21st Century.* Botanic Gardens Conservation International, London, UK.

Norstog, K. (1992). Cycad phenology. *Encephalartos* **32**: 11–13.

Norstog, K.J. and Nicholls, T.J. (1997). *The Biology of the Cycads.* Cornell University Press, Ithaca, USA.

Pavón, M. (1999). *Germinación y sobrevivencia de Dioon edule Lindley (Zamiaceae) en su habitat natural.* Bachelor thesis Facultad de Biología, Universidad Veracruzana, Xalapa, Mexico.

Pérez, M.A. (1994). *Estudio preliminar sobre germinación en semillas de espadaña Dioon merolae De Luca, Sabato & Vázq. Torres (Zamiaceae)*. Bachelor thesis, Instituto de Ciencias y Arte de Chiapas, Tuxtla Gutierrez.

Pérez-Farrera, M.A., Quintana-Ascensio, P.F., Salvatierra-Izaba, B. and Vovides, A.P. (2000). Population dynamics of *Ceratozamia matudai* Lundell (Zamiaceae) in El Triunfo Biosphere Reserve, Chiapas, Mexico. *Journal of the Torrey Botanical Club,* **127**(4): 291–299.

Pérez, M.A. and Vovides, A.P. (1997). *Manual para el cultivo y propagación de cycadas.* INE-SEMARNAP Mèxico City, Mexico.

Sánchez-Rotonda, G. (1993). *Reistro de la polinización por coleópteros en una población de Ceratozamia mexicana Brongn. (Zamiaceae).* Bachelor thesis, Facultad de Biología, Universidad Veracruzana, Xalapa, Mexico.

Sánchez-Tinoco, M.Y. (1988). *Estudio morfológico de una población de Ceratozamia mexicana Brongn. (Zamiaceae).* Bachelor thesis, Facultad de Biología, Universidad Veracruzana, Xalapa, Mexico.

Stevenson, D.W., Osborne, R. and Hill, K.D. (1995). The world list of cycads, pp. 55–64. In: P. Vorster (ed.). *Proceedings of The Third International Conference on Cycad Biology.* Cycad Society of South Africa, Stellenbosch, South Africa.

Vovides, A.P. (1981). Lista preliminar de plantas Mexicanas raras o en peligro de extinción. *Biótica* **6**: 219–228.

Vovides, A.P. (1986). Trade and habitat destruction threaten Mexican cycads. *Traffic (USA)* **6**: 13.

Vovides, A.P. (1988). Relación de plantas mexicanas raras o en peligro de extinción, pp. 289–302. In: O. Flores Villela and P. Geréz, (eds). *Conservación en México: síntesis sobre Verterbrados Terrestres, Vegetación y uso del Suelo.* Apéndice F, Instituto Nacional de Investigaciones sobre Recursos Bióticos, Xalapa, Mexico.

Vovides, A.P. (1989). Problems of endangered species conservation in Mexico: cycads, an example. *Encephalartos* **20**: 29–35.

Vovides, A.P. (1990). Spatial distribution, survival and fecundity of *Dioon edule* (Zamiaceae) in a tropical deciduous forest in Veracruz, Mexico. *American Journal of Botany* **77**: 1532–1543.

Vovides, A.P. (1993). Cycad conservation in Mexico: *Dioon edule* Lindl., a case study, pp. 365–369. In: D.W. Stevenson and K.J. Norstog (eds). *Proceedings of the Second International Conference on Cycad Biology.* Palm and Cycad Societies of Australia, Milton, Queensland, Australia.

Vovides, A.P. (1997). Propagation of Mexican cycads by peasant nurseries. *Species* **29**: 18–19.

Vovides, A.P. and Iglesias, C. (1994). An integrated conservation strategy for the cycad *Dioon edule* Lindl. *Biodiversity and Conservation* **3**: 137–141.

Vovides, A.P., Luna, V. and Medina, G. (1997). Relación de algunas plantas y hongos mexicanos raros, amenazados o en peligro de extinción y sugerencias para su conservación. *Acta Botánica Mexicana* **39**: 1–42.

Chapter 23

Sustainability of Rattan Harvesting in North Sulawesi, Indonesia

Lynn Clayton
Renewable Resources Assessment Group,
T.H. Huxley School of Environment,
Imperial College, London, UK

E.J. Milner-Gulland
Renewable Resources Assessment Group,
T.H. Huxley School of Environment,
Imperial College, London, UK

Agung Sarjono
Asosiasi Pengusaha Hutan Indonesia, Jakarta,
Indonesia

Introduction

Rattan, the pliable stems of forest palms, has attracted interest as a crop with great potential as a profitable and sustainable non-timber forest product, that might form the basis of integrated conservation and development programmes (Siebert, 1993). It is the most important of the minor forest products harvested in Indonesia (MacKinnon *et al.*, 1996) and its extraction from buffer-zone areas can be a source of income to local communities around protected areas. Indonesia's forests, where its rattan species grow, are disappearing at an alarming rate of 1.5 million ha per year (Walton and Holmes, 2000). Despite this, there is a lack of basic biological information on the taxonomy of rattans and their ecology, particularly under natural conditions. In this chapter, we describe an on-going project, funded by the Darwin Initiative, to assess the sustainability of rattan harvesting in North Sulawesi, Indonesia. As this project is on-going, we concentrate on discussing the data collection phase; the types of data collected, the reasons for collecting these data, and some of the issues involved in data collection. We present some preliminary results, and discuss the directions in which the work is continuing.

We aim to assess the sustainability of harvesting in a broad sense. On the ecological side, this includes assessing whether harvesting is causing rattan populations to decline to the point of local extinction or whether the populations can continue to survive and provide harvestable canes into the future. It also includes the effects of rattan harvesting on the forest ecosystem; for example, does it have adverse effects on species that rely on rattan fruits for food, and does it lead to a change in understory community structure that might cause a loss of biodiversity? On the economic side, we aim to assess whether current levels of rattan harvesting are likely to continue to provide a source of employment and revenues to all those involved in rattan supply, from collectors through to exporters, and whether there are any policies that could be instituted to improve the industry's prospects, whether they be controls on levels of wild harvesting or subsidies and loans to improve infrastructure. These objectives are particularly pertinent to Article 8 of the Convention on Biological Diversity (CBD) (see Appendix 1), concerning the management of biological resources important for the conservation of biological diversity within and outside protected areas, and to Article 10, concerning measures relating to the use of biological resources to minimise impacts on biological diversity.

The rattan industry in North Sulawesi grew rapidly in the mid-1980s to reach a harvest level of around 15,000 tonnes a year (Central Statistical Board of North Sulawesi, 1998). There was a dramatic drop in the amount of rattan produced in 1988–9 compared to the previous year. The Indonesian government banned the export of unfinished rattan from January 1989 in order to stimulate development of Indonesian rattan-based processing industries (Manokaran, 1990). MacKinnon *et al.* (1996) note that this led initially to falling revenues for

Indonesian collectors and cultivators, and suggest that the explanation of the large increase in production followed by a drop was that traders were selling large amounts in anticipation of the ban, following which production was cut back. Since then, rattan production in North Sulawesi province has remained high but variable. It is an important industry, employing an estimated 4,000 people in the province, and is the largest and most developed of the non-timber industries based in the province's forests (the others being the wild meat trade (Clayton and Milner-Gulland, 2000, and gold panning). As a potentially sustainable and lucrative forest-based industry, it could have an important role to play in any management plans for forest conservation, either on a large scale, or in extractive zones around reserves. This is particularly true because rattans require standing forest for support.

Rattans are spiny climbing Old World members of the family Palmae (Arecaceae). They belong to a large subfamily of the palms known as the Calamoideae (Uhl and Dranfield, 1987), all of which share a unique characteristic – the presence of overlapping reflexed scales on the fruit (Dransfield, 1992a). Some rattan species are single-stemmed, and harvest results in the death of the plant. Others are multi-stemmed ('clumped') and can be harvested continually. In clumped species, suckers are produced at the base of the original stem which develop into new stems. Some species also produce stolons up to 3 m in length (Dransfield 1992b). Stems are covered by spine-bearing leaf sheaths which detach when the stem is mature, leaving the stem bare. In most rattans aerial stems do not branch. Two whip-like organs are associated with climbing; the cirrus, an extension of the leaf apex, bears reflexed thorns and is present in many rattan species. Other species bear a flagellum, originating from the top of the leaf sheath opposite the leaf base. Grapnel-like spines anchor these whips to potential support trees. There are two main methods of flowering among rattans. In some genera, inflorescences are borne terminally, following a long period of vegetative growth; stems die after flowering although the clump continues to live (hapaxanthic flowering). In most genera however, after a juvenile period of vegetative growth maturity is reached; inflorescences are produced continually thereafter and flowering does not result in the death of the stem (pleonanthic flowering) (Dransfield, 1992b). Rattan stems do not increase in diameter with age as do dicotyledonous tree stems; rather there is a gradual increase in the diameter of the stem before it starts significantly to grow upwards; once the stem diameter has reached a maximum the stem then begins significant aerial growth (Dransfield, 1992b).

Data Collection

Assessment of sustainable practice requires consideration of a number of disciplines, most notably ecology and economics. However, it is rare for a single study to collect all these data types with the aim of integrating them into an overall assessment of sustainability. This paper focuses on a pilot study in the Paguyaman watershed, which is a major rattan-collecting area in North Sulawesi and where contacts with rattan collectors are already well established (Figure 23.1). This area is also of particular interest, as part of the Paguyaman watershed was formally gazetted as a nature reserve (31,000 ha) by the Indonesian government in July 1999, hence offering the potential for future buffer-zone rattan projects. Techniques for assessment of sustainable practice can be established based on the pilot study here, and can then be extended to other watershed areas.

Rattan identification

A key issue in the project was identification of the rattan species under investigation. The taxonomy of Sulawesi's rattans is very poorly known yet a sound taxonomic base is essential for a scientifically repeatable study, with uncritical use of local names causing considerable confusion. In this study we used local collectors' names as a first step to distinguishing species, but made extensive herbarium collections of each type including fruiting and flowering material for future taxonomic identification, following the methods of Dransfield (1986). An advantage of continuing to use local names alongside true species identifications is that the interactions between collectors and rattans are based on the types or taxa that they identify. If a type contains species that are significantly different in their biology, then similar harvesting pressure may vary significantly in its effects on the species' population dynamics.

Harvest rates in the Paguyaman watershed

Data were collected by observing the number of rafts passing a particular location on the Nantu river (the north-west branch of the Paguyaman river). Each raft is composed of bundles of harvested rattan, tied to a few wooden floats on which the collector stands. Field assistants living at the Adudu base camp collected these data every day on the Nantu river. This camp is located on the river bank; all rafts from upstream must pass this point and are clearly visible from the camp, making recording of numbers of passing rafts straightforward. A sample of rafts were stopped near the base camp and examined to give information on the type and quality of the rattans being collected, as well as on the average number of canes per raft. Collectors were

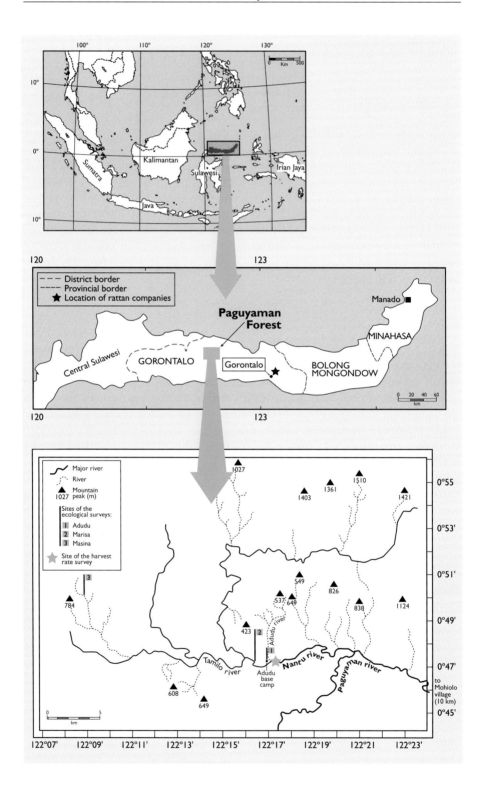

asked where they had been collecting, how long they had been in the forest for, and how many people were in their group. Each cane was examined and its diameter measured using a collector's ruler. A problem with this method is that rafts are composed of bare canes and these can only be readily identified to local names.

The number of rafts observed each day over the study period was combined with the information from the sampled rafts on the number and type of canes per raft to give an estimate of the offtake rate. This enables us to estimate the total amount of rattan of each type being harvested from the forest of the Paguyaman watershed upstream of the sampling location during the study period. The information on the location of the harvest was collected to allow us to investigate spatial differences in harvest pressure, in the types of rattan growing in each area, and in collector selectivity, when combined with ecological surveys of the same areas. Data on the time spent in the forest by each group of collectors were recorded in order to estimate the costs of rattan collection.

Industry structure and profitability

Data were collected on the costs to and prices earned by rattan collectors, in order to calculate their profits. The supply structure in the rattan trade, from collector to rattan company, was also investigated, including prices at every stage, to show how profits are distributed and how possible efficiency improvements, such as through capital provision, might be achieved. Information was gathered during informal interviews and discussions with experienced collectors acquainted with the authors for almost ten years. Companies in Gorontalo, the main rattan processing town in North Sulawesi, were contacted to investigate amounts of rattan handled there, and to gain a wider perspective on the province's rattan trade. Company owners were interviewed and asked about the problems they perceive their industry to be facing and the reasons for these. Data on volume of rattan traded were obtained from one of these companies, as well as from official government statistics (Central Statistical Board of North Sulawesi, 1998). The Central Statistical Board of North Sulawesi also provided data on general economic indicators, such as inflation rates.

Figure 23.1   (Left) The location of the Paguyaman watershed, showing the sites of the ecological surveys, the harvest rate survey and the rattan factory.

Ecological transects

Three transects were set up in the Paguyaman watershed, two in exploited areas (Adudu and Marisa) and one in an unharvested area, Masina, (see Figure 23.1). The aim was to collect data on the number and type of rattans in each area, their density, number of stems per clump and stem maturity, in particular the number of harvestable canes that could be cut from each stem. Data were also collected on environmental variables that might affect rattan populations, including physical factors such as altitude, aspect, soil type, slope, as well as biological factors such as forest cover (which affects both the availability of supports and light levels). This information was used to analyse:

- Factors affecting the distribution of rattans, both physical and biological. This allows us to extrapolate rattan distributions from the sample to the whole watershed, and gives insight into the ecological requirements of different rattans. This would be useful for any future rattan cultivation projects, as well as for increasing understanding of the dynamics of the forest ecosystem;

- Demographic data on survival, growth and reproduction of different rattans, which allows us to construct a model of rattan population ecology. The model can be used to predict optimal harvest rates, and demonstrate how the sustainability of harvesting and potential yields depends on rattan growth form and ecology. In a multi-species harvesting system such as this, it is particularly important to have this information for all harvested species, as some may be much more vulnerable than others, and may be harvested unsustainably as a component of the overall offtake;

- Types of rattan found in each area, previous harvesting levels (estimated crudely from the number of cut stems) and current availability of harvestable canes. This allows the calculation of the relative profitability of a given area, and at what point after collectors have cleared an area it becomes worthwhile to start harvesting there again.

The key type of ecological information that is currently missing, and which cannot easily be obtained from a short study such as this, is the growth rate of individual stems, so that age, length and stem maturity can be related. This is crucial for modelling the rate at which commercial yield can be produced by a rattan population, and so the sustainable offtake rate. However, it can only be obtained through experiments on individual rattans, marking them as seedlings and seeing how quickly they grow, ideally in natural conditions as well as in plantation trials. This point illustrates the weakness of the short-term data collected so far. A small sample of rattans was marked and measured in this way at the Adudu base camp and is continuing to be monitored. Data from two species of cultivated rattans (*Calamus trachycoleus*

Beccari and *C. caesius* Blume) in Kalimantan suggest that canes may be ready for harvesting approximately ten years after planting (Menon, 1980; Godoy and Tan, 1989) so clearly such trials are both urgent and in need of a long-term approach.

Ecological data collection methods

Three transects were established using a sighting compass and measuring tape, running north from the main river. Each transect was 10 m wide and of variable length. The transect was divided into 10 m long sections. Flagging tape was used to mark and number each 10 m length of the transect, as well as the transect boundaries. Every rattan plant along the transect was examined and described. Two teams carried out this work, one team of four people cutting, measuring and marking the transect and the second team (two people) examining and describing each rattan plant. Both transect establishment and rattan recording were time-consuming work and were hampered by the mountainous terrain: 300–500 m of transect could be established in one day (depending on weather and terrain), while every rattan along 150–200 m of transect could be described per day. Progress was considerably quicker (400–500 m/day) if only plants with mature stems were described, instead of every rattan plant. However, while the recording of only mature stems provides a quick estimate of harvest potential it is inadequate for demographic studies. Altitude, slope, light availability and aspect were recorded at every ten metre point along the transects (using altimeter, clinometer, light-meter and compass) and a soil sample was collected for analysis from each transect location.

Costs of fieldwork

Data from the ecological transects were costly to collect, in terms of time and physical effort. For example, 384 person hours were necessary to establish and collect all data from the 1,650 m long Masina transect (eight days work by a team of six persons, each working an eight-hour day, 7 am to 3 pm). A further seven days were necessary for this team to travel to and from Masina from the nearest village, by longboat and on foot. All supplies and equipment were transported on foot to Masina from the Adudu base camp. Location of an unexploited rattan area was also difficult, since few areas of the Paguyaman watershed remain which have not yet been accessed by collectors. This required a further three weeks ground surveys plus three days of discussions with local collectors. Counts of rafts required the continuous presence of field assistants at the Adudu base camp between August 1997 and July 1999. Examination of raft content required approximately twenty minutes per raft. Field equipment used in data collection was relatively simple and did not involve large costs.

Results

Ecological data

We present here a few of the preliminary results, to give a flavour of the data collected. Many more transects are needed to cover the Paguyaman watershed more fully, especially as our initial surveys suggest that rattan communities are highly diverse from location to location. The main feature of the rattan communities found on the ecological transects was this diversity, both within and between transects (Table 23.1). We concentrate here on comparing the results from Adudu, a lowland exploited area which has higher light levels due to some timber removal in the area, and Masina, a remote, more mountainous and unexploited area with totally undisturbed forest cover. The two areas also have different soil types: Masina is a gold-rich area while Adudu is not. The differences in composition of the rattan community between the sites are likely to be due primarily to topographic differences. This is unfortunate in terms of parameterising a harvesting model, because data on rattan population structure and dynamics for model parameterisation can only be obtained from unharvested sites where population dynamics have not been altered to an unquantifiable degree by harvesting. However, in most areas of North Sulawesi only the most remote and inaccessible (hence more mountainous) forests are unexploited – if this means that the communities are very different, parameterisation of any harvesting models will be very difficult. Further transects in unexploited parts of the Paguyaman watershed should provide additional information on this subject.

Comparing the two parts of the transects where all plants were counted, which were of the same length (350 m) in each location, Masina had 18 types of rattan with a density of 47.8 plants/100m², while Adudu had 13 types with a density of 57.1 plants/100 m². Each site had one dominant rattan type, one or two common types, and several less common rattans, although there were more uncommon rattans at Masina (Figure 23.2). The dominant types were different in each case. At Masina, Buku tinggi was the dominant type, comprising 49% of plants in the transect; the next most dominant type was Topalo (22%) followed by Susu, Beluo and Batang merah (each comprising 6%), with smaller percentages of other species. At Adudu, Batang biasa was the dominant type, comprising 44% of plants in the transect; Susu and Tohiti each made up 24% of plants here, with smaller percentages of other species. Both Buku tinggi and Batang biasa are commercially valuable, but Batang makes up the vast majority of rattan rafted from the watershed (see below). Of those types that were found in significant numbers at both sites (Tohiti, Susu, Beluo), the clump sizes were similar at both sites, as might be expected (Table 23.2).

Table 23.1   The rattan types found in each of the transect sites, with possible species attributions and information on growth form (clumped or solitary) and commercial importance

Local name	Latin name	Abbrev	Adudu	Marisa	Masina	Growth	Sold?
Batang biasa	*Calamus zollingeri?*	BatB	Yes	Yes	Yes	Clumped	Yes*
Batang merah	?	BatM	No	Yes	Yes	Clumped	Yes
Beluo	*Korthalsia celebica* H.Wendl.	Bel	Yes	Yes	Yes	Clumped	No
Bukan susu	?	BS	Yes	Yes	No	Clumped?	No
Buku tinggi	?	BT	Yes	Yes	Yes	Clumped	Yes
Dia lagi	?	DiaL	No	Yes	No	Clumped?	No
Huwulungo	?	Huw	No	No	Yes	Solitary	Minor
Malie	?	Ma	Yes	Yes	Yes	Clumped	No
Polioto	?	Po	Yes	No	Yes	Solitary	No
Ronti	?	Ron	Yes	Yes	Yes	Clumped	Minor*
Susu	*Daemonorops robusta?*	Sus	Yes	Yes	Yes	Clumped	No
Tarumpun	?	Tar	No	No	Yes	Solitary	Minor
Tarumpun daun gros	?	TDG	No	No	Yes	Clumped	Minor
Tarumpun daun gros bertunas	?	TarDGTu	No	No	Yes	?	Minor
Tarumpun daun halus/tohiti halus	?	TarDH	No	No	Yes	Solitary	Minor
Tarumpun tunggal (=tohiti topalo?)	?	TarTung	No	No	Yes	Solitary	Minor
Tohiti biasa	?	Toh	Yes	Yes	Yes	Solitary	Yes*
Tohiti bubanger	?	TohBu	No	No	Yes	Solitary	Yes
Tohiti tunas	?	TohTu	No?	Yes	Yes	Solitary	Minor
Topalo	?	Tp	Yes	No	Yes	Solitary	Minor
Topalo tunas	?	TpTu	–	Yes	Yes	Solitary?	Minor
Topeto	?	Topet	No	Yes	No	?	?*
Umbul	*Calamus symphysipus?*	Umb	Yes	Yes	Yes	Solitary	Yes*
Umbul tidak coklat	?	Umbtc	?	No	Yes	Solitary?	?

Note: Each row is a separate rattan type as recognised by collectors, with clear morphological differences from the other types. The 'Sold' column shows whether it is commercially important; *indicates that the type was found during the inspections of rattan collectors' rafts. The abbreviated name (Abbrev) is that used in the figure legends.

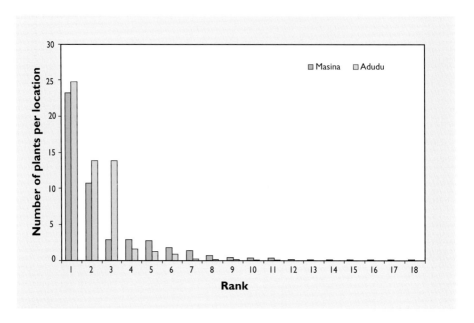

Figure 23.2   The number of plants found in 100 m² of rattan transect, ranked from the commonest to the least common type, for the two sites, Masina and Adudu.

These data give useful information on the growth forms of the different rattan types. This is important to analyse because it is a step towards modelling the growth of rattans. Models can be an important tool in understanding rattan population dynamics and in predicting the effect of different harvest rates on a population, and hence sustainable yields. Important parameters to be incorporated into a model of rattan harvesting are dependent on the precise nature of the model but would include cane growth rates, germination, survival and maturation rates. Figure 23.3 shows the distributions of numbers of stems per plant (Figure 23.3a) and maturity classes (Figure 23.3b) for several different rattan types in the unexploited transect, Masina. Those types whose distributions of clump size and maturity class were found not to be significantly different from each other (using a χ^2 test) were aggregated into the 'Others' distribution. In Figure 23.3a, types Batang merah, Susu, Tarumpun and Tarumpun daun gros were aggregated. For Figure 23.3b, Susu, Tarumpun, Tarumpun daun gros, Tohiti and Topalo were aggregated together. Such aggregation can be useful for obtaining a crude overall estimate of sustainable yields for the whole watershed, as it avoids the need for modelling each rattan type individually. There were clear differences between the types: Beluo (*Korthalsia celebica* H. Wendl. – the only member of its genus occurring in Sulawesi), in particular, stands out as having a very different growth form than the others, with more stems, and more mature stems, per plant. This may be explained by the fact that, in contrast to most other rattans, this species exhibits hapaxanthic flowering behaviour (individual stems

Table 23.2 A comparison of the clump size and number of harvestable stems for rattans found at the two transect locations (only showing types for which at least 20 plants were recorded)

Type (see Table 23.1)	Mean	Median	Plants	Harvestable %	Cut
a) Masina					
BatM*	2.55	1	38	0	0
Bel	3.68	3	28	11.7	0
BT*	3.18	1	262	9.1	0
Sus	2.19	1	48	0	0
Tar (*)	1.62	1	39	1.6	0
TDG (*)	1.69	1	29	0	0
Toh*	1.00	1	25	0	0
Tp (*)	1.01	1	385	1.0	0
Overall	2.11	1.25	854	3	0
b) Adudu					
BatB*	3.75	1	231	0.1	45
Sus	2.16	1	223	1.6	1
Toh*	1.00	1	483	2.3	0
Umb*	1.07	1	42	2.0	0
Bel	3.35	3	17	31.6	0
Ron (*)	3.67	4	9	3.0	0
Overall	2.50	1.83	1,005	7	46

Note: **Mean** = mean number of stems in the clump. **Median** = median number of stems in the clump. Solitary types have both a mean and a median number of stems of one. **Plants** = number of plants recorded in the transect (all plants, transects were of similar length). **Harvestable** = the percentage of living stems that are of harvestable length (species of major commercial value *, species of minor commercial value (*)). **Cut** = number of stems that had already been cut by collectors. As it is not possible to tell when a stem was cut, this serves only to show which types are targeted.

flowering only once before dying and being replaced by sucker shoots from the base). This species is of no commercial value, mature stems being fine and spindly (3–5 mm diameter). The solitary rattans (Tohiti, Topalo) resemble the majority of rattan types in their maturity distribution, with many juvenile stems and few mature stems (Figures 23.3a, b). Although Buku tinggi is similar to the majority of rattan types in the number of stems it produces and their maturity, it is singled out because unlike them, it produces a significant proportion of its new plants through stolons (trailing stems above ground, capable of producing roots and shoots at their nodes). The stolons observed were up to two metres in length. The plants that had produced stolons were significantly larger and more mature than those that had not done so (Figures 23.4a, b).

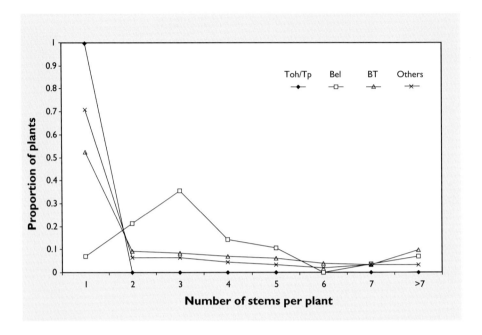

Figure 23.3a   Distributions of the number of stems per plant in the Masina transect, using data from the area in which all plants were counted. The rattan types with large enough sample sizes and distributions that were not significantly different from each other were aggregated into the 'Others' distribution; these were Batang merah, Susu, Tarumpun and Tarumpun daun gros. For species codes, see Table 23.1.

Spatial distributions

A preliminary examination of the spatial distributions of rattan types along the two transects revealed some interesting patterns, which we are now investigating further using spatial models. In Masina, all plants including seedlings were counted along a short section of the transect (the same length as the Adudu transect), and then mature plants were counted along a much longer transect on either end of the detailed transect. This allows us to compare spatial distributions of mature and immature plants, and to investigate the distribution of plants on both larger and smaller scales.

The Masina transect covered two mountains, the first from transect points –500 m to 0 m with a peak at –370 m, the second from transect points 0 m to 1,150 m with a peak at 800 m. Inspection of the distribution of rattans along the transect suggests that there is a strong influence of topography on location of rattan types. Thus Umbul (Figure 23.5a) is found on the slopes of the first mountain, decreasing in abundance as altitude decreases, while Buku tinggi

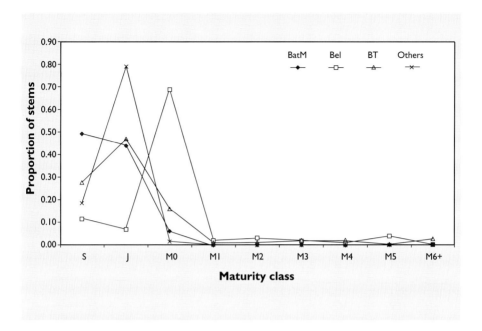

Figure 23.3b Distributions of the maturity class of the stems in the Masina transect, using data from the area in which all plants were counted. These were divided into seedlings (S), juvenile stems (J), mature stems which are hard but not yet long enough to provide a commercially harvestable cane (M0), and mature stems that could provide from one to more than six commercially harvestable canes (M1–M6+). The rattan types with large enough sample sizes and distributions that were not significantly different from each other were aggregated into the 'Others' distribution; these were Susu, Tarumpun, Tarumpun daun gros, Tohiti and Topalo. For species codes, see Table 23.1.

(Figure 23.5a) is found in large numbers between the mountains, and Susu and Tohiti (Figure 23.5b) are found mostly on the second mountain. Beluo, on the other hand, is found throughout the transect. There is also evidence of different types clustering together. Paired correlation coefficients calculated for the presence of one type of rattan in a transect location against the presence of each other type, indicated that the abundances of Susu, Batang biasa and Tohiti are positively correlated with each other, and negatively correlated (separately) with Buku tinggi and Topalo. These correlations could be simply products of similar habitat requirements, or indicate inter-specific interactions between the rattan types. Which is the case will be clearer following detailed analysis of these data.

Within a particular rattan type, the distribution of individuals through the transect may provide an indication of reproductive behaviour (distance of seed dispersal, degree of reliance on vegetative propagation). If plant distribution

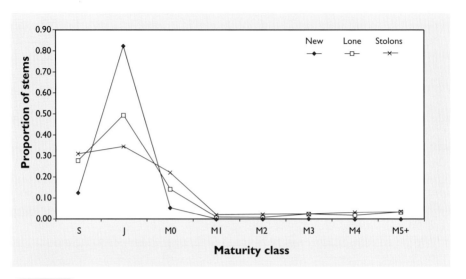

Figure 23.4a Differences between Buku tinggi plants that have formed stolons (Stolons), those that are still visibly stolons from a parent plant (New) and those that have not formed stolons (Lone), in terms of the distribution of maturity classes of their stems. All three distributions are significantly different (χ^2 test, P <0.001). For definitions of maturity classes, see Figure 23.3b.

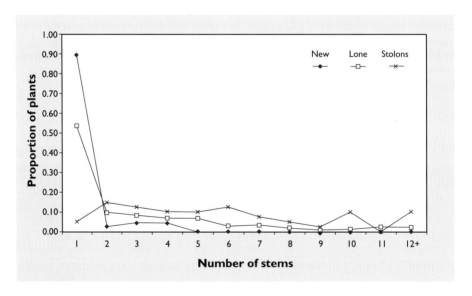

Figure 23.4b Differences between Buku tinggi plants that have formed stolons (Stolons), those that are still visibly stolons from a parent plant (New) and those that have not formed stolons (Lone), in terms of the number of stems per plant. All three distributions are significantly different (χ^2 test, P <0.001).

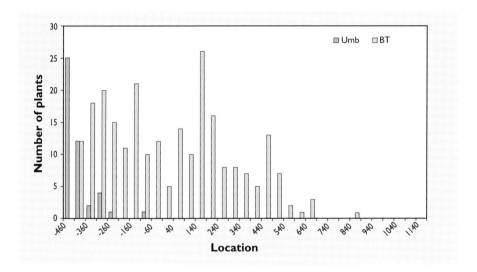

Figure 23.5a   Distribution of Umbul and Buku tinggi plants along the Masina transect (only those plants with at least one mature stem were counted). The data are aggregated so that each location represents the number of plants recorded over the previous 500 m² section of the transect (thus −460 m is the number recorded from −500 to −460 inclusive). The transect covered two mountains; the first from −500 m to 0 m with a peak at −370 m, the second from 0 m to 1,150 m with a peak at 800 m.

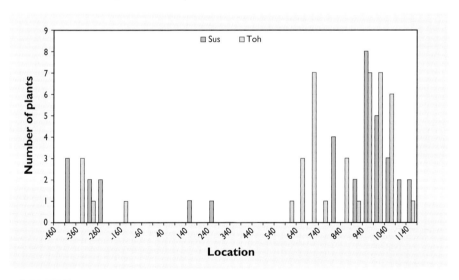

Figure 23.5b   Distribution of Tohiti and Susu plants along the Masina transect (only those plants with at least one mature stem were counted). The data are aggregated so that each location represents the number of plants recorded over the previous 500 m² section of the transect (thus −460 m is the number recorded from −500 to −460 inclusive). The transect covered two mountains; the first from −500 m to 0 m with a peak at −370 m, the second from 0 m to 1,150 m with a peak at 800 m.

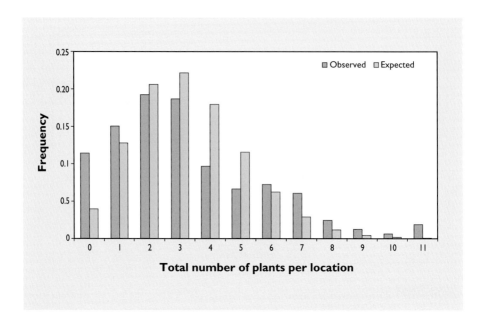

Figure 23.6   Frequency distribution of the total number of rattan plants (of all types) found in a particular 100 m² location on the Masina transect (Observed) compared to the expected frequency distribution if the plants were randomly distributed amongst transect locations (Expected). The two distributions are significantly different (χ^2 test, P <0.001), with the observed distribution having more locations that have either very few or many plants in them than would be expected.

was entirely random, then the number of individuals recorded at a particular location would be well described by a Poisson distribution. Deviations from the Poisson distribution might indicate a degree of clumping behaviour by the plants. Indeed, all the types tested (those with at least 20 plants in the sample) showed significant deviation from the Poisson distribution, with more locations having no plants and more having many plants than would be expected if they were distributed at random (Figure 23.6). Beluo, however, is an interesting case; all the other types showed significant clumping both at the original 100 m² transect location and when locations were aggregated to 500 m². However, the distribution of Beluo plants was only significantly different from random in the aggregated data. This suggests that Beluo aggregation is at a different scale to the others, possibly because it has many more stems, or because of its different flowering behaviour or hermaphroditism. This analysis assumes that plants are scattered amongst locations with no effect of neighbouring locations on each other. However in fact the transect is continuous, and there may well be interactions between neighbouring locations. Thus more sophisticated modelling is required to fully understand rattan distributions, particularly of this species.

Economics of the rattan trade

Between 18 and 25 of September 1997, the contents of 58 rafts were inspected, belonging to eight different groups of rattan collectors, each of which had come from a different collection location. The rafts were made up of an average of 155 canes each, and the predominant rattan type was Batang biasa, which comprised 84% of all canes examined. Of the remaining canes Tohiti comprised 9%, Umbul 5% and Ronti 2%. However, a note of caution is necessary as it is difficult to assign species or even exact vernacular names to bare rattan canes. Thus the Batang biasa component is known to include a small amount of Batang merah, while the Tohiti component includes some Topalo and possibly some of the Tarumpun types as well. Batang biasa is the type that was discovered to have been most heavily harvested at Adudu, suggesting it is an important type for rattan collectors in the Paguyaman watershed (Table 23.2). The number of rattan collectors and rafts passing the Adudu field station showed a steep decline from the start of observations in August 1997, stabilising at a low level from June to September 1998, and rising again thereafter (Figure 23.7). This may be connected with the economic crisis in Indonesia: our observations of the people passing the field station suggested that many rattan collectors had turned to gold panning in this period, because of an increase in the value of gold. This suggestion is supported by interviews with collectors passing the camp. Plotting the monthly change in inflation rate over the study period against the number of rattan rafts passing the camp suggests that there is a relationship between the state of the general economy and rattan collection; further analysis comparing changes in rattan and gold prices directly is required. However, our data on the offtake of rattan show how strongly linked to the general economy rattan collection is, and will enable us to estimate the profitability of rattan collection and thus predict future offtake levels.

Data obtained from one major rattan company in Gorontalo indicated that 5,907 tonnes of rattan left this company between January 1995 and December 1996, for transport to furniture factories in Java. This rattan was obtained from locations throughout North Sulawesi and 1,545 tonnes of this (26%) was collected at Paguyaman. A breakdown of this volume by trade names indicates that 50.8% of this volume was Batang, while Umbul comprised 27.5% and Tohiti 1%. Further analysis of company records covering volume traded, described by location and trade names, will give a province-wide picture of rattan purchases by this company.

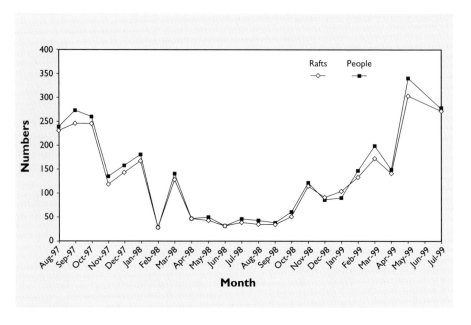

Figure 23.7   The number of rattan rafts and rattan collectors passing the Adudu field station each month from August 1997 to July 1999.

Concluding remarks

The chapter has given a broad overview of the aims of our study, and of the range of data collected. The major factor that is still missing is taxonomic identification of the herbarium specimens that have been collected, so that local names for rattan types can be related to rattan species. Our study is on-going, with work now focusing on further field data collection, data analysis and modelling.

Lessons learnt

A key point emerging from this study is the importance of establishing long-term demographic studies of rattan populations, in order to parameterise our models of the sustainability of rattan harvesting. We have also shown the wide range of biological and economic data that are required if sustainability is to be assessed, since sustainability is so multi-faceted (Gatto, 1995).

Recommendations and aims for future action

Future work will include further transects in unexploited parts of the Paguyaman watershed to help parameterise the model, plantation trials in the buffer zone of the newly created Paguyaman reserve, and collection of further data on the rattan trade from companies in Gorontalo. A further key objective is the marking and measuring of a larger sample of individual stems along the transects, in order to monitor how quickly they grow. Collectors are walking further and further into remote parts of the Paguyaman watershed to find rattan. Thus this work is becoming urgent if we are to assist in moving the rattan industry towards sustainability, while preserving the diversity of the forests on which the industry depends.

Acknowledgements

We are very grateful to the Centre for Research and Development of the Indonesian Institute of Sciences, the Directorate General for Protection and Nature Conservation of the Ministry of Forestry and Plantations, the Association of Indonesian Forest Concession Holders, and Bellanico and Saripermindo Murni rattan companies for support and assistance in Indonesia. We thank the UK Department of the Environment Transport and the Regions' (DETR) Darwin Initiative for the Survival of Species for funding this work. We are also grateful for financial support from the People's Trust for Endangered Species and British Airways Environment Programme. We especially thank: J. Komolontang, M. Stockdale, B. Wowor and many rattan collectors at Paguyaman for their assistance.

References

Central Statistical Board of North Sulawesi (1998). *North Sulawesi in Figures 1998*. Manado, Indonesia.

Clayton, L. and Milner-Gulland, E.J. (2000). The trade in wildlife in North Sulawesi, Indonesia, pp. 473–498. In: J.R. Robinson and E.L. Bennett (eds). *Hunting for sustainability in tropical forests*. Columbia University Press, New York, USA.

Dransfield, J. (1986). A guide to collecting palms. *Annals of the Missouri Botanical Garden* **73**: 166–176.

Dransfield, J. (1992a). The taxonomy of rattans, pp. 1–10. In: Wan Razali Wan Mohd, Dransfield, J. and Manokaran, N.(eds). *A Guide to the Cultivation of Rattans*. Malayan Forest Records No 35. Forest Department, Kuala Lumpur, Malaysia.

Dransfield, J. (1992*b*). Morphological considerations: the structure of rattans, pp. 11–31. In: Wan Razali Wan Mohd, J. Dransfield and N. Manokaran (eds). *A Guide to the Cultivation of Rattans.* Malayan Forest Records No 35. Forest Department, Kuala Lumpur, Malaysia.

Gatto, M. (1995). Sustainability: is it a well-defined concept? *Ecological Applications,* **5**: 1181–1183.

Godoy, R. and Tan, C.F. (1989). The profitability of smallholder rattan cultivation in southern Borneo. *Human Ecology* **17**: 347–363.

MacKinnon, K., Hatta, G., Halim, H. and Mangalik, A. (1996). *The Ecology of Kalimantan.* Periplus Editions, Singapore.

Manokaran, N. (1990). *The State of the Rattan and Bamboo Trade.* Rattan Information Centre Occasional Paper No 7. Rattan Information Centre, Forest Research Institute Malaysia, Kepong, Malaysia.

Menon, K.D. (1980). Rattans: A State of the Art Review, pp. 1–76. In: *Rattan: a report of a workshop held in Singapore, 4–6 June 1979.* International Development Research Centre, Canada.

Siebert, S.F. (1993). Rattan and extractive reserves. *Conservation Biology* **7**: 749–750.

Uhl, N.W. and Dransfield, J. (1987). *Genera Palmarum: a Classification of Palms based on the work of H.E. Moore. Jr.* L.H. Bailey Hortorium and the International Palm Society, Lawrence, Kansas, USA.

Walton, T. and Holmes, D. (2000). Indonesia's forests are vanishing faster than ever. *International Herald Tribune,* 25th January 2000.

Chapter 24

Ethnobotanical Survey Methods to Monitor and Assess the Sustainable Harvesting of Medicinal Plants in Uganda

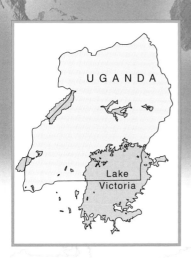

Maud Kamatenesi-Mugisha
Makerere University, Kampala, Uganda

Remigius Bukenya-Ziraba
Makerere University, Kampala, Uganda

Introduction

The sustainable use of medicinal plants is of world-wide concern. The conservation of medicinal plants in Uganda is gaining support following the realisation that important habitats containing medicinal plant resources are being degraded at an alarming rate. This degradation is a result of agricultural clearance, intensive logging, fuelwood and charcoal production, shifting cultivation and war damage. Ugandan biodiversity faces further threats from over-exploitation, habitat fragmentation, climate change, the introduction of invasive species and, possibly, the introduction of genetically modified organisms. This chapter will discuss some of the efforts being made to address the problem of declining medicinal plant resources through the involvement of the local users. As stipulated in the Convention on Biological Diversity (CBD) (Articles 1, 6, 8, 9 and 10; Glowka *et al.*; see Appendix 1), conservation and development agendas need to integrate with local community activities and promote community participation towards conserving biodiversity.

Approximately 80% of the people in Uganda use medicinal plants for their day-to-day primary healthcare needs (Kamatenesi, 1997; Ogwal, 1994; WHO, 1991). Local people have developed a store of empirical information concerning the therapeutic values of local plants. The loss of habitats not only leads to the loss of the medicinal plant resources upon which people depend for their survival, but also the extinction of unique indigenous knowledge.

The use of herbal medicine in rural areas, especially in Africa, is widespread (Sofowora, 1993; Figure 24.1). Both Agenda 21 and the CBD recognise that the different stakeholders involved in resource utilisation need to participate in order to implement proper management programmes for both long-term and short-term interventions (Glowka *et al.*, 1994). Accordingly, the local people, as well as the herbalists, will more easily accept community programmes aimed at conservation of medicinal plants.

Traditional medical practitioners (TMPs) and herbalists have come to realise that a number of medicinal plants have become scarce and are already locally reported as seriously threatened due to habitat loss and over-exploitation. For instance, in south-western Uganda, around Bwindi Impenetrable National Park, there is a heavy dependence by local communities on *Rytigynia* Blume species (Rubiaceae). Due to habitat loss, *Rytigynia* species are already extirpated from private land. Two species endemic to the Albertine Rift, *R. kigeziensis* Verdc. and *R. kiwuensis* (K. Krause) Robyns, are thought to be seriously threatened with extinction.

Article 2 of the CBD emphasises the sustainable utilisation of biodiversity. This is defined as the use of components of biological diversity in a way and at a rate that does not lead to the long-term decline of biological diversity, thereby maintaining its potential to meet the needs and aspirations of present and

Figure 24.1 Locally harvested medicinal plants remain a vital source of healthcare for millions of rural Africans, representing a direct link between sustainable development and the management of plant resources.

future generations. Therefore, the sustainable utilisation of wild medicinal plant resources must be secured if the rights of future generations are not to be jeopardised.

Government policy towards traditional medicine

In Uganda, TMP activities have been legally suppressed; under Ugandan laws and constitution they were included in the Witchcraft Act of 1957. The Act categorises the practices of traditional medicine as a crime, whether used for treating diseases or killing a person. In reaction to this, both genuine TMPs and 'quack' healers restricted their operations to hidden places and worked mostly at night. Furthermore, the Ugandan Medical Practitioners and Dental Surgeons Act of 1968 prohibits unlicensed persons from practising medicine, dentistry or surgery. However, Section 36 of the Act allows the practice of any system of therapeutics by a person recognised by the community to which they belong, with the provision that the scope of their work was limited to that community only. The 1968 Act also allows that person to be duly trained in such practice. Although this allowed the practice of traditional medicine only in the communities where these TMPs were found, the widespread use of traditional medicine in the country was ignored. Statute Supplement Number 11 of the Medical and Dental Practitioners Statute, 1996, repeals the Medical Practitioners and Dental Surgeons Act of 1968 (Statute Supplement No. 7) and makes no provision for traditional medicine. Further to this, in traditional medical systems, taxes (trading license and income tax), supposed to be paid by practising TMPs, are neither assessed nor paid. The lack of clear laws and regulations on traditional medicine has in hindsight suppressed the conservation of medicinal plants. This lack of clarity about laws on traditional medicine has led to a lack of official support for the implementation of measures necessary to conserve medicinal plants.

The Uganda Government through its own institutions, such as the National Environment Management Authority (NEMA) and Natural Chemotherapeutics Research Laboratory (NCRL), under Uganda's Ministry of Health, and in collaboration with non-governmental organisations (NGOs) have recently started programmes aimed at conserving medicinal plants. For instance, in 1991–1994, the World Health Organisation (WHO) and the United Nations Children's Fund (UNICEF) started training traditional birth attendants in rural areas who rely heavily on medicinal plants. These organisations also provide essential equipment for midwifery practice, such as examination gloves. This has promoted the role of midwives in primary health care. Further, Uganda's ratification of the CBD in 1993 has meant increased commitment to the conservation of biodiversity. This has resulted in the formulation of a National Biodiversity Strategy and Action Plan by NEMA focusing on the management of Uganda's biodiversity.

Currently, the Uganda Government is formulating a policy on traditional medicine and bioprospecting as components under the National Biodiversity Strategy and Action Plan. However, the policy formulation is still in process and at present there are no clear operational laws regarding traditional medicine in Uganda. The policy being formulated will address issues regarding access, use, and conservation/management of medicinal resources. This is in line with Article 6 of the CBD that encourages governments to formulate national biodiversity strategies that are multidisciplinary and multi-sectoral in nature, aimed at setting up general measures for conservation and sustainable use of biodiversity.

Monitoring and Assessment

Selection of methods to monitor and assess sustainable utilisation of medicinal plants

i) Participatory approaches

In the management of medicinal plants, a number of approaches for collaborative research, planning, implementation and monitoring need to be developed. Through participatory involvement, local people themselves have been encouraged to generate information, forecast potential problems and encourage practical solutions to these problems. Participatory methods are increasingly adopted by Ugandan NGOs and government organisations (Nabasa *et al.*, 1995); these include Rapid Rural Appraisal (RRA), Participatory Rural Appraisal (PRA), Market Rural Appraisal (MRA) and other similar techniques. These methods emphasise the involvement of local people in the investigations with the aim of developing practical solutions to the problems affecting communities (Farrington and Martin, 1988).

Those earning a living from medicinal plants can offer valuable information on the availability and harvesting methods of medicinal plants. Selecting and interviewing specific families intensively, through an income ranking process, provides a comparison of the relative income and assets of members of the community. This reveals the proportion of the population most dependent on medicinal resource harvesting (Martin, 1995). Aspects of gender and social stratification help in understanding and assessing who is in basic control of the medicinal plant resources. Use of floral inventories and indigenous management practices are vital tools in monitoring and sustainable utilisation of biodiversity (Mugabe and Clarke, 1998).

Kyoshabire (1998) used PRA methods to generate information on medicinal plants, namely plant habitats, growth forms, parts used and number of plants used, around Bwindi Impenetrable National Park. She employed both qualitative and quantitative methods, such as informal discussions, semi-structured interviews, field excursions and local market surveys. The quantitative methods, used were preference and pair-wise ranking. By using these methods, a wealth of information was generated, identifying the medicinal plants that the local people value. For instance, 30 commonly used medicinal plants were mentioned and among those listed are two locally threatened species: *Piper guineense* Thonn. (Piperaceae) and *R. kigeziensis*. Information generated from such studies can provide baseline data for monitoring medicinal plants under threat and can allow the sustainability of their usage to be deduced.

ii) Multiple use zones concept
Articles 9 and 10 of the CBD clearly recognise the need for local people's involvement in the management of utilised natural resources and protected areas. Communities adjacent to protected areas have often suffered from protectionist measures and have become hostile to conservation efforts as a result of reduced access to resources, employment, income, and even eviction from their own lands. Multiple Use Zones (MUZs) have been legalised by the Uganda Wildlife Authority (UWA) as a means of reducing conflict between park authorities and local communities and in instilling a sense of ownership and responsibility towards protected areas. MUZs have been introduced on a pilot basis in Bwindi Impenetrable National Park, Queen Elizabeth Biosphere Reserve, Lake Mburo and Mount Elgon National Parks.

In Bwindi, three pilot MUZs were set up in three parishes. Registered herbalists are allowed to harvest registered medicinal plants in specified amounts as stated in a Memorandum of Understanding. To prepare for this, surveys and negotiations were carried out, led by the CARE-Development Through Conservation Project involving both park authorities and the local communities. The surveys collected data on species status and identified stakeholders, and harvesting and monitoring protocols were formulated. The supplementary methods used included Rapid Vulnerability Assessment (RVA) and PRA (Wild and Mutebi, 1996). The RVA is a systematic method used to rapidly assess the vulnerability of a plant species used by people. It is used to collect both ecological and social data. Factors to consider under RVAs include: life form, habitat specificity, abundance and distribution, growth rate, response to harvesting, parts used, pattern of selection and use, demand, seasonal harvesting, traditional conservation practices, commercialisation, and possible substitutes to the affected species as a way of reducing demand on them. In using the RVA method, at each level decisions can be made as to which aspects should be emphasised to meet the management priorities and take appropriate management decisions. The beauty of RVA is that it integrates indigenous and scientific knowledge (Wild and Mutebi, 1996). Subsequently, this resulted in the signing of agreements with local community leaders in each of the three parishes, and has reduced the hostilities and resentment between villagers and the government authorities in the protected areas. Indeed, in MUZs in Bwindi Forest, the local people have been trained on how to manage medicinal plant resources and they are eager and willing to adopt new techniques, such as cultivation. Generally, in Uganda, the traditional healers, village heads and leaders are very much aware of the local resource problems. When the local people are given the opportunity, through awareness and training about the status of the medicinal plant resources under their custody, their participation can be both fast and voluntary.

iii) Monitoring the management of medicinal plants

Monitoring is the regular collection and analysis of information to determine whether activities (in this case the harvesting of medicinal plants) are meeting defined management goals (Tuxill and Nabhan, 1998). The information collected can be ecological, physical or social (Martin, 1995; Tuxill and Nabhan, 1998). Monitoring is usually done at regular intervals e.g. weekly, monthly, yearly. For efficient management of medicinal plants there is a need to have a clear monitoring system in place to detect any changes in the harvested resource and provide an early warning. Monitoring should involve the local human population and use indigenous knowledge in the recording of species occurrence, distribution, use, documentation of historical and current information, including knowledge of both generalist and specialist resource users.

Kamatenesi (1997) studied the impact of utilisation of *Rytigynia* species (*R. kigeziensis*, *R. kiwuensis* and *R. bagashwei* (S. Moore) Robyns) in the multiple use zones of Bwindi Impenetrable National Park to provide baseline data in order to monitor the level and impact of harvesting the *Rytigynia* bark. By forest sampling in the three MUZs, areas adjacent to the MUZs and in the bamboo zone, assessments were made of size class distribution, population structure and impact of use on the population structure of *Rytigynia* species. A stratified sampling strategy was adopted which took into account the elevation, topographical features and vegetation aspects. Plots of 20 m × 20 m plots were used in the studies of abundance, size class distribution, the extent of bark damage and the amount of bark available for harvesting. Plots were marked with aluminium tags and flagging tapes to allow relocation for monitoring purposes, and spaced at 15 m intervals. To estimate the extent of bark damage to *Rytigynia* species, the seven point scale of damage assessment (0 = no damage to 6 = ring-barked stem) was used as developed by Cunningham (1988). To measure bark thickness and mass, bark samples of about 2 cm × 3 cm were trimmed carefully at approximate heights of 0.25 m, 0.75 m, 1.3 m and 1.75 m. To examine the trends in the data collected, correlation and regression data analyses were performed using the Systat® programme package and Excel®.

These studies revealed over-harvesting of mature stems with diameters larger than 3 cm, the complete ring barking of the basal 2 m of some stems, and, in some areas, root collection by the local community from the MUZs. It was observed that the bark mass that was available for harvesting in the MUZs was less than its demand from the resource users, and, therefore, the harvesting of the bark was not sustainable unless other measures, such as cultivation, resource and species monitoring and resource substitution, were put in place. The recorded methods of harvesting of the medicinal plants were destructive, posing a real threat to the conservation of these species.

Field studies in the MUZs in the Bwindi forest (Kamatenesi, 1997) showed that the actual amounts of *Rytigynia* bark harvested were much higher than allowed. At the same time the number of herbalists collecting *Rytigynia* bark was increasing due to the increasing consumer demand. Wild and Mutebi (1996) report a high human population density around Bwindi Park, reaching 200–400 people per square kilometre (km^2), making it one of the most densely populated areas south of the Sahara. Therefore, the demand exceeds the available resources in the forest. Whilst this case study illustrates the value of the MUZ concept as a means to promote local people's understanding of conservation, the issues of sustainability for medicinal plants in great demand are not yet fully resolved. However, the use of *ex situ* cultivation, to increase the available supplies outside the protected areas, is a possible contribution towards long-term and sustainable utilisation.

iv) Rapid ethnobotanical survey

Bukenya-Ziraba *et al.*, (1997) used the Rapid Survey Method to assess the utilisation of *Prunus africana* (Hook. f.) Kalkman (Rosaceae) in Kasyoha-Kitomi and Kalinzu forests in Bushenyi District, Western Uganda. *P. africana* is an Afromontane hardwood tree valued locally and internationally as an economic and medicinal plant resource. The bark is the major source of an extract used to treat benign prostate hypertrophy, an increasingly common condition in older men. The commercially traded bark is taken from wild *P. africana* populations in the montane forests of Africa, and populations throughout Africa are in decline (Simons *et al.*, 1998) (see Jaenicke *et al.*; Chapter 20 and Sunderland *et al*; Chapter 21). The bark or processed extracts are used for preparation of commercial drugs. Bukenya *et al.* (1997) used 1 ha plots only in the Kasyoha-Kitomi forest for a quick assessment of the status of

P. africana in the forest. The survey revealed a major decline in wild populations. Between 1991 and 1992, there was intensive and damaging harvesting of the bark including the complete removal of the bark from tree trunks, and, at times, the felling of trees. Nearly all the large trees of *P. africana* had died leaving very few trees with a diameter at breast height (dbh) between 10–50 cm. As a result, no economically viable harvest or income from *P. africana* will be possible from this forest over the next 20 years. For sustainable use, the future conservation of *P. africana* is likely to be through cultivation and possibly restocking with some kind of recovery planning measures put in place in order to increase the production levels. Furthermore, the commercial harvesting of *Prunus* bark for export has been banned in Uganda. The major threats now to *P. africana* are pitsawing in Kasyoha-Kitomi forest and drum making in Mpanga Forest Reserve.

Recommended Action

1. Building capacity

Mugabe and Clark (1998) emphasised the role of capacity building at local levels for the management of biodiversity. Through training, the local participants (the plant users, traditional healers, and the general public) will learn the same skills employed by extension workers and protected area resource managers, and will enable them to adopt sustainable harvesting of medicinal plant resources. Articles 12 and 13 of the CBD stress the issues related to research and training, and public education and awareness respectively, especially in developing countries like Uganda.

2. Administrators and beneficiaries of herbal medicine

With the devolution of planning and monitoring to villagers, people in rural communities are no longer seen simply as informants but as teachers, extentionists, activists and monitors of change. As a result of the present trends of trying to integrate local communities in Uganda in conservation and management agendas, the perception of villagers as recipients of knowledge is ceasing to carry meaning. Instead, they provide knowledge useful to biodiversity conservation based on their long-standing experience and practical observation. In Uganda, this has been strengthened by the current form of governance and decentralisation of power to districts and to village levels through the local councils political system. Members of the society have the power to make decisions and participate in their implementation through this local council system of governance, which promotes the bottom-up approach in decision making. An emphasis on village specialists integrates marginalised groups, such as the women, the illiterate and the young. This further includes the village para-professionals, village traditional healers, experts like midwives, bone-setters, women veterinarians and so on, and allows full use of their skills and knowledge.

However, the efforts of participatory planning and development will be futile without sustained contact with potential beneficiaries. The conservation managers and policy makers at all levels need to have sustained contact with resource users in order to obtain maximum benefits from conservation efforts. This will require investing in human resource development that includes improving people's communication skills, livelihood skills and fostering multi-disciplinary and multi-sectoral working.

3. Healer to healer extension

Traditional healers have always been involved in the transfer of their expertise from generation to generation, and practitioner to practitioner. Healer to healer extension is a form of informal education (CBD, Article 13) that promotes community participation, but deals with specialist resource users. Traditional botanical knowledge has been handed down from elderly to the young through oral tradition under intra- and inter-cultural influences of sex, language, beliefs (totems) and education. These have promoted transfer and exchange of ethnobotanical knowledge especially in the area of medicine.

Healer to healer extension is a low-cost method of motivating traditional healers, groups, and organisations to change habits and practices. The process of adoption and diffusion of information and skills is very fast. It also provides crucial leadership experience to villagers and provides role models that they can reasonably aspire to emulate. When people's ideas and interests – not just their labour – are sought-after and incorporated fully into development projects and programmes, the chances of success are increased. For healers 'seeing is believing' and the best educators for traditional healers are fellow healers. For example, the NCRL, under the Ministry of Health, has an effective collaboration with traditional healers. They meet at the NCRL offices and share information, pay visits to each other, and receive training in relevant techniques, such as the cultivation of medicinal plants and proper storage of the harvested medicines.

4. Individuals and co-operatives

The successful adoption of agro-ecological technologies has largely been due to the attention paid to local social organisation (Pretty *et al.*, 1996). Success, measured in the form of change sustained over a long period of time, has been achieved through either building on existing patterns of organisations (both formal and informal) or the development of appropriate new technologies. In Uganda's case, some traditional healers are organised into groups, two such groups being:

- Buganda Nedaggala Lyaayo; and
- The National Council for Traditional Healers and Herbalists Association (NACOTHA).

The healers in these organisations have been given certificates of practice in the registered organisations. These healers also meet with the NCRL and Traditional and Modern Health Practitioners Together Against AIDS (THETA). Furthermore, in villages where individual healers are practising on their own, they belong to traditional leadership structures, such as the national political system of local councils at village level, youth groups and clan groups. These administrative (local councils) and social units in Uganda also encourage participation of local communities including healers. The present political organisation in Uganda of the 'Local Councils' incorporates everyone from village to district level in decision making. These local leaders help in the formulation of by-laws governing resource use at local levels, such as wetlands and forests, and may be vital in medicinal plants management.

5. Domestication and propagation of medicinal plants

Ex situ conservation provides excellent opportunities for research and sustainable utilisation of the components of biological diversity being conserved. In the case of rare, threatened or over-exploited plants, cultivation can provide material without further threatening the survival of wild populations. Cultivation can also reduce the possibility of mis-identification and adulteration (WHO, IUCN and WWF, 1993).

An effective way to provide medicinal plant materials is through cultivation (home gardens) by the local people and *ex situ* conservation (botanic gardens and gene banks) at a national level (Bukenya-Ziraba, 1998). The declining habitats of native plants in Uganda, such as in Kasyoha-Kitomi, Bwindi Impenetrable Forest and the surrounding vegetation, means that they can no longer supply the expanding market for medicinal plant products at both local and national levels. In Uganda, a range of government agencies including the NEMA and the National Research Organisation, could promote the production of medicinal plants for the local people to adopt as farm crops. Through these organisations and government bodies, the local people could be given propagules for cultivation. For example, important threatened medicinal plants, such as *Warburgia ugandensis* Sprague (Canellaceae), *Rytigynia* species and *P. africana* could be propagated and distributed to the farmers. Good examples of the cultivation of medicinal plants in Uganda include the Rukararwe Development Scheme and the Medicinal Plants and Biodiversity Project funded by the International Development Research Centre, co-ordinated by the NCRL. The Medicinal Plants and Biodiversity Project has five pilot botanical gardens in Uganda. The project works with traditional healers, half of whom are women. The aim is to encourage all traditional healers to have medicinal plant gardens to safeguard wild populations from extinction through over collection. A total of 15 medicinal plants are propagated including: *P. africana, W. ugandensis, Rauvolfia vomitaria* Afzel. (Apocynaceae) and *Maytenus senegalensis* (Lam.) Exell. (Celastraceae). The project also offers training in methods of propagation and harvesting of plant materials for medicinal purposes.

The Rukararwe Rural Development Scheme in Bushenyi district, Western Uganda has a medicinal garden growing over 110 medicinal plants species belonging to 40 plant families (Bukenya-Ziraba, *et al.*, 1997, Bukenya-Ziraba 1998). The project processes the herbs and packages them for sale. The project employs traditional healers and trained medical personnel to treat the patients.

6. Establishment of institutions that promote the study of traditional knowledge on medicinal plants

Traditional knowledge about medicinal plants is in itself highly threatened. For instance, the Batwa (Pygmies) in south-western Uganda, were originally nomadic forest dwellers living as food gatherers and hunters. They have now settled in communities and their knowledge of traditional medicine is now facing extinction and should be documented.

The Botany Department at Makerere has started a new course in ethnobotany for training undergraduates who will put ethnobotany on a sound scientific base for sustainable use and management of plant resources for prosperity. However, a national institute focusing on traditional medicine is required. For instance, the NCRL could be empowered to provide formally recognised courses in traditional medicine. Ethnobotanical research activities have to be broadened further rather than relying entirely on individual researchers and academicians in order to document useful traditional knowledge on herbal medicine. The Uganda Ministry of Health could incorporate approved traditional remedies into primary health care. Traditional healers need to work closely with trained health workers to satisfy the healthcare needs of Ugandans. THETA has been trying to work with both modern and traditional healers in research against chronic diarrhoea and herpes.

7. Government regulations on collection of wild medicinal plants

Due to a population growth rate of 2.5% per annum and the increasing scarcity of medicinal plants, the traditional roles played by taboos, totems and beliefs are being eroded by commercial pressures. In major cities, such as Kampala, trade in medicinals is an increasingly lucrative business. One response could be the regulation of the collection and trade of wild harvested medicinals. This would mean that commercial market vendors and traditional healers could apply for licences. Traders would pay taxes to the government units where they are collecting. The amounts to be collected, where to collect and plant parts collected would be specified in the permits. National licensing and regulations are ways to increase awareness in the local communities about the need to manage medicinal plant resources. The added value attached to these plants in terms of economic gains could yield tangible results towards domestication of these herbs. This could be complemented through international regulation, such as the Convention on International Trade in Endangered Species (CITES).

The creation of trust funds to entice local communities to conserve medicinal plants is yet another approach towards sustainable utilisation. For example, the setting up of Bwindi Impenetrable and Mgahinga Mountain Gorilla National Parks Trust Fund, a World Bank Project to help fund the surrounding community and local individual projects that are environmentally friendly, is a key example of adding monetary value to conservation efforts. The Bwindi Trust Fund is a long-term project that will last as long as these two parks remain conserved. The fund benefits local communities, researchers and conservation managers, but the former take the biggest share when allocating funds.

The intellectual property rights of traditional medical practitioners, individual herbalists and communities need to be protected (CBD Articles 1 (objective 3), 15 and 16). The receiving of benefits or sharing of returns arising from the resources and the knowledge, for which they are the custodians, would be a means of promoting the sustainable utilisation of medicinal plants. This could best be done through funding community-based projects generated by the local communities themselves. For instance, in the surrounding areas around Bwindi Impenetrable National Park, Queen Elizabeth Biosphere Reserve and Lake Mburo National Park, the Uganda Wildlife Authority and non-governmental organisations (e.g. United Nations Educational, Scientific and Cultural Organisation, Man and the Biosphere Programme (UNESCO-MAB), CARE-International) are funding a number of small-scale projects resulting from the revenue sharing scheme from tourism and other conservation activities (Kamatenesi and Origa, 1998). Such projects include building of schools, clinics/dispensaries, road construction as well as donations to women's handcraft shops, bee-keeping and youth or wildlife clubs. The communities, together with conservation managers and community leaders, initiate these projects. Although revenue sharing schemes are not fully formalised, the approach is very healthy.

Conclusion

Although the conservation of medicinal plants and biodiversity in general is gaining support and political credibility in Uganda, human population growth has placed increased pressures on the natural resource base. The cultural link between the local people and their indigenous knowledge in medicinal plant utilisation is at risk. However, due to habitat loss and over-exploitation, the TMPs and herbalists have come to realise that medicinal plants from the wild have become scarce and some plants are already locally reported to be extinct or highly threatened. Therefore, the local people, in particular the herbalists, have started to domesticate the important medicinal plants of great value. Also, local organisations aimed at mobilising traditional healers into proper management of medicinal plant resources, have been formed. For example, the Rukararwe Development Scheme, THETA, Buganda Nedaggala Lyaayo, and NACOTHA.

Furthermore, the Uganda government through its institutions, has initiated programmes aimed at conserving medicinal plants and the implementation of the CBD. The initiation of new courses in ethnobotany and the conducting of ethnobotanical and ethnopharmacological research work in higher institutions of learning like Makerere University is another effective method of promoting medicinal biodiversity conservation.

The Ugandan government is formulating a policy on traditional medicine as a component of its National Biodiversity Strategy and Action Plan. This policy will be one of the most important tools in the conservation of medicinal plants in Uganda. Allowing limited and specified resource use from the protected areas under the MUZs arrangement is yet another important policy in the conservation and monitoring of biodiversity through collaborative management.

The conservation of medicinal plants will be of great potential not only to the local population, but also to the economic development of Uganda. To gain maximum benefit from the conservation of medicinal plants, monetary value has to be attached to conservation efforts, such as the setting up of medicinal gardens, research into potent medicinal plants, and the patenting and intellectual property rights for the local communities, researchers, and professionals. This will involve setting up laboratories with adequate facilities for the assessment of the efficacy of medicinal herbs and collaborative research in Uganda. One way of adding value to conservation efforts is through schemes like revenue sharing to provide social facilities like schools, health centres and road networks according to the people's needs. For instance, the Uganda Wildlife Authority had earlier estimated that 20% of the revenue collected from the use of facilities/services in protected areas would be used in community development. Details and other formalities of sharing are being worked out and the process is already in progress in some protected areas. The implementation of such programmes involving monetary value and resource use will be by the conservation agencies (e.g. Bwindi Trust, UWA), local leaders and political leaders of the districts concerned and at the national level.

Acknowledgements

We are grateful to Dr Possy Mugyenyi and Ms Sophia Apio for providing information of Ugandan laws governing traditional medicine. Further thanks are extended to the Department of Botany, Makerere University for providing facilities during the preparation of the manuscript. Special thanks go to the Editorial team for making arrangements to review the paper and for providing useful comments on the manuscript.

References

Bukenya-Ziraba, R. (1998). Germplasm collection and conservation in Uganda: prospects and constraints, pp. 91–106. In: P. Adams and J.E. Adams (eds). *Conservation and Utilisation of African Plants. Conservation of Plant Genes III.* Missouri Botanical Garden Press, Vol. 71, St Louis, USA.

Bukenya-Ziraba, R., Doenges, P., Duez, P., Lejoly, J. and Ogwal-Okeng, J. (1997). *Medicinal Plants Sub-sector Review, Pharmacopoeia Promoting Programme.* Preparatory study, Final Report. Brussels, Belgium.

Cunningham, A.B. (1988). *An investigation of the Herbal Medicine Trade in Natal/KwaZulu.* Investigational Report No. 29. Pietermitzburg, Institute of Natural Resources, University of Natal, Republic of South Africa.

Farrington, J. and Martin, A. (1988). *Farmer Participatory Research: a review of concept practices.* Agricultural Research and Extension Network paper No. 19. IDO, London, UK.

Glowka, L., Burhenne-Guilmin, F., Synge, H., McNeely, J.A. and Gündling, L. (1994). *A Guide to the Convention on Biological Diversity.* Environmental Policy and Law paper No. 30. IUCN, Gland, Switzerland.

Kamatenesi, M.M. (1997). *The Utilisation of the Medicinal Plants* Rytigynia *spp. (Nyakibazi) in the Multiple Use Zones of Bwindi Impenetrable National Park, Uganda.* M.Sc. Thesis (MUIENR), Makerere University, Kampala, Uganda.

Kamatenesi, M.M. and Origa, H.O. (1998). Gender and Utilisation of Plant Resources of Queen Elizabeth National Park, pp. 64–76. In: M.B. Musoke (ed.). *Gender and Wildlife Management.* 3rd Uganda National UNESCO – BRAAF Seminar, Dec. 1998. Ishaka, Bushenyi District, Uganda.

Kyoshabire, M. (1998*). Medicinal Plants and Herbalists Preferences around Bwindi Impenetrable National Park, Uganda.* M.Sc. Thesis (Botany), Makerere University, Kampala, Uganda.

Martin, G.J. (1995). *Ethnobotany: a people and plants conservation manual.* Chapman and Hall, London, UK.

Mugabe, J. and Clark, N. (1998). *Managing Biodiversity: National Systems of Conservation and Innovation in Africa.* ACTS PRESS, Nairobi, Kenya.

Nabasa, J., Rutwara, G., Walker, F. and Were, C. (1995). *Participatory Rural Appraisal. Practical Experiences.* ACTIONAID, UGANDA. Natural Resources Institute, Overseas Development Administration, London, UK.

Ogwal, E.N.K. (1994). Medicinal Plants used during antenatal and post natal periods and early infant care in Busoga, pp. 768–770. In: X.M. Van de Burgt and J.M. Van Medenbach de Rooy (eds). *The Biodiversity of African Plants, Proceedings XIV AETFAT Congress.* Kluwer Academic Publishers, Dordrecht, The Netherlands.

Pretty, J., Guijt, I., Scoones, I. and Thompson, J. (1996). Regenerating agriculture: the agroecology of low-external input and community based development, pp. 125–145. In: J. Kirkby, P. O'Keefe, and J. Timberlake (eds). *The Earthscan Reader in Sustainable Development.* Earthscan Publications Ltd., London, UK.

Simons, A.J., Dawson, I.K., Duguma, B. and Tchoundjeu, Z. (1998). Passing Problems: Prostate and Prunus. African Team Works to Maintain Sustainable Supply of Pygeum Bark; International Centre for Research in Agroforestry, Nairobi, Kenya. *Herbal Medicine in Africa; HERBALGRAM* No. 43: 49.

Sofowora, A. (1993). *Medicinal Plants and Traditional Medicine in Africa.* Spectrum Books Ltd., Ibadan, Nigeria.

Tuxill, J. and Nabhan, G.P. (1998). *Plants and Protected Areas: a guide to In Situ Management. People and Plants Conservation Manual.* Stanley Thomas (Publishers) Ltd., Cheltenham, UK.

WHO (1991). *Guidelines for the assessment of herbal remedies.* Traditional Medicine Programme of the World Health Organisation, Geneva, Switzerland.

WHO, IUCN and WWF (1993). *Guidelines on the Conservation of Medicinal Plants.* WWF, Gland, Switzerland.

Wild, R.G. and Mutebi, J. (1996). *Conservation through community use of plant resources. Establishing collaborative management at Bwindi Impenetrable and Mgahinga Gorilla National Parks, Uganda.* People and Plants working paper 5, UNESCO, Paris.

Appendix 1

Convention on Biological Diversity – Text and Annexes

Introduction

The Earth's biological resources are vital to humanity's economic and social development. As a result, there is a growing recognition that biological diversity is a global asset of tremendous value to present and future generations. At the same time, the threat to species and ecosystems has never been so great as it is today. Species extinction caused by human activities continues at an alarming rate.

In response, the United National Environment Programme (UNEP) convened the Ad Hoc Working Group of Experts on Biological Diversity in November 1988 to explore the need for an international convention on biological diversity. Soon after, in May 1989, it established the Ad Hoc Working Group of Technical and Legal Experts to prepare an international legal instrument for the conservation and sustainable use of biological diversity. The experts were to take into account "the need to share costs and benefits between developed and developing countries" as well as "ways and means to support innovation by local people".

By February 1991, the Ad Hoc Working Group had become known as the Intergovernmental Negotiating Committee. Its work culminated on 22 May 1991 with the Nairobi Conference for the Adoption of the Agreed Text of the Convention on Biological Diversity.

The Convention was opened for signature on 5 June 1992 at the United Nations Conference on Environment and Development (the Rio "Earth Summit"). It remained open for signature until 4 June 1993, by which time it had received 168 signatures. The Convention entered into force on 29 December 1993, which was 90 days after the 30th ratification. The first session of the Conference of the Parties was scheduled for 28 November – 9 December 1994 in the Bahamas.

The Convention on Biological Diversity was inspired by the world community's growing commitment to sustainable development. It represents a dramatic step forward in the conservation of biological diversity, the sustainable use of its components, and the fair and equitable sharing of benefits arising from the use of genetic resources.

Convention on Biological Diversity Preamble

The Contracting Parties,

Conscious of the intrinsic value of biological diversity and of the ecological, genetic, social, economic, scientific, educational, cultural, recreational and aesthetic values of biological diversity and its components,

Conscious also of the importance of biological diversity for evolution and for maintaining life sustaining systems of the biosphere,

Affirming that the conservation of biological diversity is a common concern of humankind,

Reaffirming that States have sovereign rights over their own biological resources,

Reaffirming also that States are responsible for conserving their biological diversity and for using their biological resources in a sustainable manner,

Concerned that biological diversity is being significantly reduced by certain human activities,

Aware of the general lack of information and knowledge regarding biological diversity and of the urgent need to develop scientific, technical and institutional capacities to provide the basic understanding upon which to plan and implement appropriate measures,

Noting that it is vital to anticipate, prevent and attack the causes of significant reduction or loss of biological diversity at source,

Noting also that where there is a threat of significant reduction or loss of biological diversity, lack of full scientific certainty should not be used as a reason for postponing measures to avoid or minimize such a threat,

Noting further that the fundamental requirement for the conservation of biological diversity is the *in situ* conservation of ecosystems and natural habitats and the maintenance and recovery of viable populations of species in their natural surroundings,

Noting further that *ex situ* measures, preferably in the country of origin, also have an important role to play,

Recognizing the close and traditional dependence of many indigenous and local communities embodying traditional lifestyles on biological resources, and the desirability of sharing equitably benefits arising from the use of traditional knowledge, innovations and practices relevant to the conservation of biological diversity and the sustainable use of its components,

Recognizing also the vital role that women play in the conservation and sustainable use of biological diversity and affirming the need for the full participation of women at all levels of policy-making and implementation for biological diversity conservation,

Stressing the importance of, and the need to promote, international, regional and global cooperation among States and intergovernmental organizations and the non-governmental sector for the conservation of biological diversity and the sustainable use of its components,

Acknowledging that the provision of new and additional financial resources and appropriate access to relevant technologies can be expected to make a substantial difference in the world's ability to address the loss of biological diversity,

Acknowledging further that special provision is required to meet the needs of developing countries, including the provision of new and additional financial resources and appropriate access to relevant technologies,

Noting in this regard the special conditions of the least developed countries and small island States,

Acknowledging that substantial investments are required to conserve biological diversity and that there is the expectation of a broad range of environmental, economic and social benefits from those investments,

Recognising that economic and social development and poverty eradication are the first and overriding priorities of developing countries,

Aware that conservation and sustainable use of biological diversity is of critical importance for meeting the food, health and other needs of the growing world population, for which purpose access to and sharing of both genetic resources and technologies are essential,

Noting that, ultimately, the conservation and sustainable use of biological diversity will strengthen friendly relations among States and contribute to peace for humankind,

Desiring to enhance and complement existing international arrangements for the conservation of biological diversity and sustainable use of its components, and

Determined to conserve and sustainably use biological diversity for the benefit of present and future generations,

Have agreed as follows:

Article 1

Objectives

The objectives of this Convention, to be pursued in accordance with its relevant provisions, are the conservation of biological diversity, the sustainable use of its components and the fair and equitable sharing of the benefits arising out of the utilization of genetic resources, including by appropriate access to genetic resources and by appropriate transfer of relevant technologies, taking into account all rights over those resources and to technologies, and by appropriate funding.

Article 2

Use of Terms

For the purposes of this Convention:

"Biological diversity" means the variability among living organisms from all sources including, *inter alia*, terrestrial, marine and other aquatic ecosystems and the ecological complexes of which they are part; this includes diversity within species, between species and of ecosystems.

"Biological resources" includes genetic resources, organisms or parts thereof, populations, or any other biotic component of ecosystems with actual or potential use or value for humanity.

"Biotechnology" means any technological application that uses biological systems, living organisms, or derivatives thereof, to make or modify products or processes for specific use.

"Country of origin of genetic resources" means the country which possesses those genetic resources in *in situ* conditions.

"Country providing genetic resources" means the country supplying genetic resources collected from *in situ* sources, including populations of both wild and domesticated species, or taken from *ex situ* sources, which may or may not have originated in that country.

"Domesticated or cultivated species" means species in which the evolutionary process has been influenced by humans to meet their needs.

"Ecosystems" means a dynamic complex of plant, animal and micro-organism communities and their non-living environment interacting as a functional unit.

"Ex situ conservation" means the conservation of components of biological diversity outside their natural habitats.

"Genetic material" means any material of plant, animal, microbial or other origin containing functional units of heredity.

"Genetic resources" means genetic material of actual or potential value.

"Habitat" means the place or type of site where an organism or population naturally occurs.

"In situ conditions" means conditions where genetic resources exist within ecosystems and natural habitats, and, in the case of domesticated or cultivated species, in the surroundings where they have developed their distinctive properties.

"In situ conservation" means the conservation of ecosystems and natural habitats and the maintenance and recover of viable populations of species in their natural surroundings and, in the case of domesticated or cultivated species, in the surroundings where they have developed their distinctive properties.

"Protected area" means a geographically defined area which is designated or regulated and managed to achieve specific conservation objectives.

"Regional economic integration organization" means an organization constituted by sovereign States of a given region, to which its member States have transferred competence in respect of matters governed by this Convention and which as been duly authorized, in accordance with its internal procedures, to sign, ratify, accept, approve or accede to it.

"Sustainable use" means the use of components of biological diversity in a way and at a rate that does not lead to the long-term decline of biological diversity, thereby maintaining its potential to meet the needs and aspirations of present and future generations.

"Technology" includes biotechnology.

Article 3

Principle

States have, in accordance with the Charter of the United Nations and the principles of international law, the sovereign right to exploit their own resources pursuant to their own environmental policies, and the responsibility to ensure that activities within their jurisdiction or control do not cause damage to the environment of other States or of areas beyond the limits of national jurisdiction.

Article 4

Jurisdictional Scope

Subject to the rights of other States, and except as otherwise expressly provided in this Convention, the provisions of this Convention apply, in relation to each Contracting Party:

(a)   In the case of components of biological diversity, in areas within the limits of its national jurisdiction; and

(b)   In the case of processes and activities, regardless of where their effects occur, carried out under its jurisdiction or control, within the area of its national jurisdiction or beyond the limits of national jurisdiction.

Article 5

Cooperation

Each Contracting Party shall, as far as possible and as appropriate, cooperate with other Contracting Parties, directly or, where appropriate, through competent international organizations, in respect of areas beyond national jurisdiction and on other matters of mutual interest, for the conservation and sustainable use of biological diversity.

Article 6

General Measures for Conservation and Sustainable Use

Each Contracting Party shall, in accordance with its particular conditions and capabilities:

(a)   Develop national strategies, plans or programmes for the conservation and sustainable use of biological diversity or adapt for this purpose existing strategies, plans or programmes which shall reflect, *inter alia*, the measures set out in this Convention relevant to the Contracting Party concerned; and

(b)   Integrate, as far as possible and as appropriate, the conservation and sustainable use of biological diversity into relevant sectoral or cross-sectoral plans, programmes and policies.

Article 7

Identification and Monitoring

Each Contracting Party shall, as far as possible and as appropriate, in particular for the purposes of Articles 8 to 10:

(a)   Identify components of biological diversity important for its conservation and sustainable use having regard to the indicative list of categories set down in Annex I;

(b)   Monitor, through sampling and other techniques, the components of biological diversity identified pursuant to subparagraph (a) above, paying particular attention to those requiring urgent conservation measures and those which offer the greatest potential for sustainable use;

(c)   Identify processes and categories of activities which have or are likely to have significant adverse impacts on the conservation and sustainable use of biological diversity, and monitor their effects through sampling and other techniques; and

(d)   Maintain and organize, by any mechanism data, derived from identification and monitoring activities pursuant to subparagraphs (a), (b) and (c) above.

Article 8

In situ Conservation

Each Contracting Party shall, as far as possible and as appropriate:

(a)   Establish a system of protected areas or areas where special measures need to be taken to conserve biological diversity;

(b)   Develop, where necessary, guidelines for the selection, establishment and management of protected areas or areas where special measures need to be taken to conserve biological diversity;

(c)   Regulate or manage biological resources important for the conservation of biological diversity whether within or outside protected areas, with a view to ensuring their conservation and sustainable use;

(d)   Promote the protection of ecosystems, natural habitats and the maintenance of viable populations of species in natural surroundings;

(e)   Promote environmentally sound and sustainable development in areas adjacent to protected areas with a view to furthering protection of these areas;

(f)   Rehabilitate and restore degraded ecosystems and promote the recovery of threatened species, *inter alia*, through the development and implementation of plans or other management strategies;

(g)   Establish or maintain means to regulate, manage or control the risks associated with the use and release of living modified organisms resulting from biotechnology which are likely to have adverse environmental impacts that could affect the conservation and sustainable use of biological diversity, taking also into account the risks to human health;

(h)   Prevent the introduction of, control or eradicate those alien species which threaten ecosystems, habitats or species;

(i)   Endeavour to provide the conditions needed for compatibility between present uses and the conservation of biological diversity and the sustainable use of its components;

(j)   Subject to its national legislation, respect, preserve and maintain knowledge, innovations and practices of indigenous and local communities embodying traditional lifestyles relevant for the conservation and sustainable use of biological diversity and promote their wider application with the approval and involvement of the holders of such knowledge, innovations and practices and encourage the equitable sharing of the benefits arising from the utilization of such knowledge, innovations and practices;

(k)   Develop or maintain necessary legislation and/or other regulatory provisions for the protection of threatened species and populations;

(l)   Where a significant adverse effect on biological diversity has been determined pursuant to Article 7, regulate or manage the relevant processes and categories of activities; and

(m)   Cooperate in providing financial and other support for *in situ* conservation outlined in subparagraphs (a) to (l) above, particularly to developing countries.

Article 9

Ex situ Conservation

Each Contracting Party shall, as far as possible and as appropriate, and predominantly for the purpose of complementing *in situ* measures:

(a)   Adopt measures for the *ex situ* conservation of components of biological diversity, preferably in the country of origin of such components;

(b)   Establish and maintain facilities for *ex situ* conservation of and research on plants, animals and micro-organisms, preferably in the country of origin of genetic resources;

(c)   Adopt measures for the recovery and rehabilitation of threatened species and for their reintroduction into their natural habitats under appropriate conditions;

(d)   Regulate and manage collections of biological resources from natural habitats for *ex situ* conservation purposes so as not to threaten ecosystems and *in situ* populations of species, except where special temporary *ex situ* measures are required under subparagraph (c) above; and

(e)   Cooperate in providing financial and other support for *ex situ* conservation outlined in subparagraphs (a) to (d) above and in the establishment and maintenance of *ex situ* conservation facilities in developing countries.

Article 10

Sustainable use of Components of Biological Diversity

Each Contracting Party shall, as far as possible and as appropriate:

(a) Integrate consideration of the conservation and sustainable use of biological resources into national decision making;

(b) Adopt measures relating to the use of biological resources to avoid or minimize adverse impacts on biological diversity;

(c) Protect and encourage customary use of biological resources in accordance with traditional cultural practices that are compatible with conservation or sustainable use requirements;

(d) Support local populations to develop and implement remedial action in degraded areas where biological diversity has been reduced; and

(e) Encourage cooperation between its governmental authorities and its private sector in developing methods for sustainable use of biological resources.

Article 11

Incentive Measures

Each Contracting Party shall, as far as possible and as appropriate, adopt economically and socially sound measures that act as incentives for the conservation and sustainable use of components of biological diversity.

Article 12

Research and Training

The Contracting Parties, taking into account the special needs of developing countries, shall:

(a) Establish and maintain programmes for scientific and technical education and training in measures for the identification, conservation and sustainable use of biological diversity and its components and provide support for such education and training for the specific needs of developing countries;

(b)   Promote and encourage research which contributes to the conservation and sustainable use of biological diversity, particularly in developing countries, *inter alia*, in accordance with decisions of the Conference of the Parties taken in consequence of recommendations of the Subsidiary Body on Scientific, Technical and Technological Advice; and

(c)   In keeping with the provisions of Articles 16, 18 and 20, promote and cooperate in the use of scientific advances in biological diversity research in developing methods for conservation and sustainable use of biological resources.

Article 13
Public Education and Awareness

The Contracting Parties shall:

(a)   Promote and encourage understanding of the importance of, and the measures required for, the conservation of biological diversity, as well as its propagation through media, and the inclusion of these topics in educational programmes; and

(b)   Cooperate, as appropriate, with other States and international organizations in developing educational and public awareness programmes, with respect to conservation and sustainable use of biological diversity.

Article 14
Impact Assessment and Minimizing Adverse Impacts

1. Each Contracting Party, as far as possible and as appropriate, shall:

(a)   Introduce appropriate procedures requiring environmental impact assessment of its proposed projects that are likely to have significant adverse effects on biological diversity with a view to avoiding or minimizing such effects and, where appropriate, allow for public participation in such procedures;

(b)   Introduce appropriate arrangements to ensure that the environmental consequences of its programmes and policies that are likely to have significant impacts on biological diversity are duly taken into account;

(c)   Promote, on the basis of reciprocity, notification, exchange of information and consultation on activities under their jurisdiction or control which are likely to significantly affect adversely the biological diversity of other States or areas beyond the limits of national jurisdiction, by encouraging the conclusion of bilateral, regional or multilateral arrangements, as appropriate;

(d)   In the case of imminent or grave danger or damage, originating under its jurisdiction or control, to biological diversity within the area under jurisdiction of other States or in areas beyond the limits of national jurisdiction, notify immediately the potentially affected States of such danger or damage, as well as initiate action to prevent or minimize such danger or damage; and

(e)   Promote national arrangements for emergency responses to activities or events, whether caused naturally or otherwise, which present a grave and imminent danger to biological diversity and encourage international cooperation to supplement such national efforts and, where appropriate and agreed by the States or regional economic integration organizations concerned, to establish joint contingency plans.

2. The Conference of the Parties shall examine, on the basis of studies to be carried out, the issue of liability and redress, including restoration and compensation, for damage to biological diversity, except where such liability is a purely internal matter.

Article 15
Access to Genetic Resources

1. Recognizing the sovereign rights of States over their natural resources, the authority to determine access to genetic resources rests with the national governments and is subject to national legislation.

2. Each Contracting Party shall endeavour to create conditions to facilitate access to genetic resources for environmentally sound uses by other Contracting Parties and not to impose restrictions that run counter to the objectives of this Convention.

3. For the purpose of this Convention, the genetic resources being provided by a Contracting Party, as referred to in this Article and Articles 16 and 19, are only those that are provided by Contracting Parties that are countries of origin of such resources or by the Parties that have acquired the genetic resources in accordance with this Convention.

4. Access, where granted, shall be on mutually agreed terms and subject to the provisions of this Article.

5. Access to genetic resources shall be subject to prior informed consent of the Contracting Party providing such resources, unless otherwise determined by the Party.

6. Each Contracting Party shall endeavour to develop and carry out scientific research based on genetic resources provided by other Contracting Parties with the full participation of, and where possible in, such Contracting Parties.

7. Each Contracting Party shall take legislative, administrative or policy measures as appropriate, and in accordance with Articles 16 and 19 and, where necessary, through the financial mechanism established by Articles 20 and 21 with the aim of sharing in a fair and equitable way the results of research and development and the benefits arising from the commercial and other utilization of genetic resources with the Contracting Party providing such resources. Such sharing shall be upon mutually agreed terms.

Article 16

Access to and Transfer of Technology

1. Each Contracting Party, recognizing that technology includes biotechnology, and that both access to the transfer of technology among Contracting Parties are essential elements for the attainment of the objectives of this Convention, undertakes subject to the provisions of this Article to provide and/or facilitate access for and transfer to other Contracting Parties of technologies that are relevant to the conservation and sustainable use of biological diversity or make use of genetic resources and do not cause significant damage to the environment.

2. Access to and transfer of technology referred to in paragraph 1 above to developing countries shall be provided and/or facilitates under fair and most favourable terms, including on concessional and preferential terms where mutually agreed, and, where necessary, in accordance with the financial mechanism established by Articles 20 and 21. In the case of technology subject to patents and other intellectual property rights, such access and transfer shall be provided on terms which recognize and are consistent with the adequate and effective protection of intellectual property rights. The application of this paragraph shall be consistent with paragraphs 3, 4 and 5 below.

3. Each Contracting Party shall take legislative, administrative or policy measures, as appropriate, with the aim that Contracting Parties, in particular those that are developing countries, which provide genetic resources are provided access to and transfer of technology which makes use of those resources, on mutually agreed terms, including technology protected by patents and other intellectual property rights, where necessary, through the provisions of Articles 20 and 21 and in accordance with international law and consistent with paragraphs 4 and 5 below.

4. Each Contracting Party shall take legislative, administrative or policy measures, as appropriate, with the aim that the private sector facilitates access to, joint development and transfer of technology referred to in paragraph 1 above for the benefit of both governmental institutions and the private sector of developing countries and in this regard shall abide by the obligations included in paragraphs 1, 2 and 3 above.

5. The Contracting Parties, recognizing that patents and other intellectual property rights may have an influence on the implementation of this Convention, shall cooperate in this regard subject to national legislation and international law in order to ensure that such rights are supportive of and do not run counter to its objectives.

Article 17
Exchange of Information

1. The Contracting Parties shall facilitate the exchange of information, from all publicly available sources, relevant to the conservation and sustainable use of biological diversity, taking into account the special needs of developing countries.

2. Such exchange of information shall include exchange of results of technical, scientific and socio-economic research, as well as information on training and surveying programmes, specialized knowledge, indigenous and traditional knowledge as such and in combination with the technologies referred to an Article 16, paragraph 1. It shall also, where feasible, include repatriation of information.

Article 18
Technical and Scientific Cooperation

1. The Contracting Parties shall promote international technical and scientific cooperation in the field of conservation and sustainable use of biological diversity, where necessary, through the appropriate international and national institutions.

2. Each Contracting Party shall promote technical and scientific cooperation with other Contracting Parties, in particular developing countries, in implementing this Convention, *inter alia*, through the development and implementation of national policies. In promoting such cooperation, special attention should be given to the development and strengthening of national capabilities, by means of human resources development and institution building.

3. The Conference of the Parties, at its first meeting, shall determine how to establish a clearing-house mechanism to promote and facilitate technical and scientific cooperation.

4. The Contracting Parties shall, in accordance with national legislation and policies, encourage and develop methods of cooperation for the development and use of technologies, including indigenous and traditional technologies, in pursuance of the objectives of this Convention. For this purpose, the Contracting Parties shall also promote cooperation in the training of personnel and exchange of experts.

5. The Contracting Parties shall, subject to mutual agreement, promote the establishment of joint research programmes and joint ventures for the development of technologies relevant to the objectives of this Convention.

Article 19
Handling of Biotechnology and Distribution of its Benefits

1. Each Contracting Party shall take legislative, administrative or policy measures, as appropriate, to provide for the effective participation in biotechnological research activities by those Contracting Parties, especially developing countries, which provide the genetic resources for such research, and where feasible in such Contracting Parties.

2. Each Contracting Party shall take all practicable measures to promote and advance priority access on a fair and equitable basis by Contracting Parties, especially developing countries, to the results and benefits arising from biotechnologies based upon genetic resources provided by those Contracting Parties. Such access shall be on mutually agreed terms.

3. The Parties shall consider the need for and modalities of a protocol setting out appropriate procedures, including, in particular, advance informed agreement, in the field of the safe transfer, handling and use of any living modified organism resulting from biotechnology that may have adverse effect on the conservation and sustainable use of biological diversity.

4. Each Contracting Party shall, directly or by requiring any natural or legal person under its jurisdiction providing the organisms referred to in paragraph 3 above, provide any available information about the use and safety regulations required by that Contracting Party in handling such organisms, as well as any available information of the potential adverse impact of the specific organisms concerned to the Contracting Party into which those organisms are to be introduced.

Article 20

Financial Resources

1. Each Contracting Party undertakes to provide, in accordance with its capabilities, financial support and incentives in respect of those national activities which are intended to achieve the objectives of this Convention, in accordance with its national plans, priorities and programmes.

2. The developed country Parties shall provide new and additional financial resources to enable developing country Parties to meet the agreed full incremental costs to them of implementing measures which fulfil the obligations of this Convention and to benefit from its provisions and which costs are agreed between a developing country Party and the institutional structure referred to in Article 21, in accordance with policy, strategy, programme priorities and eligibility criteria and an indicative list of incremental costs established by the Conference of the Parties. Other Parties, including countries undergoing the process of transition to a market economy, may voluntarily assume the obligations of the developed country Parties. For the purpose of this Article, the Conference of the Parties, shall at its first meeting establish a list of developed country Parties and other Parties which voluntarily assume the obligations of the developed country Parties. The Conference of the Parties shall periodically review and if necessary amend the list. Contributions from other countries and sources on a voluntary basis would also be encouraged. The implementation of these commitments shall take into account the need for adequacy, predictability and timely flow of funds and the importance of burden-sharing among the contributing Parties included in the list.

3. The developed country Parties may also provide, and developing country Parties avail themselves of, financial resources related to the implementation of this Convention through bilateral, regional and other multilateral channels.

4. The extent to which developing country Parties will effectively implement their commitments under this Convention will depend on the effective implementation by developed country Parties of their commitments under this Convention related to financial resources and transfer of technology and will take fully into account the fact that economic and social development and eradication of poverty are the first and overriding priorities of the developing country Parties.

5. The Parties shall take full account of the specific needs and special situation of least developed countries in their actions with regard to funding and transfer of technology.

6. The Contracting Parties shall also take into consideration the special conditions resulting from the dependence on, distribution and location of, biological diversity within developing country Parties, in particular small island States.

7. Consideration shall also be given to the special situation of developing countries, including those that are most environmentally vulnerable, such as those with arid and semi-arid zones, coastal and mountainous areas.

Article 21

Financial Mechanism

1. There shall be a mechanism for the provision of financial resources to developing country Parties for purposes of this Convention on a grant or concessional basis the essential elements of which are described in this Article. The mechanism shall function under the authority and guidance of, and be accountable to, the Conference of the Parties for purposes of this Convention. The operations of the mechanism shall be carried out by such institutional structure as may be decided upon by the Conference of the Parties at its first meeting. For purposes of this Convention, The Conference of the Parties shall determine the policy, strategy, programme priorities and eligibility criteria relating to the access to and utilization of such resources. The contributions shall be such as to take into account the need for predictability, adequacy and timely flow of funds referred to in Article 20 in accordance with the amount of resources needed to be decided periodically by the Conference of the Parties and the importance of burden-sharing among the contributing Parties included in the list referred to in Article 20, paragraph 2. Voluntary contributions may also be made by the developed country Parties and by other countries and sources. The mechanism shall operate within a democratic and transparent system of governance.

2. Pursuant to the objectives of this Convention, the Conference of the Parties shall at its first meeting determine the policy, strategy and programme priorities, as well as detailed criteria and guidelines for eligibility for access to and utilization of the financial resources including monitoring and evaluation on a regular basis of such utilization. The Conference of the Parties shall decide on the arrangements to give effect to paragraph 1 above after consultation with the institutional structure entrusted with the operation of the financial mechanism.

3. The Conference of the Parties shall review the effectiveness of the mechanism established under this Article, including the criteria and guidelines referred to in paragraph 2 above, not less than two years after the entry into force of this Convention and thereafter on a regular basis. Based on such review, it shall take appropriate action to improve the effectiveness of the mechanism if necessary.

4. The Contracting Parties shall consider strengthening existing financial institutions to provide financial resources for the conservation and sustainable use of biological diversity.

Article 22

Relationship with other International Conventions

1. The provisions of this Convention shall not affect the rights and obligations of any Contracting Party deriving from any existing international agreement, except where the exercise of those rights and obligations would cause a serious damage or threat to biological diversity.

2. Contracting Parties shall implement this Convention with respect to the marine environment consistently with the rights and obligations of States under the law of the sea.

Article 23

Conference of the Parties

1. A Conference of the Parties is hereby established. The first meeting of the Conference of the Parties shall be convened by the Executive Director of the United National Environment Programme not later than one year after the entry into force of this Convention. Thereafter, ordinary meetings of the Conference of the Parties shall be held at regular intervals to be determined by the Conference at its first meeting.

2. Extraordinary meetings of the Conference of the Parties shall be held at such other times as may be deemed necessary by the Conference, or at the written request of any Party, provided that, within six months of the request being communicated to them by the Secretariat, it is supported by at least one third of the Parties.

3. The Conference of the Parties shall by consensus agree upon and adopt rules of procedure for itself and for any subsidiary body it may establish, as well as financial rules governing the funding of the Secretariat. At each ordinary meeting, it shall adopt a budget for the financial period until the next ordinary meeting.

4. The Conference of the Parties shall keep under review the implementation of this Convention, and, for this purpose, shall:

(a) Establish the form and the intervals for transmitting the information to be submitted in accordance with Article 26 and consider such information as well as reports submitted by any subsidiary body;

(b) Review scientific, technical and technological advice on biological diversity provided in accordance with Article 25;

(c) Consider and adopt, as required, protocols in accordance with Article 28;

(d)   Consider and adopt, as required, in accordance with Articles 29 and 30, amendments to this Convention and its annexes;

(e)   Consider amendments to any protocol, as well as to any annexes thereto, and, if so decided, recommend their adoption to the parties to the protocol concerned;

(f)   Consider and adopt, as required, in accordance with Article 30, additional annexes to this Convention;

(g)   Establish such subsidiary bodies, particularly to provide scientific and technical advice, as are deemed necessary for the implementation of this Convention;

(h)   Contact, through the Secretariat, the executive bodies of conventions dealing with the matters covered by this Convention with a view to establishing appropriate forms of cooperation with them; and

(i)   Consider and undertake any additional action that may be required for the achievement of the purposes of this Convention in the light of experience gained in its operation.

5. The United Nations, its specialized agencies and the International Atomic Energy Agency, as well as any State not Party to this Convention, may be represented as observers at meetings of the Conference of the Parties. Any other body or agency, whether governmental or non-governmental, qualified in fields relating to conservation and sustainable use of biological diversity, which has informed the Secretariat of its wish to be represented as an observer at a meeting of the Conference of the Parties, may be admitted unless at least one third of the Parties present object. The admission and participation of observers shall be subject to the rules of procedure adopted by the Conference of the Parties.

Article 24

Secretariat

1. A secretariat is hereby established. Its functions shall be:

(a)   To arrange for and service meetings of the Conference of the Parties provided for in Article 23;

(b)   To perform the functions assigned to it by any protocol;

(c)   To prepare reports on the execution of its functions under this Convention and present them to the Conference of the Parties;

(d)   To coordinate with other relevant international bodies and, in particular to enter into such administrative and contractual arrangements as may be required for the effective discharge of its functions; and

(e)   To perform such other functions as may be determined by the Conference of the Parties.

2. At its first ordinary meeting, the Conference of the Parties shall designate the secretariat from amongst those existing competent international organizations which have signified their willingness to carry out the secretariat functions under this Convention.

Article 25

Subsidiary Body on Scientific, Technical and Technological Advice

1. A subsidiary body for the provision of scientific, technical and technological advice is hereby established to provide the Conference of the Parties and, as appropriate, its other subsidiary bodies with timely advice relating to the implementation of this Convention. This body shall be open to participation by all Parties and shall be multidisciplinary. It shall comprise government representatives competent in the relevant field of expertise. It shall report regularly to the Conference of the Parties on all aspects of its work.

2. Under authority of and in accordance with guidelines laid down by the Conference of the Parties, and upon its request, this body shall:

(a)   Provide scientific and technical assessment of the status of biological diversity;

(b)   Prepare scientific and technical assessments of the effects of types of measures taken in accordance with the provisions of this Convention;

(c)   Identify innovative, efficient and state of the art technologies and know-how relating to the conservation and sustainable use of biological diversity and advise on the ways and means of promoting development and/or transferring such technologies;

(d)   Provide advice on scientific programmes and international cooperation in research and development related to conservation and sustainable use of biological diversity; and

(e)   Respond to scientific, technical, technological and methodological questions that the Conference of the Parties and its subsidiary bodies may put to the body.

3. The functions, terms of reference, organization and operation of this body may be further elaborated by the Conference of the Parties.

Article 26

Reports

Each Contracting Party shall, at intervals to be determined by the Conference of the Parties, present to the Conference of the Parties, reports on measures which it has taken for the implementation of the provisions of this Convention and their effectiveness in meeting the objectives of this Convention.

Article 27

Settlement of Disputes

1. In the event of a dispute between Contracting Parties concerning the interpretation or application of this Convention, the parties concerned shall seek solution by negotiation.

2. If the parties concerned cannot reach agreement by negotiation, they may jointly seek the good offices of, or request mediation by, a third party.

3. When ratifying, accepting, approving or acceding to this Convention, or at any time thereafter, a State or regional economic integration organization may declare in writing to the Depositary that for a dispute not resolved in accordance with paragraph 1 or paragraph 2 above, it accepts one or both of the following means of dispute settlement as compulsory:

(a)   Arbitration in accordance with the procedure laid down in Part 1 of Annex II;

(b)   Submission of the dispute to the International Court of Justice.

4. If the parties to the dispute have not, in accordance with paragraph 3 above, accepted the same or any procedure, the dispute shall be submitted to conciliation in accordance with Part 2 of Annex II unless the parties otherwise agree.

5. The provisions of this Article shall apply with respect to any protocol except as otherwise provided in the protocol concerned.

Article 28

Adoption of Protocols

1. The Contracting Parties shall cooperate in the formulation and adoption of protocols to this Convention.

2. Protocols shall be adopted at a meeting of the Conference of the Parties.

3. The text of any proposed protocol shall be communicated to the Contracting Parties by the Secretariat at least six months before such a meeting.

Article 29

Amendment of the Convention or Protocols

1. Amendments to this Convention may be proposed by any Contracting Party. Amendments to any protocol may be proposed by any Party to that protocol.

2. Amendments to this Convention shall be adopted at a meeting of the Conference of the Parties. Amendments to any protocol shall be adopted at a meeting of the Parties to the Protocol in question. The text of any proposed amendment to this Convention or to any protocol, except as may otherwise be provided in such protocol, shall be communicated to the Parties to the instrument in question by the secretariat at least six months before the meeting at which it is proposed for adoption. The secretariat shall also communicate proposed amendments to the signatories to the Convention for information.

3. The Parties shall make every effort to reach agreement on any proposed amendment to this Convention or to any protocol by consensus. If all efforts at consensus have been exhausted, and no agreement reached, the amendment shall as a last resort be adopted by a two-third majority vote of the Parties to the instrument in question present and voting at the meeting, and shall be submitted by the Depositary to all Parties for ratification, acceptance or approval.

4. Ratification, acceptance or approval of amendments shall be notified to the Depositary in writing. Amendments adopted in accordance with paragraph 3 above shall enter into force among Parties having accepted them on the nineteenth day after the deposit of instruments of ratification, acceptance or approval by at least two thirds of the Contracting Parties to this Convention or of the Parties to the protocol concerned, except as may otherwise be provided in such protocol. Thereafter the amendments shall enter into force for any other Party on the nineteenth day after that Party deposits its instrument of ratification, acceptance or approval of the amendments.

5. For the purposes of this Article, "Parties present and voting" means Parties present and casting an affirmative or negative vote.

Article 30

Adoption and Amendment of Annexes

1. The annexes to this Convention or to any protocol shall form an integral part of the Convention or of such protocol, as the case may be, and, unless expressly provided otherwise, a reference to this Convention or its protocols constitutes at the same time a reference to any annexes thereto. Such annexes shall be restricted to procedural, scientific, technical and administrative matters.

2. Except as may be otherwise provided in any protocol with respect to its annexes, the following procedure shall apply to the proposal, adoption and entry into force of additional annexes to this Convention or of annexes to any protocol:

(a)   Annexes to this Convention or to any protocol shall be proposed and adopted according to the procedure laid down in Article 29;

(b)   Any Party that is unable to approve an additional annex to this Convention or an annex to any protocol to which it is Party shall so notify the Depositary, in writing, within one year from the date of the communication of the adoption by the Depositary. The Depositary shall without delay notify all Parties of any such notification received. A Party may at any time withdraw a previous declaration of objection and the annexes shall thereupon enter into force for that Party subject to subparagraph (c) below;

(c)   On the expiry of one year from the date of the communication of the adoption by the Depositary, the annex shall enter into force for all Parties to this Convention or to any protocol concerned which have not submitted a notification in accordance with the provisions of subparagraph (b) above.

3. The proposal, adoption and entry into force of amendments to annexes to this Convention or to any protocol shall be subject to the same procedure as for the proposal, adoption and entry into force of annexes to the Convention or annexes to any protocol.

4. If an additional annex or an amendment to an annex is related to an amendment to this Convention or to any protocol, the additional annex or amendment shall not enter into force until such time as the amendment to the Convention or to the protocol concerned enters into force.

Article 31

Right to Vote

1. Except as provided for in paragraph 2 below, each Contracting Party to this Convention or to any protocol shall have one vote.

2. Regional economic integration organizations, in matters within their competence, shall exercise their right to vote with a number of votes equal to the number of their member States which are Contracting Parties to this Convention or the relevant protocol. Such organizations shall not exercise their right to vote if their member States exercise theirs, and vice versa.

Article 32

Relationship between this Convention and its Protocols

1. A State or a regional economic integration organization may not become a Party to a protocol unless it is, or becomes at the same time, a Contracting Party to this Convention.

2. Decisions under any protocol shall be taken only by the Parties to the protocol concerned. Any Contracting Party that has not ratified, accepted or approved a protocol may participate as an observer in any meeting of the parties to that protocol.

Article 33

Signature

This Convention shall be open for signature at Rio de Janeiro by all States and any regional economic integration organization from 5 June 1992 until 14 June 1992, and at the United Nations Headquarters in New York from 15 June 1992 to 4 June 1993.

Article 34

Ratification, Acceptance or Approval

1. This Convention and any protocol shall be subject to ratification, acceptance or approval by States and by regional economic integration organizations. Instruments of ratification, acceptance or approval shall be deposited with the Depositary.

2. Any organization referred to in paragraph 1 above which becomes a Contracting Party to this Convention or any protocol without any of its member States being a Contracting Party shall be bound by all the obligations under the Convention or the protocol, as the case may be. In the case of such organizations, one or more of whose member States is a Contracting Party to this Convention or relevant protocol, the organization and its member States shall decide on their respective responsibilities for the performance of their obligations under the Convention or protocol, as the case may be. In such cases, the organization and the member States shall not be entitled to exercise rights under the Convention or relevant protocol concurrently.

3. In their instruments of ratification, acceptance or approval, the organizations referred to in paragraph 1 above shall declare the extend of their competence with respect to the matters governed by the Convention or the relevant protocol. These organizations shall also inform the Depositary of any relevant modification in the extend of their competence.

Article 35

Accession

1. This Convention and any protocol shall be open for accession by States and by regional economic integration organizations from the date on which the Convention or the protocol concerned is closed for signature. The instruments of accession shall be deposited with the Depositary.

2. In their instruments of accessions, the organizations referred to in paragraph 1 above shall declare the extent of their competence with respect to the matters governed by the Convention or the relevant protocol. These organizations shall also inform the Depositary of any relevant modification in the extent of their competence.

3. The provisions of Article 34, paragraph 2, shall apply to regional economic integration organizations which accede to this Convention or any protocol.

Article 36

Entry into Force

1. This Convention shall enter into force on the ninetieth day after the date of deposit of the thirtieth instrument of ratification, acceptance, approval or accession.

2. Any protocol shall enter into force on the ninetieth day after the date of deposit of the number of instruments of ratification, acceptance, approval or accession, specified in that protocol, has been deposited.

3. For each Contracting Party which ratifies, accepts or approves this Convention or accedes thereto after the deposit of the thirtieth instrument of ratification, acceptance, approval or accession, it shall enter into force on the ninetieth day after the date of deposit by such Contracting Party of its instrument of ratification, acceptance, approval or accession.

4. Any protocol, except as otherwise provided in such protocol, shall enter into force for a Contracting Party that ratifies, accepts or approves that protocol or accedes thereto after its entry into force pursuant to paragraph 2 above, on the ninetieth day after the date on which that Contracting Party deposits its instrument of ratification, acceptance, approval or accession, or on the date on which this Convention enters into force for that Contracting Party, whichever shall be the later.

5. For the purposes of paragraphs 1 and 2 above, any instrument deposited by a regional economic integration organization shall not be counted as additional to those deposited by member States of such organization.

Article 37

Reservations

No reservations may be made to this Convention.

Article 38
Withdrawals

1. At any time after two years from the date on which this Convention has entered into force for a Contracting Party, that Contracting Party may withdraw from the Convention by giving written notification to the Depositary.

2. Any such withdrawal shall take place upon expiry of one year after the date of its receipt by the Depositary, or on such later date as may be specified in the notification of the withdrawal.

3. Any Contracting Party which withdraws from this Convention shall be considered as also having withdrawn from any protocol to which it is party.

Article 39
Financial Interim Arrangements

Provided that it has been fully restructured in accordance with the requirements of Article 21, the Global Environment Facility of the United Nations Development Programme, the United Nations Environment Programme and the International Bank for Reconstruction and Development shall be the institutional structure referred to in Article 21 on an interim basis, for the period between the entry into force of this Convention and the first meeting of the Conference of the Parties or until the Conference of the Parties decided which institutional structure will be designated in accordance with Article 21.

Article 40
Secretariat Interim Arrangements

The secretariat to be provided by the Executive Director of the United Nations Environment Programme shall be the secretariat referred to in Article 24, paragraph 2, on an interim basis for the period between the entry into force of this Convention and the first meeting of the Conference of the Parties.

Article 41

Depositary

The Secretary-General of the United Nations shall assume the functions of Depositary of this Convention and any protocols.

Article 42

Authentic Texts

The original of this Convention, of which the Arabic, Chinese, English, French, Russian and Spanish texts are equally authentic, shall be deposited with the Secretary-General of the United Nations.

IN WITNESS WHEREOF the undersigned, being duly authorized to that effect, have signed this Convention.

Done at Rio de Janeiro on this fifth day of June, one thousand nine hundred and ninety-two.

Annex I

Identification and Monitoring

1. Ecosystems and habitats: containing high diversity, large numbers of endemic or threatened species, or wilderness; required by migratory species; of social, economic, cultural or scientific importance; or, which are representative, unique or associated with key evolutionary or other biological processes;

2. Species and communities which are: threatened; wild relatives of domesticated or cultivated species; of medicinal, agricultural or other economic value; or social, scientific or cultural importance; or importance for research into the conservation and sustainable use of biological diversity, such as indicator species; and

3. Described genomes and genes of social, scientific or economic importance.

Annex II

Part 1 – Arbitration

Article 1

The claimant party shall notify the secretariat that the parties are referring a dispute to arbitration pursuant to Article 27. The notification shall state the subject-matter of arbitration and include, in particular, the articles of the Convention or the protocol, the interpretation or application of which are at issue. If the parties do not agree on the subject matter of the dispute before the President of the tribunal is designated, the arbitral tribunal shall determine the subject matter. The secretariat shall forward the information thus received to all Contracting Parties to this Convention or to the protocol concerned.

Article 2

1. In disputes between two parties, the arbitral tribunal shall consist of three members. Each of the parties to the dispute shall appoint an arbitrator and the two arbitrators so appointed shall designate by common agreement the third arbitrator who shall be the President of the tribunal. The latter shall not be a national of one of the parties to the dispute, nor have his or her usual place of residence in the territory of one of these parties, nor be employed by any of them, nor have dealt with the case in any other capacity.

2. In disputes between more than two parties, parties in the same interest shall appoint one arbitrator jointly by agreement.

3. Any vacancy shall be filled in the manner prescribed for the initial appointment

Article 3

1. If the President of the arbitral tribunal has not been designated within two months of the appointment of the second arbitrator, the Secretary-General of the United Nations shall, at the request of a party, designate the President within a further two-month period.

2. If one of the parties to the dispute does not appoint an arbitrator within two months of receipt of the request, the other party may inform the Secretary-General who shall make the designation within a further two-month period.

Article 4

The arbitral tribunal shall render its decisions in accordance with the provisions of this Convention any protocols concerned, and international law.

Article 5

Unless the parties to the dispute otherwise agree, the arbitral tribunal shall determine its own rules of procedure.

Article 6

The arbitral tribunal may, at the request of one of the parties, recommend essential interim measures of protection.

Article 7

The parties to the dispute shall facilitate the work of the arbitral tribunal and, in particular, using all means at their disposal, shall:

(a)   Provide it with all relevant documents, information and facilities; and

(b)   Enable it, when necessary, to call witnesses or experts and receive their evidence.

Article 8

The parties and the arbitrators are under an obligation to protect the confidentiality of any information they receive in confidence during the proceedings of the arbitral tribunal.

Article 9

Unless the arbitral tribunal determines otherwise because of the particular circumstances of the case, the costs of the tribunal shall be borne by the parties to the dispute in equal shares. The tribunal shall keep a record of all its costs, and shall furnish a final statement thereof to the parties.

Article 10

Any Contracting Party that has an interest of a legal nature in the subject-matter of the dispute which may be affected by the decision in the case, may intervene in the proceedings with the consent of the tribunal.

Article 11

The tribunal may hear and determine counterclaims arising directly out of the subject-matter of the dispute.

Article 12

Decisions both on procedure and substance of the arbitral tribunal shall be taken by a majority vote of its members.

Article 13

If one of the parties to the dispute does not appear before the arbitral tribunal or fails to defend its case, the other party may request the tribunal to continue the proceedings and to make its award. Absence of a party or a failure of a party to defend its case shall not constitute a bar to the proceedings. Before rendering its final decision, the arbitral tribunal must satisfy itself that the claim is well founded in fact and law.

Article 14

The tribunal shall render its final decision within five months of the date on which it is fully constituted unless it finds it necessary to extend the time-limit for a period which should not exceed five more months.

Article 15

The final decision of the arbitral tribunal shall be confined to the subject-matter of the dispute and shall state the reasons on which it is based. It shall contain the names of the members who have participated and the date of the final decision. Any member of the tribunal may attach a separate or dissenting opinion to the final decision.

Article 16

The award shall be binding on the parties to the dispute. It shall be without appeal unless the parties to the dispute have agreed in advance to an appellate procedure.

Article 17

Any controversy which may arise between the parties to the dispute as regards the interpretation or manner of implementation of the final decision may be submitted by either party for decision to the arbitral tribunal which rendered it.

Part 2 – Conciliation

Article 1

A conciliation commission shall be created upon the request of one of the parties to the dispute. The commission shall, unless the parties otherwise agree, be composed of five members, two appointed by each Party concerned and a President chosen jointly by those members.

Article 2

In disputes between more than two parties, parties in the same interest shall appoint their members of the commission jointly by agreement. Where two or more parties have separate interests or there is a disagreement as to whether they are of the same interest, they shall appoint their members separately.

Article 3

If any appointments by the parties are not made within two months of the date of the request to create a conciliation commission, the Secretary-General of the United Nations shall, if asked to do so by the party that made the request, make those appointments within a further two-month period.

Article 4

If a President of the conciliation commission has not been chosen within two months of the last of the members of the commission being appointed, the Secretary-General of the United Nations shall, if asked to do so by a party, designate a President within a further two-month period.

Article 5

The conciliation commission shall take its decisions by majority vote of its members. It shall, unless the parties to the dispute otherwise agree, determine its own procedure. It shall render a proposal for resolution of the dispute, which the parties hall consider in good faith.

Article 6

A disagreement as to whether the conciliation commission has competence shall be decided by the commission.

For more information about the Convention on Biological Diversity, its implementation and activities of the Secretariat and the Contacting Parties see: www.biodiv.org

Appendix 2

IUCN Red List Categories – Version 3.1

Prepared by the IUCN Species Survival Commission

(As approved by the 51st meeting of the IUCN Council, Gland, Switzerland – 9 February 2000)

SPECIES SURVIVAL COMMISSION

I. Introduction

1. The IUCN Red List Categories are intended to be an easily and widely understood system for classifying species at high risk of global extinction. The general aim of the system is to provide an explicit, objective framework for the classification of the broadest range of species according to their extinction risk. However, while the Red List may focus attention on those taxa at the highest risk it is not the sole means of setting priorities for conservation measures for their protection.

Extensive consultation and testing in the development of the system strongly suggests that it is robust across most organisms. However, it should be noted that although the system places species into the threatened categories with a high degree of consistency, the criteria do not take into account the life histories of every species. Hence, in certain individual cases, the risk of extinction may be under- or over-estimated.

2. Before 1994 the more subjective threatened species categories used in IUCN Red Data Books and Red Lists had been in place, with some modification, for almost 30 years. Although the need to revise the categories had long been recognised (Fitter & Fitter, 1987), the current phase of development only began in 1989 following a request from the IUCN Species Survival Commission (SSC) Steering Committee to develop a more objective approach. The IUCN Council adopted the new Red List system in 1994.

The IUCN Red List Categories and Criteria have several specific aims:

- to provide a system that can be applied consistently by different people;
- to improve objectivity by providing users with clear guidance on how to evaluate different factors which affect risk of extinction;
- to provide a system which will facilitate comparisons across widely different taxa;
- to give people using threatened species lists a better understanding of how individual species were classified.

3. Since their adoption by IUCN Council in 1994, the IUCN Red List Categories have become widely recognised internationally and they are now used in a range of publications and listings produced by IUCN, as well as by numerous governmental and non-governmental organisations. Such broad and extensive use revealed the need for a number of improvements and SSC was mandated by the 1996 World Conservation Congress (WCC Res. 1.4) to conduct a review of the system (IUCN, 1996). This document presents the revisions accepted by the IUCN Council.

The proposals presented in this document result from a continuing process of drafting, consultation and validation. The production of a large number of draft proposals has led to some confusion, especially as each draft has been used for classifying some set of species for conservation purposes. To clarify matters, and to open the way for modifications as and when they become necessary, a system for version numbering is as follows:

Version 1.0: Mace & Lande (1991)

The first paper discussing a new basis for the categories, and presenting numerical criteria especially relevant for large vertebrates.

Version 2.0: Mace *et al.* (1992)

A major revision of Version 1.0, including numerical criteria appropriate to all organisms and introducing the non-threatened categories.

Version 2.1: IUCN (1993)

Following an extensive consultation process within SSC, a number of changes were made to the details of the criteria, and fuller explanation of basic principles was included. A more explicit structure clarified the significance of the non-threatened categories.

Version 2.2: Mace & Stuart (1994)

Following further comments received and additional validation exercises, some minor changes to the criteria were made. In addition, the Susceptible category present in Versions 2.0 and 2.1 was subsumed into the Vulnerable category. A precautionary application of the system was emphasised.

Version 2.3: IUCN (1994)

IUCN Council adopted this version, which incorporated changes as a result of comments from IUCN members, in December 1994. The initial version of this document was published without the necessary bibliographic details, such as date of publication and ISBN number, but these were included in the subsequent reprints in 1998 and 1999. This version was used for the *1996 IUCN Red List of Threatened Animals* (Baillie and Groombridge, 1996), *The World List of Threatened Trees* (Oldfield *et al.*,1998) and the *2000 IUCN Red List of Threatened Species* (Hilton-Taylor, 2000).

Version 3.0: IUCN/SSC Criteria Review Working Group (1999)

Following comments received, a series of workshops were convened to look at the IUCN Red List Criteria following which, changes were proposed affecting the criteria, the definitions of some key terms and the handling of uncertainty.

Version 3.1: IUCN (2001)

The IUCN Council adopted this latest document, which incorporated changes as a result of comments from the IUCN and SSC memberships and from a final meeting of the Criteria Review Working Group, in February 2000.

All new assessments from January 2001 should use the latest adopted version and cite the year of publication and version number.

4. In the rest of this document the proposed system is outlined in several sections. Section II, the Preamble, presents basic information about the context and structure of the system, and the procedures that are to be followed in applying the criteria to species. Section III provides definitions of key terms used. Section IV presents the categories, while Section V details the quantitative criteria used for classification within the threatened categories. Annex I provides guidance on how to deal with uncertainty when applying the criteria; Annex II suggests a standard format for citing the Red List Categories and Criteria; and Annex III outlines the documentation requirements for taxa to be included on IUCN's global Red Lists. It is important for the effective functioning of the system that all sections are read and understood to ensure that the definitions and rules are followed (**Note:** Annexes I, II and III will be updated on a regular basis).

II. Preamble

The information in this section is intended to direct and facilitate the use and interpretation of the categories (Critically Endangered, Endangered, etc.), criteria (A to E), and subcriteria (1, 2, etc.; a, b, etc.; i, ii, etc.):

1.   Taxonomic level and scope of the categorisation process

The criteria can be applied to any taxonomic unit at or below the species level. In the following information, definitions and criteria, the term 'taxon' is used for convenience, and may represent species or lower taxonomic levels, including forms that are not yet formally described. There is sufficient range among the different criteria to enable the appropriate listing of taxa from the complete taxonomic spectrum, with the exception of micro-organisms. The criteria may also be applied within any specified geographical or political area, although in such cases special notice should be taken of point 14. In presenting the results of applying the criteria, the taxonomic unit and area under consideration should be specified in accordance with the documentation guidelines. The categorisation process should only be applied to wild populations inside their natural range, and to populations resulting from benign introductions. The latter are defined in the IUCN *Guidelines for Re-introductions* (IUCN, 1998) as '...an attempt to establish a species, for the purpose of conservation, outside its recorded distribution, but within an appropriate habitat and eco-geographical area. This is a feasible conservation tool only when there is no remaining area left within a species' historic range'.

2. Nature of the categories

Extinction is a chance process. Thus, a listing in a higher extinction risk category implies a higher expectation of extinction, and over the time-frames specified, more taxa listed in a higher category are expected to go extinct than those in a lower one (without effective conservation action). However, the persistence of some taxa in high-risk categories does not necessarily mean their initial assessment was inaccurate.

All taxa listed as Critically Endangered qualify for Vulnerable and Endangered, and all listed as Endangered qualify for Vulnerable. Together these categories are described as 'threatened'. The threatened categories form a part of the overall scheme. It will be possible to place all taxa into one of the categories (see Figure 1).

3. Role of the different criteria

For listing as Critically Endangered, Endangered or Vulnerable there is a range of quantitative criteria; meeting any one of these criteria qualifies a taxon for listing at that level of threat. Each taxon should be evaluated against all the criteria. Even though some criteria will be inappropriate for certain taxa (some taxa will never qualify under these however close to extinction they come), there should be criteria appropriate for assessing threat levels for any taxon. The relevant factor is whether *any one* criterion is met, not whether all are appropriate or all are met. Because it will never be clear which criteria are appropriate for a particular taxon in advance, each taxon should be evaluated against all the criteria, and any criterion met should be listed.

4. Derivation of quantitative criteria

The different criteria (A–E) are derived from a wide review aimed at detecting risk factors across the broad range of organisms and the diverse life histories they exhibit. The quantitative values presented in the various criteria associated with threatened categories were developed through wide consultation and they are set at what are generally judged to be appropriate levels, even if no formal justification for these values exists. The levels for different criteria within categories were set independently but against a common standard. Broad consistency between them was sought.

5. Conservation actions in the listing process

The criteria for the threatened categories are to be applied to a taxon whatever the level of conservation action affecting it. It is important to emphasise here that a taxon may require conservation action even if it is not listed as threatened. Conservation actions which may benefit the taxon are included as part of the documentation requirements (see Annex 3, point 6).

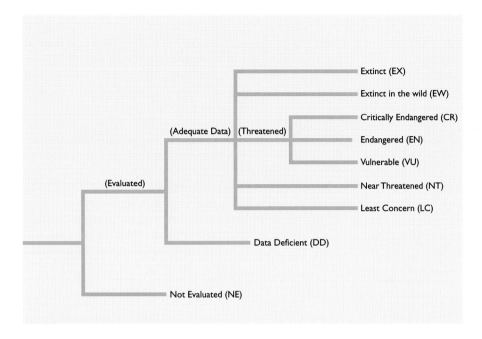

Extinct (EX)

Extinct in the wild (EW)

Critically Endangered (CR)

(Adequate Data)   (Threatened)

Endangered (EN)

Vulnerable (VU)

(Evaluated)

Near Threatened (NT)

Least Concern (LC)

Data Deficient (DD)

Not Evaluated (NE)

Figure 1   Structure of the categories

6.   Data quality and the importance of inference and projection

The criteria are clearly quantitative in nature. However, the absence of high-quality data should not deter attempts at applying the criteria, as methods involving estimation, inference and projection are emphasised as being acceptable throughout. Inference and projection may be based on extrapolation of current or potential threats into the future (including their rate of change), or of factors related to population abundance or distribution (including dependence on other taxa), so long as these can reasonably be supported. Suspected or inferred patterns in the recent past, present or near future can be based on any of a series of related factors, and these factors should be specified as part of the documentation.

Taxa at risk from threats posed by future events of low probability but with severe consequences (catastrophes) should be identified by the criteria (e.g., small distributions, few locations). Some threats need to be identified particularly early, and appropriate actions taken, because their effects are irreversible, or nearly so (e.g., pathogens, invasive organisms, hybridisation).

7. Problems of scale

Classification based on the sizes of geographic ranges or the patterns of habitat occupancy is complicated by problems of spatial scale. The finer the scale at which the distributions or habitats of taxa are mapped, the smaller the area will be that they are found to occupy, and the less likely it will be that range estimates (at least for 'area of occupancy': see Definitions, point 10) exceed the thresholds specified in the criteria. Mapping at finer scales reveals more areas in which the taxon is unrecorded. Conversely, coarse-scale mapping reveals fewer unoccupied areas resulting in range estimates that are more likely to exceed the thresholds for the threatened categories. The choice of scale at which range is estimated may thus, itself, influence the outcome of Red List assessments and could be a source of inconsistency and bias. It is impossible to provide any strict but general rules for mapping taxa or habitats; the most appropriate scale will depend on the taxa in question, and the origin and comprehensiveness of the distribution data.

8. Uncertainty

The data used to evaluate taxa against the criteria are often estimated with considerable uncertainty. Such uncertainty can arise from any one or all of the following three factors: natural variation, vagueness in the terms and definitions used, and measurement error. The way in which this uncertainty is handled can have a strong influence on the results from an evaluation. Details of methods recommended for handling uncertainty are included in Annex 1, and assessors are encouraged to read and follow these principles.

In general, when uncertainty leads to wide variation in the results of assessments the range of possible outcomes should be specified. A single category must be chosen and the basis for the decision should be documented; it should be both precautionary and credible.

When data are very uncertain, the category of 'Data Deficient' may be assigned. However, in this case the assessor must provide documentation showing that this category indicates that this category has been assigned because data are inadequate to determine a threat category. It is important to recognise that taxa that are poorly known can often be assigned a threat category on the basis of background information concerning the deterioration of their habitat and/or other causal factors; therefore the liberal use of 'Data Deficient' is discouraged.

9. Implications of listing

Listing in the categories of Not Evaluated and Data Deficient indicates that no assessment of extinction risk has been made, though for different reasons. Until such time as an assessment is made, taxa listed in these categories should not be treated as if they were non-threatened. It may be appropriate (especially for Data Deficient forms) to give them the same degree of protection as threatened taxa, at least until their status can be assessed.

10. Documentation

All assessments should be documented. Threatened classifications should state the criteria and subcriteria that were met. No assessment can be accepted for the IUCN Red List as valid unless at least one criterion is given. Therefore, if more than one criterion or subcriterion is met, then each should be listed. Therefore, if a re-evaluation indicates that the documented criterion is no longer met, this should not result in automatic reassignment to a lower category of threat (downlisting). Instead, the taxon should be re-evaluated against all the criteria to clarify its status. The factors responsible for qualifying the taxon against the criteria, especially where inference and projection are used, should be documented (see Annexes 2 and 3). The documentation requirements for other categories are also specified in Annex 3.

11. Threats and priorities

The category of threat is not necessarily sufficient to determine priorities for conservation action. The category of threat simply provides an assessment of the extinction risk under current circumstances, whereas a system for assessing priorities for action will include numerous other factors concerning conservation action such as costs, logistics, chances of success, and other biological characteristics of the subject.

12. Re-evaluation

Re-evaluation of taxa against the criteria should be carried out at appropriate intervals. This is especially important for taxa listed under Near Threatened, Data Deficient and for threatened taxa whose status is known or suspected to be deteriorating.

13. Transfer between categories

The following rules govern the movement of taxa between categories:

(A) A taxon may be moved from a category of higher threat to a category of lower threat if none of the criteria of the higher category has been met for five years or more.

(B) If the original classification is found to have been erroneous, the taxon may be transferred to the appropriate category or removed from the threatened categories altogether, without delay (but see Section 10).

(C) Transfer from categories of lower to higher risk should be made without delay.

14. Use at regional level

The IUCN Red List Categories and Criteria were designed for global taxon assessments. However, many people are interested in applying them to subsets of global data, especially at regional, national or local levels. To do this it is important to refer to guidelines prepared by the IUCN/SSC Regional Applications Working Group (Gärdenfors *et al.*, 1999). When applied at national or regional levels it must be recognised that a global category may not be the same as a national or regional category for a particular taxon. For example, taxa classified as Least Concern globally might be Critically Endangered within a particular region where numbers are very small or declining, perhaps only because they are at the margins of their global range. Conversely, taxa classified as Vulnerable on the basis of their global declines in numbers or range might be Least Concern within a particular region where their populations are stable. It is also important to note that taxa endemic to regions or nations will be assessed globally in any regional or national applications of the criteria, and in these cases great care must be taken to check that an assessment has not already been undertaken by a Red List Authority (RLA), and that the categorisation is agreed with the relevant RLA (e.g., an SSC Specialist Group known to cover the taxon).

III. Definitions

1. Population and Population Size (Criteria A, C and D)

The term 'population' is used in a specific sense in the Red List Criteria that is different to its common biological usage. Population is here defined as the total number of individuals of the taxon. For functional reasons, primarily owing to differences between life forms, population size is measured as numbers of mature individuals only. In the case of taxa obligately dependent on other taxa for all or part of their life cycles, biologically appropriate values for the host taxon should be used.

2. Subpopulations (Criteria B and C)

Subpopulations are defined as geographically or otherwise distinct groups in the population between which there is little demographic or genetic exchange (typically one successful migrant individual or gamete per year or less).

3. Mature individuals (Criteria A, B, C and D)

The number of mature individuals is the number of individuals known, estimated or inferred to be capable of reproduction. When estimating this quantity the following points should be borne in mind:

- Mature individuals that will never produce new recruits should not be counted (e.g. densities are too low for fertilisation).
- In the case of populations with biased adult or breeding sex ratios it is appropriate to use lower estimates for the number of mature individuals which take this into account.
- Where the population size fluctuates use a lower estimate. In most cases this will be much less than the mean
- Reproducing units within a clone should be counted as individuals, except where such units are unable to survive alone (e.g. corals).
- In the case of taxa that naturally lose all or a subset of mature individuals at some point in their life cycle, the estimate should be made at the appropriate time, when mature individuals are available for breeding.
- Re-introduced individuals must have produced viable offspring before they are counted as mature individuals.

4. Generation (Criteria A, C and E)

Generation length is the average age of parents of the current cohort (i.e. newborn individuals in the population). Generation length therefore reflects the turnover rate of breeding individuals in a population. Generation length is greater than the age at first breeding and less than the age of the oldest breeding individual, except in taxa that breed only once. Where generation length varies under threat, the more natural, i.e. pre-disturbance, generation length should be used.

5. Reduction (Criterion A)

A reduction is a decline in the number of mature individuals of at least the amount (%) stated under the criterion over the time period (years) specified, although the decline need not still be continuing. A reduction should not be interpreted as part of a fluctuation unless there is good evidence for this. The downward phase of a fluctuation will not normally count as a reduction.

6. Continuing decline (Criteria B and C)

A continuing decline is a recent, current or projected future decline (which may be smooth, irregular or sporadic) which is liable to continue unless remedial measures are taken. Fluctuations will not normally count as continuing declines, but an observed decline should not be considered as a fluctuation unless there is evidence for this.

7. Extreme fluctuations (Criteria B and C)

Extreme fluctuations can be said to occur in a number of taxa where population size or distribution area varies widely, rapidly and frequently, typically with a variation greater than one order of magnitude (i.e., a tenfold increase or decrease).

8. Severely fragmented (Criterion B)

The term 'severely fragmented' refers to the situation which increased extinction risk to the taxon results from the fact that most of its individuals are found in small and relatively isolated subpopulations (in certain circumstances this may be inferred from habitat information). These small subpopulations may go extinct, with a reduced probability of recolonisation.

9. Extent of occurrence (Criteria A and B)

Extent of occurrence is defined as the area contained within the shortest continuous imaginary boundary which can be drawn to encompass all the known, inferred or projected sites of present occurrence of a taxon, excluding cases of vagrancy (see Figure 2). This measure may exclude discontinuities or disjunctions within the overall distributions of taxa (e.g., large areas of obviously unsuitable habitat) (but see 'area of occupancy'). Extent of occurrence can often be measured by a minimum convex polygon (the smallest polygon in which no internal angle exceeds 180 degrees and which contains all the sites of occurrence).

10. Area of occupancy (Criteria A, B and D)

Area of occupancy is defined as the area within its 'extent of occurrence' (see point 9 above) which is occupied by a taxon, excluding cases of vagrancy. The measure reflects the fact that a taxon will not usually occur throughout the area of its extent of occurrence, which may contain unsuitable or unoccupied habitats. In some cases (e.g. irreplaceable colonial nesting sites, crucial feeding sites for migratory taxa) the area of occupancy is the smallest area essential at any stage to the survival of existing populations of a taxon. The size of the area of occupancy will be a function of the scale at which it is measured, and should be at a scale appropriate to relevant biological aspects of the taxon,

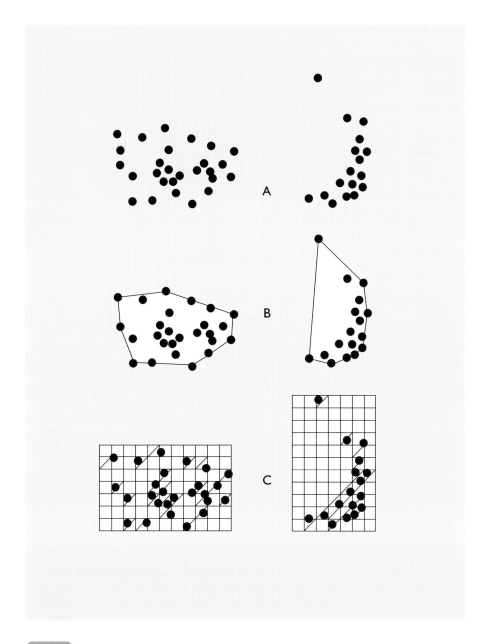

Figure 2   Two examples of the distinction between extent of occurrence and area of occupancy. (A) is the spatial distribution of known, inferred or projected sites of occurrence. (B) shows one possible boundary to the extent of occurrence, which is the measured area within this boundary. (C) shows one measure of area of occupancy which can be measured by the sum of the occupied grid squares.

the nature of threats and the available data (see point 7 in the Preamble). To avoid inconsistencies and bias in assessments caused by estimating area of occupancy at different scales, it may be necessary to standardise estimates by applying a scale-correction factor. It is difficult to give strict guidance on how standardisation should be done because different types of taxa have different scale-area relationships.

11. Location (Criteria B and D)

The term 'location' defines a geographically or ecologically distinct area in which a single threatening event can rapidly affect all individuals of the taxon present. The size of the location depends on the area covered by the threatening event and may include part of one or many subpopulations. Where a taxon is affected by more than one threatening event, location should be defined by considering the most serious plausible threat.

12. Quantitative analysis (Criterion E)

A quantitative analysis is defined here as any form of analysis which estimates the extinction probability of a taxon based on known life history, habitat requirements, threats and any specified management options. Population viability analysis (PVA) is one such technique. Quantitative analyses should make full use of all relevant available data. In a situation in which there is limited information, such data as are available can be used to provide an estimate of extinction risk (for instance, estimating the impact of stochastic events on habitat). In presenting the results of quantitative analyses, the assumptions (which must be appropriate and defensible), the data used and uncertainty in the data or quantitative model must be documented.

IV. The Categories[1]

A representation of the relationships between the categories is shown in Figure 1.

EXTINCT (EX)

A taxon is Extinct when there is no reasonable doubt that the last individual has died. A taxon is presumed Extinct when exhaustive surveys in known and/or expected habitat, at appropriate times (diurnal, seasonal, annual),

---

[1] Note: As in previous IUCN categories, the abbreviation of each category (in parenthesis) follows the English denominations when translated into other languages (see Annex II).

throughout its historic range have failed to record an individual. Surveys should be over a time frame appropriate to the taxon's life cycle and life form.

EXTINCT IN THE WILD (EW)

A taxon is Extinct in the Wild when it is known only to survive in cultivation, in captivity or as a naturalised population (or populations) well outside the past range. A taxon is presumed Extinct in the Wild when exhaustive surveys in known and/or expected habitat, at appropriate times (diurnal, seasonal, annual), throughout its historic range have failed to record an individual. Surveys should be over a time frame appropriate to the taxon's life cycle and life form.

CRITICALLY ENDANGERED (CR)

A taxon is Critically Endangered when the best available evidence indicates that it meets any of the Criteria A to E for Critically Endangered (see Section V), and it is therefore considered to be facing an extremely high risk of extinction in the wild.

ENDANGERED (EN)

A taxon is Endangered when the best available evidence indicates that it meets any of the Criteria A to E for Endangered (see Section V), and it is therefore considered to be facing a very high risk of extinction in the wild.

VULNERABLE (VU)

A taxon is Vulnerable when the best available evidence indicates that it meets any of the Criteria A to E for Vulnerable (see Section V), and it is therefore considered to be facing a high risk of extinction in the wild.

NEAR THREATENED (NT)

A taxon is Near Threatened when it has been evaluated against the criteria but does not qualify for Critically Endangered, Endangered or Vulnerable now, but is close to qualifying for or is likely to qualify for a threatened category in the near future.

LEAST CONCERN (LC)

A taxon is Least Concern when it has been evaluated against the criteria and does not qualify for Critically Endangered, Endangered, Vulnerable or Near Threatened. Widespread and abundant taxa are included in this category.

DATA DEFICIENT (DD)

A taxon is Data Deficient when there is inadequate information to make a direct, or indirect, assessment of its risk of extinction based on its distribution and/or population status. A taxon in this category may be well studied, and its biology well known, but appropriate data on abundance and/or distribution are lacking. Data Deficient is therefore not a category of threat. Listing of taxa in this category indicates that more information is required and acknowledges the possibility that future research will show that threatened classification is appropriate. It is important to make positive use of whatever data are available. In many cases great care should be exercised in choosing between DD and a threatened status. If the range of a taxon is suspected to be relatively circumscribed, and a considerable period of time has elapsed since the last record of the taxon, threatened status may well be justified.

NOT EVALUATED (NE)

A taxon is Not Evaluated when it is has not yet been evaluated against the criteria.

V. The Criteria for Critically Endangered,

Endangered and Vulnerable

CRITICALLY ENDANGERED (CR)

A taxon is Critically Endangered when the best available evidence indicates that it meets any of the following criteria (A to E), and it is therefore considered to be facing an extremely high risk of extinction in the wild:

A. Reduction in population size based on any of the following:

1. An observed, estimated, inferred or suspected population size reduction of ≥90% over the last 10 years or three generations, whichever is the longer, where the causes of the reduction are clearly reversible AND understood AND ceased, based on (and specifying) any of the following:

    (a) direct observation
    (b) an index of abundance appropriate for the taxon
    (c) a decline in area of occupancy, extent of occurrence and/or quality of habitat
    (d) actual or potential levels of exploitation
    (e) the effects of introduced taxa, hybridisation, pathogens, pollutants, competitors or parasites.

2. An observed, estimated, inferred or suspected population size reduction of ≥80% over the last 10 years or three generations, whichever is the longer, where the reduction or its causes may not have ceased OR be understood OR be reversible, based on (and specifying) any of (a) to (e) under A1.

3. A population size reduction of ≥80%, projected or suspected to be met within the next 10 years or three generations, whichever is the longer (up to a maximum of 100 years), based on (and specifying) any of (b) to (e) under A1.

4. An observed, estimated, inferred, projected or suspected population size reduction of ≥80% over any 10 year or three generation period, whichever is longer (up to a maximum of 100 years in the future), where the time period must include both the past and the future, and where the reduction or its causes may not have ceased OR may not be understood OR may not be reversible, based on (and specifying) any of (a) to (e) under A1.

B. Geographic range in the form of either B1 (extent of occurrence) OR B2 (area of occupancy) OR both:

1. Extent of occurrence estimated to be less than 100 km², and estimates indicating at least two of a–c:

    (a) Severely fragmented or known to exist at only a single location.

    (b) Continuing decline, observed, inferred or projected, in any of the following:
        (i)   extent of occurrence
        (ii) area of occupancy
        (iii) area, extent and/or quality of habitat
        (iv) number of locations or subpopulations
        (v)   number of mature individuals.

    (c) Extreme fluctuations in any of the following:
        (i)   extent of occurrence
        (ii) area of occupancy
        (iii) number of locations or subpopulations
        (iv) number of mature individuals.

2. Area of occupancy estimated to be less than 10 km², and estimates indicating at least two of a–c:

    (a) Severely fragmented or known to exist at only a single location.

    (b) Continuing decline, observed, inferred or projected, in any of the following:

(i) extent of occurrence

(ii) area of occupancy

(iii) area, extent and/or quality of habitat

(iv) number of locations or subpopulations

(v) number of mature individuals.

(c) Extreme fluctuations in any of the following:

(i) extent of occurrence

(ii) area of occupancy

(iii) number of locations or subpopulations

(iv) number of mature individuals.

C. Population size estimated to number fewer than 250 mature individuals and either:

1. An estimated continuing decline of at least 25% within three years or one generation, whichever is longer (up to a maximum of 100 years in the future) OR

2. A continuing decline, observed, projected, or inferred, in numbers of mature individuals AND at least one of the following (a–b):

(a) Population structure in the form of one of the following:

(i) no subpopulation estimated to contain more than 50 mature individuals, OR

(ii) at least 90% of mature individuals are in one subpopulation.

(b) Extreme fluctuations in number of mature individuals.

D. Population size estimated to number less than 50 mature individuals.

E. Quantitative analysis showing the probability of extinction in the wild is at least 50% within 10 years or three generations, whichever is the longer (up to a maximum of 100 years).

ENDANGERED (EN)

A taxon is Endangered when best available evidence indicates that it meets any of the following criteria (A to E), and it is therefore considered to be facing a very high risk of extinction in the wild:

A. Reduction in population size based on any of the following:

1. An observed, estimated, inferred or suspected population size reduction of ≥70% over the last 10 years or three generations, whichever is the longer, where the causes of the reduction are clearly reversible AND understood AND ceased, based on (and specifying) any of the following:

(a) direct observation
(b) an index of abundance appropriate for the taxon
(c) a decline in area of occupancy, extent of occurrence and/or quality of habitat
(d) actual or potential levels of exploitation
(e) the effects of introduced taxa, hybridisation, pathogens, pollutants, competitors or parasites.

2. An observed, estimated, inferred or suspected population size reduction of ≥50% over the last 10 years or three generations, whichever is the longer, where the reduction or its causes may not have ceased OR may not be understood OR may not be reversible, based on (and specifying) any of (a) to (e) under A1.

3. A population size reduction of ≥50%, projected or suspected to be met within the next 10 years or three generations, whichever is the longer (up to a maximum of 100 years), based on (and specifying) any of (b) to (e) under A1.

4. An observed, estimated, inferred, projected or suspected population size reduction of ≥50% over any 10 year or three generation period, whichever is longer (up to a maximum of 100 years in the future), where the time period must include both the past and the future, AND where the reduction or its causes may not have ceased OR may not be understood OR may not be reversible, based on (and specifying) any of the (a) to (e) under A1.

B. Geographic range in the form of either B1 (extent of occurrence) OR B2 (area of occupancy) OR both:

1. Extent of occurrence estimated to be less than 5000 km², and estimates indicating at least two of a–c:

    (a) Severely fragmented or known to exist at no more than five locations.

    (b) Continuing decline, observed, inferred or projected, in any of the following:

        (i) extent of occurrence
        (ii) area of occupancy
        (iii) area, extent and/or quality of habitat
        (iv) number of locations or subpopulations
        (v) number of mature individuals.

    (c) Extreme fluctuations in any of the following:

        (i) extent of occurrence
        (ii) area of occupancy
        (iii) number of locations or subpopulations
        (iv) number of mature individuals.

2. Area of occupancy estimated to be less than 500 km², and estimates indicating at least two of a–c:

   (a) Severely fragmented or known to exist at no more than five locations.

   (b) Continuing decline, observed, inferred or projected, in any of the following:

      (i)   extent of occurrence
      (ii) area of occupancy
      (iii) area, extent and/or quality of habitat
      (iv) number of locations or subpopulations
      (v)   number of mature individuals.

   (c) Extreme fluctuations in any of the following:

      (i)   extent of occurrence
      (ii) area of occupancy
      (iii) number of locations or subpopulations
      (iv) number of mature individuals.

C. Population size estimated to number less than 2500 mature individuals and either:

   1. An estimated continuing decline of at least 20% within five years or two generations, whichever is longer, (up to a maximum of 100 years in the future) OR

   2. A continuing decline, observed, projected, or inferred, in numbers of mature individuals AND at least one of the following (a–b):

      (a) Population structure in the form of one of the following:

         (i) no subpopulation estimated to contain more than 250 mature individuals, OR
         (ii) at least 95% of mature individuals are in one subpopulation.

      (b) Extreme fluctuations in number of mature individuals.

D. Population size estimated to number fewer than 250 mature individuals.

E. Quantitative analysis showing the probability of extinction in the wild is at least 20% within 20 years or five generations, whichever is the longer (up to a maximum of 100 years).

VULNERABLE (VU)

A taxon is Vulnerable when best available evidence indicates that it meets any of the following criteria (A to E), and it is therefore considered to be facing a high risk of extinction in the wild:

A. Reduction in population size based on any of the following:

1. An observed, estimated, inferred or suspected population size reduction of ≥50% over the last 10 years or three generations, whichever is the longer, where the causes of the reduction are: clearly reversible AND understood AND ceased, based on (and specifying) any of the following:

(a) direct observation
(b) an index of abundance appropriate for the taxon
(c) a decline in area of occupancy, extent of occurrence and/or quality of habitat
(d) actual or potential levels of exploitation
(e) the effects of introduced taxa, hybridisation, pathogens, pollutants, competitors or parasites.

2. An observed, estimated, inferred or suspected population size reduction of ≥30% over the last 10 years or three generations, whichever is the longer, where the reduction or its causes may not have ceased OR may not be understood OR may not be reversible, based on (and specifying) any of (a) to (e) under A1.

3. A population size reduction of ≥30%, projected or suspected to be met within the next ten years or three generations, whichever is the longer (up to a maximum of 100 years), based on (and specifying) any of (b) to (e) under A1.

4. An observed, estimated, inferred, projected or suspected population size reduction of ≥30% over any 10 year or three generation period, whichever is longer (up to a maximum of 100 years), where the time period includes both the past and the future, AND where the reduction or its causes may not have ceased OR may not be understood OR may not be reversible, based on (and specifying) any of the (a) to (e) under A1.

B. Geographic range in the form of either B1 (extent of occurrence) OR B2 (area of occupancy) OR both:

1. Extent of occurrence estimated to be less than 20,000 km2, and estimates indicating at least two of a–c:

(a) Severely fragmented or known to exist at no more than 10 locations.

(b) Continuing decline, observed, inferred or projected, in any of the following:

    (i)   extent of occurrence
    (ii) area of occupancy
    (iii) area, extent and/or quality of habitat
    (iv) number of locations or subpopulations
    (v)   number of mature individuals.

  (c) Extreme fluctuations in any of the following:

    (i)   extent of occurrence
    (ii) area of occupancy
    (iii) number of locations or subpopulations
    (iv) number of mature individuals.

2. Area of occupancy estimated to be less than 2000 km2, and estimates indicating at least two of a–c:

  (a) Severely fragmented or known to exist at no more than 10 locations.

  (b) Continuing decline, observed, inferred or projected, in any of the following:

    (i)   extent of occurrence
    (ii) area of occupancy
    (iii) area, extent and/or quality of habitat
    (iv) number of locations or subpopulations
    (v)   number of mature individuals.

  (c) Extreme fluctuations in any of the following:

    (i)   extent of occurrence
    (ii) area of occupancy
    (iii) number of locations or subpopulations
    (iv) number of mature individuals.

C. Population size estimated to number fewer than 10,000 mature individuals and either:

  1. An estimated continuing decline of at least 10% within 10 years or three generations, whichever is longer, (up to a maximum of 100 years in the future) OR

  2. A continuing decline, observed, projected, or inferred, in numbers of mature individuals AND at least one of the following (a–b):

    (a) Population structure in the form of one of the following:

      (i) no subpopulation estimated to contain more than 1000 mature individuals, OR
      (ii) all mature individuals are in one subpopulation.

    (b) Extreme fluctuations in number of mature individuals.

D. Population very small or restricted in the form of either of the following:

1. Population size estimated to number fewer than 1000 mature individuals.

2. Population with a very restricted area of occupancy (typically less than 20km²) or number of locations (typically 5 or fewer) such that it is prone to the effects of human activities or stochastic events within a very short time period in an uncertain future, and is thus capable of becoming Critically Endangered or even Extinct in a very short time period.

E. Quantitative analysis showing the probability of extinction in the wild is at least 10% within 100 years.

Annex 1: Uncertainty

The Red List Criteria should be applied to a taxon based on the available evidence concerning its numbers, trend and distribution. In cases where there are evident threats to a taxon through, for example, deterioration of its only known habitat, a threatened listing may be justified, even though there may be little direct information on the biological status of the taxon itself. In all these instances there are uncertainties associated with the available information and how it was obtained. These uncertainties may be categorised as natural variability, semantic uncertainty and measurement error (Akçakaya *et al.*, 2000). This section provides guidance on how to recognise and deal with these uncertainties when using the criteria.

Natural variability results from the fact that species' life histories and the environments in which they live change over time and space. The effect of this variation on the criteria is limited, because each parameter refers to a specific time or spatial scale. Semantic uncertainty arises from vagueness in the definition of terms or a lack of consistency in different assessors' usage of them. Despite attempts to make the definitions of the terms used in the criteria exact, in some cases this is not possible without the loss of generality. Measurement error is often the largest source of uncertainty; it arises from the lack of precise information about the parameters used in the criteria. This may be due to inaccuracies in estimating the values or a lack of knowledge. Measurement error may be reduced or eliminated by acquiring additional data. For further details, see Akçakaya *et al.* (2000) and Burgman *et al.* (1999).

One of the simplest ways to represent uncertainty is to specify a best estimate and a range of plausible values. The best estimate itself might be a range, but in any case the best estimate should always be included in the range of plausible values. When data are very uncertain, the range for the best estimate might be the range of plausible values. There are various methods that can be used to establish the plausible range. It may be based on confidence intervals, the opinion of a single expert, or the consensus opinion of a group of experts. Whichever method is used should be justified in the documentation.

When interpreting and using uncertain data, preferences and attitudes toward risk and uncertainty may play an important role. Attitudes have two components. First, assessors need to consider whether they will include the full range of plausible values in assessments, or whether they will exclude extreme values from consideration (known as dispute tolerance). An assessor with a low dispute tolerance would include all values, thereby increasing the uncertainty, whereas an assessor with a high dispute tolerance would exclude extremes, reducing the uncertainty. Second, assessors need to consider whether they have a precautionary or evidentiary attitude to risk (known as risk tolerance). A precautionary attitude will classify a taxon as threatened unless it is certain that it is not threatened, whereas an evidentiary attitude will classify a taxon as threatened only when there is strong evidence to support a threatened classification. Assessors should resist an evidentiary attitude and adopt a precautionary but realistic attitude to uncertainty when applying the criteria, for example, by using plausible lower bounds, rather than best estimates, in determining population size, especially if it is fluctuating. All preferences and attitudes should be explicitly documented.

The assessment using a point estimate (i.e. single numerical value) will lead to a single Red List Category. However, when a plausible range for each parameter is used to evaluate the criteria, a range of categories may be obtained reflecting the uncertainties in the data. A single category, based on a specific attitude to uncertainty, should always be listed along with the criteria met while the range of plausible categories should be indicated in the documentation (see Annex 3).

Where data are so uncertain that any category is plausible, the category of 'Data Deficient' should be assigned. However, it is important to recognise that this category indicates that the data are inadequate to determine the degree of threat faced by a taxon, not necessarily that the taxon is poorly known or indeed not threatened. Although Data Deficient is not a threatened category, it indicates a need to obtain more information on a taxon to determine the appropriate listing; moreover it requires documentation with whatever available information there is.

Annex 2: Citation of the IUCN Red List Categories and Criteria

In order to promote the use of a standard format for citing the Red List Categories and Criteria the following forms of citation are recommended:

1.   The Red List Category may be written out in full or abbreviated as follows (when translated into other languages, the abbreviations should follow the English denominations):

> Extinct or EX
>
> Extinct in the Wild or EW
>
> Critically Endangered or CR
>
> Endangered or EN
>
> Vulnerable or VU
>
> Near Threatened or NT
>
> Least Concern or LC
>
> Data Deficient or DD
>
> Not Evaluated or NE

2.   Under Section V (the criteria for Critically Endangered, Endangered and Vulnerable) there is a hierarchical alphanumeric numbering system of criteria and sub-criteria. These criteria and subcriteria (all three levels) form an integral part of the Red List assessment and all those that result in the

assignment of a threatened category must be specified after the Category. Under the criteria A to C and D under Vulnerable, first level in the hierarchy is indicated by the use of numbers (1–4) and if more thane one is met, they are separated by means of the '+' symbol. The second level is indicated by the use of the lower case alphabet characters (a–e). These are listed without any punctuation. A third level of the hierarchy under Criteria B and C involves the use of lower case roman numerals (i–v). These are placed in parentheses (with no space between the preceeding alphabet character and start of the parenthesis) and separated by the use of commas if more than one is listed. The following are examples of such usage:

EX

CR A1cd

VU A2c+3c

EN B1ac(i, ii, iii)

EN A2c; D

VU D1+2

CR A2c+3c; B1ab(iii)

CR D

VU D2

EN B2ab(i, ii, iii)

VU C2a(ii)

EN A1c; B1ab(iii); C2a(i)

EN B2b(iii)c(ii)

EN B1ab(i, ii, v)c(iii, iv); B2b(i)c(ii, v)

VU B1ab(iii)+2ab(iii)

EN A2abc+3bc+4abc; B1b(iii, iv, v)c(ii, iii, iv)+2b(iii, iv, v)c(ii, iii, iv)

Annex 3: Documentation Requirements for Taxa Included on the IUCN Red List

The following is the minimum set of information, which should accompany every assessment submitted for incorporation into the *IUCN Red List of Threatened Species*™:

- Scientific name including authority details.
- English common name(s) and any other widely used common names (specify the language of each name supplied).
- Red List Category and Criteria.
- Countries of occurrence (including country subdivisions for large nations, e.g., states within the USA, and overseas territories, e.g., islands far from the mainland country).

- For marine species, the Fisheries Areas in which they occur should be recorded (see www.iucn.org/themes/ssc/sis/faomap.htm for the Fisheries Areas as delimited by The Food and Agriculture Organisation of the United Nations (FAO).

- For inland water species, the names of the river systems, lakes, etc. to which they are confined.

- A map showing the geographic distribution (extent of occurrence).

- A rationale for the listing (including any numerical data, inferences or uncertainty that relate to the criteria and their thresholds).

- Current population trends (increasing, decreasing, stable or unknown).

- Habitat preferences (using a modified version of the Global Land Cover Characterisation (GLCC) classification which is available electronically from www.iucn.org/themes/ssc/sis/authority.htm or on request from redlist@ssc-uk.org).

- Major threats (indicating past, current and future threats using a standard classification which is available from the SSC web site or e-mail address as shown above).

- Conservation measures, (indicating both current and proposed measures using a standard classification which is available from the SSC web site or e-mail address as shown above).

- Information on any changes in the Red List status of the taxon, and why the status has changed.

- Data sources (cited in full, including unpublished sources and personal communications).

- Name(s) and contact details of the assessor(s).

- Before inclusion on the IUCN Red List, all assessments will be evaluated by at least two members of a Red List Authority. The Red List Authority is appointed by the Chair of the IUCN Species Survival Commission and is usually a sub-group of a Specialist Group. The names of the evaluators will appear with each assessment.

In addition to the minimum documentation, the following information should also be supplied where appropriate:

- If a quantitative analysis is used for the assessment (i.e. Criterion E), the data, assumptions and structural equations (e.g., in the case of a Population Viability Analysis) should be included as part of the documentation.

- For Extinct or Extinct from the Wild taxa, extra documentation is required indicating the effective data of extinction, possible causes of the extinction and the details of surveys which have been conducted to search for the taxon.
- For taxa listed as Near Threatened, the rationale for listing the taxon (e.g., they are dependent on ongoing conservation measures).
- For taxa listed as Data Deficient, the documentation should include what little information is available.

Assessments may be made using version 2.0 of the software package RAMAS® Red List (Akçakaya and Ferson, 2001). This program assigns taxa to the Red List Categories according to the rules of the IUCN Red List Criteria and has the advantage of being able to explicitly handle uncertainty in the data. The software captures most of the information required for the documentation above, but in some cases the information will be reported differently. The following points should be noted:

- If RAMAS® Red List is used to obtain a listing, this should be stated.
- Uncertain values should be entered into the program as a best estimate and a plausible range, or as an interval (see RAMAS® Red List manual or help files for further details).
- The settings for attitude towards risk and uncertainty (i.e. dispute tolerance, risk tolerance and burden of proof) are all pre-set at a mid-point. If any of these settings are changed this should be documented and fully justified, especially if a less precautionary position is adopted.
- Depending on the uncertainties, the resulting classification can be a single category and/or a range of plausible categories. In such instances, the following approach should be adopted (the program will usually indicate this automatically in the Results window):
  - If the range of plausible categories extends across two or more of the threatened categories (e.g. Critically Endangered to Vulnerable) and no preferred category is indicated, the precautionary approach is to take the highest category shown, i.e. CR in the above example. In such cases, the range of plausible categories should be documented under the rationale including a note that a precautionary approach was followed in order to distinguish it from the situation in the next point. The following notation has been suggested e.g., CR* (CR-VU).
  - If a range of plausible categories is given and a preferred category is indicated, the rationale should indicate the range of plausible categories met e.g., EN (CR-VU).

- The program specifies the criteria that contributed to the listing (see Status window). However, when data are uncertain, the listing criteria are approximate, and in some cases may not be determined at all. In such cases, the assessors should use the Text results to determine or verify the criteria and sub-criteria met. Listing criteria derived in this way must be clearly indicated in the rationale (refer to the RAMAS® Red List Help menu for further guidance on this issue).

- If the preferred category is indicated as Least Concern, but the plausible range extends into the threatened categories, a listing of 'Near Threatened' (NT) should be used. The criteria, which triggered the extension into the threatened range, should be recorded under the rationale.

- Any assessments made using this software must be submitted with the RAMAS® Red List input files (i.e. the *.Red files).

New global assessments or reassessments of taxa currently on the IUCN Red List, may be submitted to the IUCN/SSC Red List Programme Officer for incorporation (subject to peer review) in a future edition of the *IUCN Red List of Threatened Species*™. Submissions from within the SCC network should preferable be made using the Species Information Service (SIS) database. Other submissions may be submitted electronically; these should preferably be as files produced using RAMAS® Red List or any other programmes in Microsoft Office 97 (or earlier versions) e.g., Word, Excel or Access. Submissions should be sent to:

> IUCN/SSC Red List Programme,IUCN/SSC UK Office,
> 219c Huntingdon Road,
> Cambridge, CB3 0DL, United Kingdom.
> Fax: +44-(0)1223-277845; e-mail: redlist@ssc-uk.org

For further clarification or information about the IUCN Red List Criteria, documentation requirements (including the standards used) or submission of assessments, please contact the IUCN/SSC Red List Programme Officer at the address shown above.

References

Akçakaya, H.R. and Ferson, S. (2001). *RAMAS® Red List: Threatened Species Classifications under Uncertainty.* Version 2.0. Applied Biomathematics, New York, USA.

Akçakaya, H.R., Ferson, S., Burgman, M.A., Keith, D.A., Mace, G.M. and Todd, C.R. (2000). Making consistent IUCN classifications under uncertainty. *Conservation Biology* **14**: 1001–1013.

Baillie, J. and Groombridge, B. (eds.). (1996). *1996 IUCN Red List of Threatened Animals.* IUCN, Gland, Switzerland.

Burgman, M.A., Keith, D.A. and Walshe, T.V. (1999). Uncertainty in comparative risk analysis of threatened Australian plant species. *Risk Analysis* **19**: 585–598.

Fitter, R. and Fitter, M. (eds.). (1987). *The Road to Extinction.* IUCN, Gland, Switzerland.

Gärdenfors, U., Rodríguez, J.P., Hilton-Taylor, C., Hyslop, C., Mace, G., Molur, S. and Poss, S. (1999). Draft Guidelines for the Application of IUCN Red List Criteria at national and regional Levels. *Species* **31–32**: 58–70.

Hilton-Taylor, C. (Compiler) (2000). 2000 IUCN Red List of Threatened Species. IUCN, Gland, Switzerland and Cambidge, UK.

IUCN (1993). *Draft IUCN Red List Categories.* IUCN, Gland, Switzerland.

IUCN (1994). *IUCN Red List Categories.* Prepared by the IUCN Species Survival Commission. IUCN, Gland, Switzerland.

IUCN (1996). Resolution 1.4. Species Survival Commission. *Resolutions and Recommendations*, pp. 7–8. World Conservation Congress, 13–23 October 1996, Montreal, Canada. IUCN, Gland, Switzerland.

IUCN (1998). *Guidelines for Re-introductions.* Prepared by the IUCN/SSC Re-introduction Specialist Group. IUCN, Gland, Switzerland.

IUCN/SSC Criteria Review Working Group (1999). IUCN Red List Criteria review provisional report: draft of the proposed changes and recommendations. *Species* **31–32**: 43–57.

Mace, G.M., Collar, N., Cooke, J., Gaston, K.J., Ginsberg, J.R., Leader-Williams, N., Maunder, M. and Milner-Gulland, E.J. (1992). The development of new criteria for listing species on the IUCN Red List. *Species* **19**: 16–22.

Mace, G.M. and Lande, R. (1991). Assessing extinction threats: toward a re-evaluation of IUCN threatened species categories. *Conservation Biology* **5**: 148–157.

Mace, G.M. and Stuart, S.N. (1994). Draft IUCN Red List Categories, Version 2.2. *Species* **21–22**: 13–24.

Oldfield, S., Lusty, C. and MacKinven, A. (1998). *The World List of Threatened Trees.* World Conservation Press, Cambridge, UK.

Appendix 3

The CAMP Workshop: tools for the assessment and management of threatened plant species

Onnie Byers
Conservation Breeding Specialist Group of the
IUCN Species Survival Commission,
Minnesota, USA

M. Maunder[1]
Royal Botanic Gardens, Kew, UK

[1] Now at: The National Tropical Botanical Garden Kauai,
Hawaii, USA

SPECIES SURVIVAL COMMISSION

Introduction

The Conservation Assessment and Management Plan (CAMP) protocol provides two valuable services:

- A vehicle for involving the entire conservation community in biodiversity activities; and
- A rapid assessment of conservation status of any plant or animal groups, e.g., providing local or global 'prioritization'.

CAMPs can help organizations establish their strategic plans and actions for conservation. India, for example, has used the CAMP process to carry out biodiversity inventories of its endemic species (plants and animals) and to prioritize conservation activities in fulfillment of its commitment to the Convention on Biological Diversity (CBD).

The Conservation Breeding Specialist Group (CBSG) of IUCN's Species Survival Commission (SSC) has ten years of experience in developing, testing and applying a series of scientifically based tools and processes to assist risk characterization and species management decision making. These tools, based on conservation biology (biological and physical factors), human demography, and the dynamics of social learning, are employed in intensive, problem-solving workshops to produce realistic and achievable recommendations for conservation management.

Since 1992, CAMPs have been carried out for a wide spectrum of plants, invertebrates, birds, reptiles and mammals. Examples of CAMP workshop configurations include:

- Biological – a complete taxon group (e.g. all Indian bryophytes);
- Geographical or biogeographical – taxon in a geographical region (e.g., Southern Indian medicinal plants; selected orchids of Costa Rica; non-timber forest produce of Central India);
- Political – taxon in a particular political region (e.g., endemic species of St. Helena Island; threatened plants of Costa Rica, plants of Himachal Pradesh, India);
- Review of Status – pre-evaluated or pre-prioritized requiring reduction and refinement (e.g., international, national, or institutional list; export/import lists);
- Species of particular concern (e.g., species known to be in trade; high profile species; palms; cacti, orchids); and
- Ecological (e.g., Indian mangrove species)

The CAMP process utilizes information from action plans and management plans as well as additional data from experts on the taxa, both published and unpublished. With this information, the Workshop participants evaluate status of threat of all taxa in a broad group, e.g., an order or family, country, or geographic region, using the IUCN Red List Criteria to assign categories. Based on threat status, CAMPs provide a rational and comprehensive way to determine priorities for conservation needs and identification of required information gathering and conservation management actions.

Whilst effective conservation action is built upon the utilisation of available information, it is also dependent on both the actions of humans living within the range of the threatened species and established national and international interests. However, tools to evaluate and integrate the interaction of biological, physical, and social factors on the population dynamics are scarce.

CAMP processes

CBSG Workshops bring together a broad spectrum of expertise on the management of the threatened taxa under review. Invitees include representatives of the full range of stakeholder groups with an interest in conserving and managing the species in its habitat or the consequences of such management. Ten to sixty experts come together during the Workshop to determine the status of threat for the taxa under review and to make broad-based recommendations for research and management activities. The experts have different backgrounds, e.g., biodiversity managers, field biologists, scientists from the academic community and/or the private sector, land owners, IUCN Specialist Group members and *ex situ* managers.

The CAMP Workshop is an intensive and interactive process taking place over a 3 to 5 day period, evenings included, depending on the scope of the workshop and the number of species to be assessed. The Workshop is unique in its ability to facilitate objective and systematic prioritisation of practical management and research actions that are needed for species conservation. Participants develop the assessments of risks and formulate recommendations for action using a systematically designed Taxon Data Sheet (Fig. 1). This sheet facilitates recording of detailed information about each taxon, including data on the status of populations, their habitat in the wild and recommendations for intensive conservation action. An accompanying computerized CAMP data entry program aides the collection of information, facilitates production of the report and allows information for all CAMP workshops to be accessed and queried by any interested parties.

The CAMP process utilizes information from grey literature, action plans and management plans as well as unpublished data from experts on the taxa. With this information, the Workshop participants evaluate status of threat of the listed taxa in a broad group (e.g., an order or family), country, or geographic region, using the IUCN Red List Criteria to assign categories. Based on threat status, CAMPs provide a rational and comprehensive way to determine priorities for integrated conservation actions and for required information gathering.

| Figure 1 | **CAMP Summary Information Sheet for Darwin Workshop 1998** |

1A. Group: 2	**1B. Group Members**: Hamisi Mududu, Daniel Sitoni, Donatus Bayona, Alsen Oduwo, Aggrey Rwetsiba, Jane Nyafuno, Ahmed Mndolwa, Perpetua Ipulet, Clare Hankamer	**Date Completed**

2A. Scientific name w. author citation of the taxon for which this sheet is being filled:

Gigasiphon macrosiphon (Harms) Brenan

2B. Synonyms (if any): *Bauhinia macrosiphon* Harms, *Gigasiphon humblotianum sensu* KTS

2C. Common/vernacular name(s) (specify language): Mnyanza (Digo, Kenya)

2D. Trade name(s) (specify language):

2E. Family: Leguminosae (sub-family Caesalpiniaceae)

2F. Is the taxon protected by National and/or International legislation:

☐ Yes   ☑ No

If, Yes please specify:

3A. Geographic distribution: Kenya – K7, Tanzania – T8

Gongoni FR (K7), Kaya Muhaka National Monument (K7), Rondo Plateau – Michinjiri (T8) and Buda Mafisini (Fr)

? Amani Nature Reserve (Kiuhui) (T3) [Initial search carried out by Luke, 2/3/98 but not found]

? Reported by Greenway (1951?) to have been planted in Amani Botanic Garden, but unknown location.

[Extensive surveys carried out of Mrima FR (K7) and Marenje FR (K7) (Luke, pers. com.) but *G. macrosiphon* not found].

3B. Habit of taxon: tree

3C. Habitat of the taxon: Moist semi-decidous/evergreen lowland forest.

3D. Habitat specificity (niche): lowland (100–810m), moist evergreen (100–250m).

4. ESTIMATED EXTENT OF OCCURRENCE of the taxon. (Extent of occurrence is defined as the area contained within the shortest continuous imaginary boundary encompassing all known, inferred or projected sites of present occurrence of the taxon): (tick one)

| ☐ < 100 sq. km. | ☐ 101–5,000 sq. km. | ☐ 5,001–20,000 sq. km. | ☑ > 20,001 sq. km. | ☐ unknown |

5. ESTIMATED AREA OF OCCUPANCY of the taxon (Area of occupancy is defined as the area occupied by the taxon within the 'extent of occurrence'): (tick one)

| ☑ < 10 sq. km. | ☐ 11–500 sq. km. | ☐ 501–2,000 sq. km. | ☐ > 2,001 sq. km. | ☐ unknown |

6A. Number of known locations or populations in which the taxon is distributed: _4_____

6B. Number in Protected Areas: _3_____

Note: protected areas noted in K7 subject to illegal logging and land-grabbing.

7A. Are the locations or populations:

| ☐ **Contiguous** | ☑ **Fragmented** |

8. Number of mature individuals:

| ☐ < 50 | ☐ < 2,500 | ☑ **Unknown** |
| ☐ < 250 | ☐ > 2,500 | Mbinda (1996): 44 mature individuals surveyed in transects in Gongoni FR, Buda Mafisini FR and Kaya Muwaka FR. |

9. Habitat quality:

9A. Is there any change in the habitat where the taxon occurs:

| ☑ **Yes** | ☐ **No** |

9B. If, Yes:

| ☑ **Decrease** | ☐ **Stable** |
| ☐ **Increase** | ☐ **Unknown** |

9C. If decreasing, what has been the decrease in habitat (approximately in percent) over years?:

☐ **Unknown**	☐ < 20 %	☐ > 50 %
	☑ > 20%	☐ > 80 %
		in the last __1.25___ years (Luke, pers. com. 1998; Mbinda, 1997)

9D. If stable or unknown, do you predict a decline in habitat (approximately, in percent) over years ?:

☐ Unknown	☐ < 20 %	☐ > 50 %
	☐ > 20%	☐ > 80 %
		in the next _____ years

10. Threats:

☑ Past	☐ Present	☐ Future (predicted)

10A. What are the threats to the taxon ? Indicate which are Past [P], Present [Pr], and/or Future [F]:

☐ Disease/Pathogens	☐ Edaphic factors	☐ Pollution
☐ Flooding/Scouring	☑ Loss of habitat [P, Pr, F]	☑ Habitat fragmentation [P, Pr, F]
☑ Harvest [P]	☑ Predation [P, Pr, F]	☑ Successional changes [P, Pr, F]
☐ Trade	☐ Competition from exotics	☐ Other, please specify:
☐ Catastrophic events	☐ Pesticides	
☐ Hybridisation		

10B. Are these threats resulting in population decline?:

☑ Yes	☐ No

11. Trade/harvesting:

11A. Is the taxon harvested or in trade?:

☑ Yes	☐ No	☐ Unknown

If Yes, is it:

☑ Local harvesting (non-commercial)	☐ National trade (Commercial)
☐ Local trade (Commercial)	☐ International trade (Commercial)

11B. Parts in trade/harvested:

☐ Roots	☐ Bark	☐ Products (gums, resin, etc.)
☐ Leaves	☐ Stem/twigs/branch	☐ Whole plant
☐ Fruits	☑ Wood/timber	☐ Others (please specify):
☐ Seeds	(Mbinda 1996)	

11C. Is the harvesting:

☑ Destructive to individual plants	☐ Non-destructive to individual plants

11D: Is Trade/harvesting (in any form) resulting in a perceived or inferred population decline:

☐ Yes	☑ No

(If you have more information regarding trade, harvest of this taxon, please explain in Section 19 (other comments).

12. Population trends:

12A. Is the population size/numbers of the taxon:

☐ **Stable**	☐ **Declining**
☐ **Increasing**	☑ **Unknown**

12B. If declining, what has been the decline in population (perceived or inferred in percents due to habitat loss, threats, trade, etc.) over years:

☐ **Unknown**	☐ **50 %**
☐ **< 20 %**	☐ **> 80 %**
☐ **> 20 %**	in the last _____ years

12C. If Stable or Unknown, do you predict a decline in population (due to factors such as habitat loss, threats, trade, etc.) over years:

☑ **Unknown**	☐ **> 50 %**
☐ **< 20 %**	☐ **> 80 %**
☐ **> 20 %**	in the next _____ years

13. Data quality. Are the above perceived, inferred, predicted educated/qualified estimates based on:

☐ **Census or monitoring**	☐ **Living collections/records/literature**	☑ **Herbarium/records/literature**
☑ **General field study**	☐ **Indirect information such as from trade, etc.**	☐ **Hearsay/popular belief**
☐ **Informal field sighting**		

14. IUCN Category of Threat (state criteria used): Endangered; B1, 2 b

15A. Research recommendations necessary for the taxon:

☑ **Survey**	☐ **Taxonomic studies**	☑ **Limiting factor management**
☑ **Monitoring**	☑ **Life history studies**	☑ **Habitat management**
☐ **Genetic studies**	☐ **Limiting factor research**	☑ **PHVA**
		☐ **Others** (taxon specific)

15B. Management recommendations:

– Surveys required to establish the extent of occurrence of the populations in Tanzania.

– *Ex situ* conservation. It has attractive flowers hence has a high potential as an ornamental.

– Establish nurseries for commercial use (currently grown for sale to the hotel industry by the Coastal Forest Conservation Unit, NMK, Kenya South Coast).

– The seeds are orthodox therefore it is easy to multiply/grow.

– It is a legume and therefore needs research on possibilities on multiple use in farms.

---

16. Is cultivation required:

☑ Yes   ☐ No

If yes, is it for:

☑ Conservation research   ☑ Support for wild populations e.g. reintroduction

☑ Horticultural research   ☑ Sustainable utilisation

☐ Education   ☐ Others (please specify):

---

17. Does an *ex situ* conservation programme already exist:

☑ Yes   ☐ No

If yes, give details:

Coastal Forest Conservation Unit (NMK), Ukunda, Kenya. Grown from seed, for enrichment planting and for sale to hotels. (Hamisi J Mududu, pers. com., Mbinda, 1997).

---

18. Level of understanding of cultivation of the taxon:

☑ Techniques known for taxon or similar taxa   ☐ Horticultural techniques need to be established

☐ Some techniques known for taxon or similar taxa

---

19. Other comments related to status and conservation of the species:

Harvested for timber. Seeds eaten by bush pig (Mbinda, 1996).

Mbinda, J.M. (1996) survey for the conservation of *G. macrosiphon* (Harms) Brenan (Caesalpiniaceae) at Muhaka, Gongoni and Buda Mafisini. Project in part fulfilment of Plant Conservation Techniques Course East Africa International certificate.

Modified from CBSG/SSC by Amani Darwin Workshop Team, Amani Nature Reserve, Tanzania, 1998.

Three kinds of assessments/recommendations are made for each taxon reviewed:

1. Assigning taxa to the New IUCN Red List Categories of Threat: The New IUCN Red List Categories are based both on population and distribution criteria. They provide a system that facilitates comparisons across widely different taxa. These criteria can be applied to any taxonomic unit at or below the species level, with the exception of micro-organisms (see Appendix 2).

2. Making recommendations for research and management activities: These recommendations aim to integrate research and practical management actions with known threats to contribute to the taxon's conservation. Research management can be defined as an interactive management program including a strong feedback loop between management activities, evaluation of their effectiveness, and the response of the species.

3. Making recommendations for appropriate *ex situ* programs: These recommendations are only proposed if they contribute to the conservation of the taxon either directly (*ex situ* propagation or storage) or indirectly (for research or public education). These recommendations form the foundation for development of regional strategic collection plans for the zoo and aquarium community and are now being used by and for botanical agencies.

The CAMP Process documents can be used as guidelines by national and regional biodiversity agencies as well as regional *ex situ* programs as they develop their own conservation action plans.

The CBSG Workshop processes provide an objective environment, expert knowledge, and a neutral facilitation process that supports sharing of available information across institutions and stakeholder groups, reaching agreement on the issues and available information, and then making practical management recommendations for the taxon and habitat under consideration. The process has been successful in unearthing and integrating previously unpublished information for the decision making process.

CBSG Resources

Expertise and Costs: The problems facing threatened species everywhere are complex and require information from a diverse range of specialists. No one individual or agency encompasses all of the relevant knowledge. Thus, there is a need to include a wide range of people as resources and analysts. It is important that the invited participants have reputations for expertise, political and scientific objectivity, openness, and for active transfer of wanted skills. CBSG has a volunteer network of more than 700 experts. More than 3,000 people from 400 organizations have assisted CBSG on projects and participated in workshops on a volunteer basis contributing tens of thousands of hours of time.

Indirect cost contributions to support: Use of CBSG resources and the contribution of participating experts provides a matching contribution more than equalling the proposed budget request for projects. CBSG workshop processes provide an objective environment, expert knowledge, and a neutral facilitation process that supports sharing of available information across institutions and stakeholder groups, reaching agreement on the issues and available information, and then making useful and practical management recommendations for the taxon and habitat system under consideration. The process has been remarkably successful in unearthing and integrating previously unpublished information for the decision making process.

Integration of Science, Management, and Stakeholders

There are patterns of human behavior that are cross-disciplinary and cross-cultural which affect the processes of communication, problem-solving, and collaboration:

1) in the acquisition, sharing, and analysis of information;

2) in the perception and characterization of risk;

3) in the development of trust among individuals; and

4) in 'territoriality' (personal, institutional, local, national).

Each of these has strong emotional components that shape our interactions. Recognition of these patterns has been essential in the development of processes to assist people in Working Groups to reach agreement on needed conservation actions, collaboration needed, and to establish new working relationships. Frequently, local management agencies, external consultants, and local experts have identified management actions. However, an isolated narrow professional approach which focuses primarily on the perceived biological problems seems to have little effect on the needed political and social changes (social learning) for collaboration, effective management and conservation of habitat fragments or protected areas and their species components. CBSG workshops are organized to bring together the full range of groups with a strong interest in conserving and managing the species in its habitat or the consequences of such management. One goal in all workshops is to reach a common understanding of the state of scientific knowledge available and its possible application to the decision-making process and to needed management actions. We have found that decision-making driven workshop processes with risk characterization tools, and deliberation among stakeholders are powerful tools for extracting, assembling, and exploring information. These tools also support building of working agreements and instill local ownership of the problems, the decisions required, and their management during the workshop process. The process encourages developing a shared understanding across wide boundaries of training and expertise. As participants appreciate the complexity of the problems as a group, they take more ownership of the process as well as the ultimate recommendations made to achieve workable solutions. This is essential if the management recommendations generated by the workshops are to succeed.

One of the keys to a successful CAMP process are the skills provided by a Workshop facilitator. Trained and/or experienced facilitators are available from CBSG, regional CBSG networks, and other institutions. It is not necessary to use these facilitators if you have people in your own network with facilitation skills and who know and have experience with the CAMP methodology. Facilitation is a skill that is developed over time and requires neutrality, objectivity, respect, and sensitivity to participants and issues.

Technical facilitators experienced in using the IUCN Red List Criteria are highly desirable. Anyone familiar with Red List guidelines, procedure, and background can provide the needed guidance to participants, and anyone with an understanding of basic principles of conservation biology and an open mind can learn the process.

CBSG Workshop Toolkit

The basic CBSG set of tools for workshops includes small group dynamic skills, explicit use in small groups of problem restatement, divergent thinking sessions, identification of the history and chronology of the problem, causal flow diagramming (elementary systems analysis), matrix methods for qualitative data and expert judgements, paired and weighted ranking for making comparisons between sites, criteria, and options, utility analysis, stochastic simulation modeling for single populations and metapopulation and deterministic and stochastic modeling of local human populations. Several computer packages are used to assist collection and analysis of information with these tools. We provide training in these tools and undertake training workshops for people wishing to organize their own workshops.

The CBSG CAMP Database

CAMPs are intended to provide strategic guidance for application of appropriate management and information collection techniques to threatened taxa. CAMPs provide a rational and comprehensive means of assessing priorities for intensive management, sometimes including *ex situ* management, within the context of the broader conservation needs of threatened taxa.

During the CAMP process, workshop participants utilize a data sheet to assess the risks and formulate recommendations for action. This data sheet allows participants to systematically record data on the status of populations and habitat in the wild, and permits entry of recommendations for intensive action. The Taxon Data Sheets provide documentation of reasoning behind recommendations, and may include data that does not fit into spreadsheet format. The CAMP process attempts to be as quantitative or numerate as possible. The CAMP Taxon Database provides a systematic way of recording data and providing for reassessments as the status of species change and as new information becomes available. This program is both specific and flexible allowing detailed information to be entered and a variety of queries to be made.

We also envision the data being accessed and queries being made by people in a variety of disciplines with a wide range of questions. The CAMP will feed directly into the red listing process through the simultaneous development of this program and the Ramas Red List expert system and the SSC Specialist Group database, Species Information System (SIS).

Appendix: Outline of Workshop Process

CAMP Preparation and Documentation for Process Organisers

Dates and Location

CAMP processes generally take place over a full three- or three-and-one-half day period, including evenings. CBSG's schedule is usually filled up to one year in advance, accordingly, it is essential that the host/organizer contact the CBSG Chairman or Office far in advance to make arrangements for the process dates. Each CAMP is assigned to a specific CBSG Program Officer who will be the organizer's primary contact for the process.

Appropriate Specialist Group Chairs should be consulted and involved. Often, there are special interest group meetings or conferences, which many potential participants may plan to attend. It is advisable to take advantage of these opportunities by arranging CAMP process dates in close proximity and/or in the same location so that attendance and participation can be maximized. Participants should plan on arriving the day before the CAMP begins and departing on the fourth day.

Living quarters and food for the three days should be arranged at a location that minimizes outside distractions. Arrangements should promote the equality of all participants. Participants are usually responsible for their own lodging expenses and meals.

Invitations to the CAMP Process

Invitations are generally prepared and mailed by the host/organizer unless other arrangements have been made with CBSG. Ideally, invitations should be in the mail at least six to twelve months prior to the process. The list of invitees to a CAMP is generated by the host/organizer in collaboration with the appropriate SSC Specialist Group and CBSG. Appropriate invitees may include biologists, Specialist Group members, policy level managers, NGOs that have participated in conservation efforts, botanic garden managers, academics, and other interested parties. Generally, a list of 30-40 individuals is compiled, on the assumption that approximately 15-30 will be able to attend. The host/organizer is responsible for securing commitments to participate and for all communication with invitees prior to the CAMP.

Taxonomic lists

A critical piece of information that must be provided to the CBSG Office as soon as plans are firm is a list of the taxa to be assessed at the CAMP. This list should be at the subspecies level if possible. The most widely accepted taxonomy for the group of concern should be used and should cross reference to listings in national and international legislation. For regional CAMPs, lists are needed of all endemic taxa, as well as all taxa listed as threatened within the region.

Meeting agenda

The meeting agenda is put together by the host/organizer, with input from the CBSG office and/or the appropriate wildlife agency or Specialist Group Chair. The agenda should be mailed to participants at least 30 days prior to the CAMP. Usually, there are a few brief presentations on the first morning providing an overview of the general status of the taxonomic group or region (re: conservation status and general threats), as well as a few specific presentations chosen by the organizing Specialist Group, when applicable. Presenters at these sessions should be given plenty of notice concerning the presentations so that they have adequate time to prepare. It is useful to request a written copy of presentations to be included in the final report. These presentations are followed by general overviews by CBSG staff on the CAMP process. Next, working groups are organized to review the taxonomic groups or regions. Working Groups report to the larger group of participants several times during the course of the process; participants work to reach consensus on assessments and recommendations prior to the process' end.

Equipment and meeting room needs

Meeting facilities should include a meeting room for the group, with breakout areas or rooms that also can be used during evenings. The core resources are: a slide projector and carousel, and an overhead projector (generally for the first day only), a parallel port IBM compatible laser printer (preferably in the meeting room or in an adjacent location), black or whiteboard, 4-8 flip charts and pens, tape, access to a photocopier (for up to 400 pages per day), and as many IBM-compatible laptops or desktop computers as possible. CBSG uses Microsoft Word® 6.0 or 7.0 for documents. CBSG staff bring at least one computer to CAMPs, but it is best if a total of 4 or 5 are available for participants to use during working group sessions. Adequate electrical outlets also should be provided.

After the CAMP

CBSG staff generally plan to stay on-site for at least one day following a CAMP. This time is spent drafting the preliminary participants' report so that distribution of the meeting results is expedited. Access to computers, a laser printer, and copying facilities is essential during this period. It is important that two or three participants, with a good sense of the overall scope of the problem facing the taxon or region, work with CBSG staff on this draft at this time. The host/organizer should identify these individuals, in collaboration with the appropriate wildlife agency or Specialist Group Chair (when applicable).

CBSG generally will take responsibility for printing and distribution of the participants' first draft. This is distributed to a number of volunteer editors who work with the CBSG Program staff to refine the document. These editors usually are given up to four weeks to make comments on this initial draft. After these comments are incorporated into a working draft version, CAMP documents are distributed to a broader audience including all participants, field biologists, academics, wildlife managers, *ex situ* managers, and other interested parties. Subsequent comments are incorporated into revised drafts as they are printed.

Funding

Funding is needed primarily for travel and per diem during the CAMP, preparation of the briefing document and the CAMP report, communications and some personnel costs. CBSG costs are for preparation of the documents, completion and distribution of the reports after the meeting, travel of 2-4 people, and their per diem. We estimate that CBSG's costs for each CAMP are $10,000 to $18,000 depending on the amount of work required beforehand as well as after the process in completion of the report.

The following organizational outline is a suggested set of steps only and should be modified to local circumstances. Each workshop will have its own requirements and personality. These guidelines are given as a checklist of essential steps in the process. This set of suggestions is for a single CAMP Workshop. Guidelines and suggestions for organizing a series of workshops in a national Biodiversity Inventory are also available.

Preparation: planning and organization

1. *What kind of CAMP Workshop?*

Available funding and support will both determine and be determined by the type of workshop you select and its relevance to national biodiversity needs. The nature and goal of the workshop should be influenced by the management need of the host organisation and its collaborators. A "one off" workshop may be needed to answer a particular managerial or legislative objective for a small group of species. A series may be required to assess a large number of species or a larger geographic area, a series may be used to assess changes in species status over time. The structure may be influenced by larger processes such as planning regimes for protected areas or national responses to international conventions e.g. a series of CAMPs to directly feed into a national biodiversity inventory and assessment as has happened in India.

When reviewing a group of species, a large number can be processed as initial assessments with subsequent more detailed reviews directed by the initial workshop findings. Or smaller numbers per workshop can be processed in more detail. To establish the process and generate confidence a review of endemic species can be useful. This has the advantage of being "non-controversial" – always "global" and therefore fitting within IUCN Global Red List Guidelines.

2. *Identification of key participants*

  a. Ensure representation from all relevant disciplines/stakeholder groups.
  b. Consider how diverse stakeholder participation should be. For instance, exporters and manufacturers need to be represented in trade related workshops such as CAMPs for medicinal plants. Local stakeholders (users, collectors, local traders, and landowners) may have a lot of relevant information if a means can be found to encourage them to participate.

3. *Locating specialists*

It is essential to have participation and support from the implementing agency. CBSG will not conduct a workshop without an invitation from the host country's "wildlife agency". Forest officers and wildlife managers can contribute information that is useful and current about potential threats, habitat and tell the ground realities of local politics in a protected area. Names of potential participants can be collected from other workshops, conferences, symposia, etc. The SSC can help in identifying local members of SSC Specialist Groups. Local and national research agencies and universities can play a vital role in such meetings.

4. *Convincing the relevant experts to come to your workshop and share unpublished information*

The CAMP organisers should send a series of welcoming and convincing invitation letters promoting the mutual gains to be achieved from the workshop. These should be endorsed by the relevant national agencies and international bodies e.g. SSC/IUCN. Your own organizational credibility as a fair-minded, objective, dedicated, cooperative organization can be enhanced through CAMP workshops. However the initial contacts are very important, any letter should reflect this as not everyone may know you or your organization.

5. *Organizational preparation*

A CAMP workshop is logistically complex and requires effective planning. The CBSG Process Manual is an excellent guide, it can be copied on computer and your own workshop details incorporated into it. Fund raising will be required; work with local organizations to obtain support by sponsored dinners, lunches, venue etc. Keep chart/calendar of dates, activities to be done/completed, materials sent. Check proposed dates with essential participants.

Send invitation letters to participants. Some facilitators have found that sending a series of letters helps to build anticipation and generate interest in the meeting. Invitations should include description and some details of the CAMP process; request for early confirmation of participation; offer to write letter to institution head requesting their presence officially (this is extremely important), the working language, health responsibilities (particularly for malarial regions) and clear statements on the financial responsibilities of both parties. Follow-up letters are valuable, providing more detailed information and Biological Information Sheets to be returned (ideally) prior to CAMP process (even if they can't attend). You may also want to send a copy of the CAMP Manual with this second invitation to provide some instruction and background for participants. This creates interest as it gives potential participants something they can do to prepare for the workshop. Letters should also explain that there will be no time for tours, families, time off during the workshop. A final letter should request final confirmation of attendance and provide final instructions to bring field notes, references, and other data. Personal letters may be required for experts who have crucial information. Lastly, courtesy invitations to VIPs are often necessary, an opportunity to promote the event and obtain sanction from senior decision-makers, even if they are unable to attend.

6. *Briefing Book*

Every participant must get a Briefing Book. Make extra copies because the host organization may want to give them away to special guests or the press and often additional participants join the workshop on the first day. To avoid giving away expensive Briefing Books to the press, make up a special Press Packet using the cover of the Briefing Book and a Press Release with a few other essential facts about the workshop process and the species to be assessed.

7. *Venue requirements*

Larger, formal halls are useful for opening and closing ceremonies, particularly for large and politically important workshops. To allow the workshop to proceed effectively space is required for frequent and informal plenaries where the whole group can sit comfortably and discuss as a group. Where possible avoid huge auditoriums; a medium sized room where a microphone is not required and where informal plenary sessions can be held at a moment's notice is ideal. Smaller spaces for small working groups (from about 6 - 12 persons) are essential. Tables are very important for the comfort of the group, as they will be filling out forms and using books and other reference materials. Tables should be arranged for easy and informal discussion (e.g. in a square instead of rows). Tables for Working Groups can be fixed in one large room, but sufficient space should be between tables so that disturbance from others is not a factor.

8. *Meals and coffee*

Time is of the essence at a CAMP Workshop. Starting at the same time and not spending lots of time at coffee breaks is important. Therefore all meals should be provided at the Workshop venue only. Similarly on site accommodation with provision of breakfast will get participants there on time. Dinner will ensure (usually) that they stay a couple of hours extra. Rather than take formal breaks, provide coffee and tea all the time, or have it served around working group tables.

The CAMP process itself creates excitement and dedication. Participants are usually willing to work during breaks and into the evening, particularly if asked to do so with good humor and a "pep talk" now and then about the "cause." In any case, evening sessions should be presented as optional on the agenda and when you explain ground rules and get the group's agreement about how to run the workshop.

9. *Inaugural exercise*

The Inauguration (opening ceremony) and Valedictory (closing ceremony) is usually 100% the host's responsibility, and can be useful to them for their public relations' and political needs. Organizers and facilitators may be asked to sit with dignitaries and to say a few words about the process. After the Inauguration the actual CAMP Workshop begins.

Briefing Materials

For each CAMP meeting, a Briefing Book for participants will be prepared. It is up to the organizer/host to solicit information for this Book. A useful approach is to request materials for the briefing book in the original CAMP invitation letter. Information to be included in the briefing book includes:

- Overview material on the taxon or region in question, particularly with reference to conservation or population biology;
- Information on wild populations. Should include a regional overview;
- Maps of ranges of species within the group;
- Overview material on regional habitat problems that are affecting the taxon or region in question;
- Disease/pathogen problems facing this taxonomic group;
- Taxonomic problems or questions within the group;
- Any specific environmental parameters that seem to be affecting the group (e.g., pollution, predation, harvesting);
- Information on *ex situ* populations;
- Information on *ex situ* cultivation problems;
- A bibliography, preferably as complete as possible and either on disk or in clean copy that we can scan into a computer file (CBSG can sometimes assist with this);
- Introductory material on CBSG, SSC and the CAMP process;
- A list of invitees;
- A copy of the invitation to the meeting;
- A copy of the meeting agenda;
- Logos from sponsoring organizations; and
- Photos.

Basic Workshop Agenda

Day One

- Welcome and introductions (participants introduce themselves, their institution and main area of interest). This period sets the scene with regard to the conservation and legislative context for the workshop. Usually, there are several overview presentations on the first morning which discuss the general status of the taxonomic group or region being considered at the CAMP. The specific subject of these presentations will vary but examples include such topics as: 1) conservation status; 2) general threats; 3) pest/pathogen/invasive issues; 4) reproduction; and 5) genetics. Description of CAMP process and IUCN categories (this presentation will include: 1) the role of IUCN, SSC, CBSG; 2) history of the CAMP process; 3) history of Red Data books and IUCN categories; and 4) discussion of problem solving tools and inherent human mental traits;

- It is generally the host's responsibility to identify appropriate topics and speakers for this morning session. Request written copies of presentations for inclusion in final CAMP report. An introductory presentation on the CAMP process and the IUCN categories of threat is given by CBSG facilitators;

- Fill in a sample Taxon Data Sheet using overhead projector with whole group (this can take a long time, but it is an excellent exercise as everyone can come to understand the process together and in the same format);

- Presentation of Ground Rules, and Rules for Working in Small Groups, e.g. role of facilitators, recorders, researchers. You may want to use overheads in this section to explain Ground Rules and group roles. These guidelines should also be included in the briefing book. We find it useful to get a show of hands on the Ground Rules and schedule to be sure that all participants agree to abide by these rules;

- Form Working Groups – Working Groups get together. The Working Groups will form and meet, even if it is just to introduce themselves and decide one species to start with; and

- Plenary for those species assessed before closing for Day One.

Day Two

- Working groups meet and continue to review taxa and fill in Taxon Data Sheets;

- Plenary session: After assessing several species, the Working Groups assemble for plenary so that all species information can be read out and agreed upon by the Workshop as a whole. This is extremely important as different group members will definitely have information on species other than those that their own group is assessing. Ideally, Taxon Data Sheets should be circulated to all groups before reading but sometimes this is a logistical impossibility. In addition, it is definitely NOT a good idea to wait until the last day to read out Taxon Data Sheet information, it is tedious, boring and people just stop listening. Doing this a bit at a time also gives people a needed break from the concentrated attention of the Working Group;

- Working Groups;

- Plenary session.

Day Three

- Working Groups meet and continue to review taxa;
- Plenary session;
- Formation of Special Issue Working Groups: Special Issue Working Groups can take up various issues that emerge during the Workshop. These can vary from taxonomic or nomenclatural problems to general conservation issues such as Trade or Education. Each group will produce a written report to be read out to the Plenary. The Workshop participants comment on the Report and agree to allow it as part of the Workshop Document; and
- Working Groups meet and continue to review taxa.

Day Four

- Plenary session: The Working Groups assemble for Plenary so that all information for the remaining species can be read out and agreed by the Workshop as a whole;
- End of Workshop: Collect hard and disk copies of the Taxon Data Sheets and working group reports. Explain the procedure to be used in reviewing and correcting the Draft Report and distributing the Final Report and establish consensus from participants that they agree with this procedure. Essentially what will happen is that the CBSG facilitators will get the Taxon Data Sheet information typed up into a simple format and standardize Working Group reports. This draft document will be sent to every participant, or to a small cohort of process participants agreeing to serve as voluntary editors, and he/she will get a chance to fill in gaps of sources and other information not accessible during the Workshop. However, the facilitators will not entertain major changes to the Draft which were not agreed upon by the entire Workshop. In addition, other specialists and experts will not be permitted to correct or comment on the Draft. What was done at the Workshop is the output of the participants who came and worked;
- Try to give every participant a draft list of species covered, their status as derived by the Workshop, institutions represented at the workshop and a list of participants. These can be shown to senior management when staff return from workshops thus illustrating the value of the event. It is also important to have at least a few key people stay after the workshop has ended and help tie up loose ends regarding the species assessed, organize and count the Taxon Data Sheets, and complete the Draft Report; and
- Closing ceremony, this is an important opportunity to thank those people who were integral to the success of the Workshop and to place the work in context of current national and international conservation activities.

Guidelines for Group Dynamics

There are several ground rules that should be made explicit at the beginning of every CAMP process. These ground rules should be read to the Workshop and agreed upon by all participants at the start of the Workshop.

- Every idea, plan or belief about the Taxon or Region can be examined and discussed;
- Everyone participates in discussions and no one individual or agency dominates;
- Set aside (temporarily) all special agendas except conserving the Taxon or Region in question;
- Assume good intent of all participants. Treat other participants with respect;
- Stick to the schedule – begin and end promptly;
- The primary work will be conducted in sub-groups;
- Facilitators of Plenary sessions or Working Groups can call 'time out' when discussions reach an impasse or stray far off the topic at hand;
- Agreements or recommendations are reached by consensus;
- Plan to complete and review a draft report by the end of the meeting; and
- Flexibility is key. We will adjust our process and schedule as needed to achieve our goals.

Draft and Final Reports

A Draft Report consists of an Executive Summary, all Taxon Data Sheets, the Special Issue Working Group Reports and a list of participants. Drafts are posted to all participants, or to a small cohort of process participants agreeing to serve as voluntary editors, as soon as possible after the workshop. Facilitators/organizers get the Taxon Data Sheets typed and standardized. A Red List specialist checks the category to make sure that the conservation status has been derived in accordance with the information provided and the official Red List guidelines. Participants are asked to correct the Draft and return it by a specified date (within 4 weeks) to have their corrections and additions incorporated into the Final Report.

The Final Report consists of an Executive Summary, a more lengthy Report which includes analysis of the data and perhaps incorporation of the output in relation to some of the issues raised in the workshop Special Issue Working Groups. Summary Data Tables are compiled in different formats for ease of use and output Summary Charts are included. Usually a copy of the IUCN Red List Criteria and Guidelines is included in the Report as well as a description of the CAMP Process.

When budgeting your Workshop, REMEMBER, it costs a LOT more (in time and money) to make and post a Draft and Final Report for 500 species than it does for 30 or 40 species. For large exercises, it may be worthwhile to explore alternatives to paper drafts, e.g. computer disks.

Distribution: Reports of Workshops in which large numbers of species are evaluated are expensive to produce and it is not possible to circulate them widely. All participants, workshop sponsors, and the national government agencies receive a copy of the Final Report. Others requesting a Final Report can be charged a fee. The Final Report should be listed on the CBSG publication list (hard copy and on the Web site) so that people know that a report is available.

Review Process for CAMPs

Essentially what will happen is that the CBSG facilitators will get the Taxon Data Sheet information typed up into a simple format and standardize Working Group reports. This draft document will be sent to every participant or to a small cohort of process participants agreeing to serve as voluntary editors and he/she will get a chance to fill in gaps of sources and other information not accessible during the workshop. However, the facilitators will not entertain major changes to the Draft which were not agreed upon by the entire workshop. In addition, other specialists and experts will not be permitted to correct or comment on the Draft. What was done at the Workshop is the output of the participants who came and worked.

Once the comments of the participants have been incorporated into the document, the Final CAMP Report is reviewed: 1) by distribution to a broader audience which includes wildlife managers and regional captive programs worldwide; and 2) at regional review sessions at various CBSG meetings and processes, taking advantage of local expertise with the taxonomic group in question. Thus CAMPs are not single event. Instead, they are part of a continuing and evolving process of developing conservation and recovery plans for the taxa involved. The CAMP review process allows extraction of information from experts worldwide and prioritization of actions based on levels of threat. In nearly all cases, follow-up meetings are required to consider particular issues in greater depth or on a regional basis. Moreover, some form of follow-up will always be necessary to monitor the implementation and effectiveness of the recommendations resulting from the process. In many cases a range of Population and Habitat Viability Assessment (PHVA) workshops result from the CAMPs.

CAMPs are "living" documents that will be continually reassessed and revised as new information becomes available and as global and regional situations and priorities shift. The current CAMP process will continue both by its application to new groups of taxa and regions and the refinement of the ones already under way.

For more information on the IUCN SSC Conservation Breeding Specialist Group see: www.cbsg.org/

IUCN Guidelines for Re-introduction

Prepared by the IUCN SSC Re-introduction Specialist Group – as approved by 41st Meeting of Council, May 1995

SPECIES SURVIVAL COMMISSION

Introduction

These policy guidelines have been drafted by the Re-introduction Specialist Group of the IUCN's Species Survival Commission[1], in response to the increasing occurrence of re-introduction projects worldwide, and consequently, to the growing need for specific policy guidelines to help ensure that the re-introductions achieve their intended conservation benefit, and do not cause adverse side-effects of greater impact. Although IUCN developed a Position Statement on the Translocation of Living Organisms in 1987, more detailed guidelines were felt to be essential in providing more comprehensive coverage of the various factors involved in re-introduction exercises.

These Guidelines are intended to act as a guide for procedures useful to re-introduction programmes and do not represent an inflexible code of conduct. Many of the points are more relevant to re-introductions using captive-bred individuals than to translocations of wild species. Others are especially relevant to globally endangered species with limited numbers of founders. Each re-introduction proposal should be rigorously reviewed on its individual merits. It should be noted that re-introduction is always a very lengthy, complex and expensive process.

Re-introductions or translocations of species for short-term, sporting or commercial purposes – where there is no intention to establish a viable population - are a different issue and beyond the scope of these guidelines. These include fishing and hunting activities.

This document has been written to encompass the full range of plant and animal taxa and is therefore general. It will be regularly revised. Handbooks for re-introducing individual groups of animals and plants will be developed in future.

Context

The increasing number of re-introductions and translocations led to the establishment of the IUCN Species Survival Commission's Re-introduction Specialist Group. A priority of the Group has been to update IUCN's 1987 Position Statement on the Translocation of Living Organisms, in consultation with IUCN's other Commissions.

---

[1] Guidelines for determining procedures for disposal of species confiscated in trade are being developed separately by IUCN.

It is important that the Guidelines are implemented in the context of IUCN's broader policies pertaining to biodiversity conservation and sustainable management of natural resources. The philosophy for environmental conservation and management of IUCN and other conservation bodies is stated in key documents such as "Caring for the Earth" and the "Global Biodiversity Strategy," which cover the broad themes of the need for approaches with community involvement and participation in sustainable natural resource conservation, an overall enhanced quality of human life and the need to conserve and, where necessary, restore ecosystems. With regard to the latter, the re-introduction of a species is one specific instance of restoration where, in general, only this species is missing. Full restoration of an array of plant and animal species has rarely been tried to date.

Restoration of single species of plants and animals is becoming more frequent around the world. Some succeed, many fail. As this form of ecological management is increasingly common, it is a priority for the Species Survival Commission's Re-introduction Specialist Group to develop guidelines so that re-introductions are both justifiable and likely to succeed, and that the conservation world can learn from each initiative, whether successful or not. It is hoped that these Guidelines, based on extensive review of case-histories and wide consultation across a range of disciplines will introduce more rigour into the concepts, design, feasibility and implementation of re-introduction despite the wide diversity of species and conditions involved.

Thus, the priority has been to develop guidelines that are of direct, practical assistance to those planning, approving or carrying out re-introductions. The primary audience of these Guidelines is, therefore, the practitioners (usually managers or scientists), rather than decision-makers in governments. Guidelines directed towards the latter group would inevitably have to go into greater depth on legal and policy issues.

1. Definition of Terms

a) **"Re-introduction":** an attempt to establish a species[2] in an area which was once part of its historical range, but from which it has been **extirpated** or become extinct[3]. ("Re-establishment" is a synonym, but implies that the re-introduction has been successful).

[2] The taxonomic unit referred to throughout the document is species; it may be a lower taxonomic unit (e.g. sub-species or race) as long as it can be unambiguously defined.

[3] A taxon is Extinct when there is no reasonable doubt that the last individual has died.

b) **"Translocation":** deliberate and mediated movement of wild individuals to an existing population of conspecifics.

c) **"Re-enforcement/Supplementation":** addition of individuals to an existing population of conspecifics.

d) **"Conservation/Benign Introductions":** an attempt to establish a species, for the purpose of conservation, outside its recorded distribution but within an appropriate habitat and eco-geographical area. **This is a feasible conservation tool only when there is no remaining area left within a species' historic range.**

2. Aims and Objectives of Re-introduction

a) **Aims:** The principal aim of any re-introduction should be to establish a viable, free-ranging population in the wild, of a species, subspecies or race, which has become globally or locally extinct, or extirpated, in the wild. It should be re-introduced within the species' former natural habitat and range and should require minimal long-term management.

b) **Objectives:** The objectives of a re-introduction may include: to enhance the long-term survival of a species; to re-establish a keystone species (in the ecological or cultural sense) in an ecosystem; to maintain and/or restore natural biodiversity; to provide long-term economic benefits to the local and/or national economy; to promote conservation awareness, or a combination of these.

3. Multidisciplinary Approach

A re-introduction requires a multidisciplinary approach involving a team of persons drawn from a variety of backgrounds. As well as government personnel, they may include persons from governmental natural resource management agencies, non-governmental organizations, funding bodies, universities, veterinary institutions, zoos (and private animal breeders) and/or botanic gardens, with a full range of suitable expertise. Team leaders should be responsible for coordination between the various bodies and provision should be made for publicity and public education about the project.

4. Pre-Project Activities

4a. BIOLOGICAL

(i)   **Feasibility study and background research**

  – An assessment should be made of the taxonomic status of individuals to be re-introduced. They should preferably be of the same subspecies or race as those which were extirpated, unless adequate numbers are not available. An investigation of historical information about the loss and fate of individuals from the re-introduction area, as well as molecular genetic studies, should be undertaken in case of doubt as to individuals' taxonomic status. A study of genetic variation within and between populations of this and related taxa can also be helpful. Special care is needed when the population has long been extinct.

  – Detailed studies should be made of the status and biology of wild populations (if they exist) to determine the species' critical needs. For animals, this would include descriptions of habitat preferences, intraspecific variation and adaptations to local ecological conditions, social behaviour, group composition, home range size, shelter and food requirements, foraging and feeding behaviour, predators and diseases. For migratory species, studies should include the potential migratory areas. For plants, it would include biotic and abiotic habitat requirements, dispersal mechanisms, reproductive biology, symbiotic relationships (e.g. with mycorrhizae, pollinators), insect pests and diseases. Overall, a firm knowledge of the natural history of the species in question is crucial to the entire re-introduction scheme.

  – The species, if any, that has filled the void created by the loss of the species concerned, should be determined; an understanding of the effect the re-introduced species will have on the ecosystem is important for ascertaining the success of the re-introduced population.

  – The build-up of the released population should be modelled under various sets of conditions, in order to specify the optimal number and composition of individuals to be released per year and the numbers of years necessary to promote establishment of a viable population.

  – A Population and Habitat Viability Analysis (PHVA) will aid in identifying significant environmental and population variables and assessing their potential interactions, which would guide long-term population management.

(ii) **Previous Re-introductions**

– Thorough research into previous re-introductions of the same or similar species and wide-ranging contacts with persons having relevant expertise should be conducted prior to and while developing the re-introduction protocol.

(iii) **Choice of release site and type**

– The site should be within the historic range of the species. For an initial re-enforcement there should be few remnant wild individuals. For a re-introduction, there should be no remnant population to prevent disease spread, social disruption and introduction of alien genes. In some circumstances, a re-introduction or re-enforcement may have to be made into an area which is fenced or otherwise delimited, but it should be within the species' former natural habitat and range.

– A conservation/benign introduction should be undertaken only as a last resort when no opportunities for re-introduction into the original site or range exist and only when a significant contribution to the conservation of the species will result.

– The re-introduction area should have assured, long-term protection (whether formal or otherwise).

(iv) **Evaluation of re-introduction site**

– Availability of suitable habitat: re-introductions should only take place where the habitat and landscape requirements of the species are satisfied, and likely to be sustained for the foreseeable future. The possibility of natural habitat change since extirpation must be considered. Likewise, a change in the legal/political or cultural environment since the species' extirpation needs to be ascertained and evaluated as a possible constraint. The area should have sufficient carrying capacity to sustain growth of the re-introduced population and support a viable (self-sustaining) population in the long run.

– Identification and elimination, or reduction to a sufficient level, of previous causes of decline: could include disease; over-hunting; over-collection; pollution; poisoning; competition with or predation by introduced species; habitat loss; adverse effects of earlier research or management programmes; competition with domestic livestock, which may be seasonal.

- Where the release site has undergone substantial degradation caused by human activity, a habitat restoration programme should be initiated before the re-introduction is carried out.

(v) **Availability of suitable release stock**

- It is desirable that source animals come from wild populations. If there is a choice of wild populations to supply founder stock for translocation, the source population should ideally be closely related genetically to the original native stock and show similar ecological characteristics (morphology, physiology, behaviour, habitat preference) to the original sub-population.

- Removal of individuals for re-introduction must not endanger the captive stock population or the wild source population. Stock must be guaranteed available on a regular and predictable basis, meeting specifications of the project protocol.

- Individuals should only be removed from a wild population after the effects of translocation on the donor population have been assessed, and after it is guaranteed that these effects will not be negative.

- If captive or artificially propagated stock is to be used, it must be from a population which has been soundly managed both demographically and genetically, according to the principles of contemporary conservation biology.

- Re-introductions should not be carried out merely because captive stocks exist, nor solely as a means of disposing of surplus stock.

- Prospective release stock, including stock that is a gift between governments, must be subjected to a thorough veterinary screening process before shipment from original source. Any animals found to be infected or which test positive for non-endemic or contagious pathogens with a potential impact on population levels, must be removed from the consignment, and the uninfected, negative remainder must be placed in strict quarantine for a suitable period before retest. If clear after retesting, the animals may be placed for shipment.

- Since infection with serious disease can be acquired during shipment, especially if this is intercontinental, great care must be taken to minimise this risk.

- Stock must meet all health regulations prescribed by the veterinary authorities of the recipient country and adequate provisions must be made for quarantine if necessary.

vi) **Release of captive stock**

- Most species of mammals and birds rely heavily on individual experience and learning as juveniles for their survival; they should be given the opportunity to acquire the necessary information to enable survival in the wild through training in their captive environment; a captive bred individual's probability of survival should approximate that of a wild counterpart.

- Care should be taken to ensure that potentially dangerous captive-bred animals (such as large carnivores or primates) are not so confident in the presence of humans that they might be a danger to local inhabitants and/or their livestock.

4b. SOCIO-ECONOMIC AND LEGAL REQUIREMENTS

- Re-introductions are generally long-term projects that require the commitment of long-term financial and political support.

- Socio-economic studies should be made to assess impacts, costs and benefits of the re-introduction programme to local human populations.

- A thorough assessment of attitudes of local people to the proposed project is necessary to ensure long-term protection of the re-introduced population, especially if the cause of species' decline was due to human factors (e.g. over-hunting, over-collection, loss or alteration of habitat). The programme should be fully understood, accepted and supported by local communities.

- Where the security of the re-introduced population is at risk from human activities, measures should be taken to minimise these in the re-introduction area. If these measures are inadequate, the re-introduction should be abandoned or alternative release areas sought.

- The policy of the country to re-introductions and to the species concerned should be assessed. This might include checking existing provincial, national and international legislation and regulations, and provision of new measures and required permits as necessary.

- Re-introduction must take place with the full permission and involvement of all relevant government agencies of the recipient or host country. This is particularly important in re-introductions in border areas, or involving more than one state or when a re-introduced population can expand into other states, provinces or territories.

- If the species poses potential risk to life or property, these risks should be minimised and adequate provision made for compensation where necessary; where all other solutions fail, removal or destruction of the released individual should be considered. In the case of migratory/mobile species, provisions should be made for crossing of international/state boundaries.

5.   PLANNING, PREPARATION AND RELEASE STAGES

- Approval of relevant government agencies and land owners, and coordination with national and international conservation organizations.
- Construction of a multidisciplinary team with access to expert technical advice for all phases of the programme.
- Identification of short- and long-term success indicators and prediction of programme duration, in the context of agreed aims and objectives.
- Securing adequate funding for all programme phases.
- Design of pre- and post-release monitoring programme so that each re-introduction is a carefully designed experiment, with the capability to test methodology with scientifically collected data. Monitoring the health of individuals, as well as the survival, is important; intervention may be necessary if the situation proves unforeseeably favourable.
- Appropriate health and genetic screening of release stock, including stock that is a gift between governments. Health screening of closely related species in the re-introduction area.
- If release stock is wild-caught, care must be taken to ensure that: a) the stock is free from infectious or contagious pathogens and parasites before shipment and b) the stock will not be exposed to vectors of disease agents which may be present at the release site (and absent at the source site) and to which it may have no acquired immunity.
- If vaccination prior to release, against local endemic or epidemic diseases of wild stock or domestic livestock at the release site, is deemed appropriate, this must be carried out during the "Preparation Stage" so as to allow sufficient time for the development of the required immunity.
- Appropriate veterinary or horticultural measures as required to ensure health of released stock throughout the programme. This is to include adequate quarantine arrangements, especially where founder stock travels far or crosses international boundaries to the release site.

– Development of transport plans for delivery of stock to the country and site of re-introduction, with special emphasis on ways to minimise stress on the individuals during transport.

– Determination of release strategy (acclimatization of release stock to release area; behavioural training - including hunting and feeding; group composition, number, release patterns and techniques; timing).

– Establishment of policies on interventions (see below).

– Development of conservation education for long-term support; professional training of individuals involved in the long-term programme; public relations through the mass media and in local community; involvement where possible of local people in the programme.

– The welfare of animals for release is of paramount concern through all these stages.

6. POST-RELEASE ACTIVITIES

– Post-release monitoring is required of all (or a sample of) individuals. This most vital aspect may be by direct (e.g. tagging, telemetry) or indirect (e.g. spoor, informants) methods as suitable.

– Demographic, ecological and behavioural studies of released stock must be undertaken.

– Study of processes of long-term adaptation by individuals and the population.

– Collection and investigation of mortalities.

– Interventions (e.g. supplemental feeding; veterinary aid; horticultural aid) when necessary.

– Decisions for revision, rescheduling, or discontinuation of programme where necessary.

– Habitat protection or restoration to continue where necessary.

– Continuing public relations activities, including education and mass media coverage.

– Evaluation of cost-effectiveness and success of re-introduction techniques.

– Regular publication in scientific and popular literature.

For more information on the IUCN SSC Reintroduction Specialist Group see: www.iucn.org/themes/ssc/programs/rsg.htm.

Appendix 5

IUCN Guidelines for the Prevention of Biodiversity Loss caused by Alien Invasive Species

(as approved by 51ˢᵗ Meeting of the IUCN Council, Gland, Switzerland, February 2000)

SPECIES SURVIVAL COMMISSION

1. Background[1]

Biological diversity faces many threats throughout the world. One of the major threats to native biological diversity is now acknowledged by scientists and governments to be biological invasions caused by alien invasive species. The impacts of alien invasive species are immense, insidious, and usually irreversible. They may be as damaging to native species and ecosystems on a global scale as the loss and degradation of habitats.

For millennia, the natural barriers of oceans, mountains, rivers and deserts provided the isolation essential for unique species and ecosystems to evolve. In just a few hundred years these barriers have been rendered ineffective by major global forces that combined to help alien species travel vast distances to new habitats and become alien invasive species. The globalisation and growth in the volume of trade and tourism, coupled with the emphasis on free trade, provide more opportunities than ever before for species to be spread accidentally or deliberately. Customs and quarantine practices, developed in an earlier time to guard against human and economic diseases and pests, are often inadequate safeguards against species that threaten native biodiversity. Thus the inadvertent ending of millions of years of biological isolation has created major ongoing problems that affect developed and developing countries.

The scope and cost of biological alien invasions is global and enormous, in both ecological and economic terms. Alien invasive species are found in all taxonomic groups: they include introduced viruses, fungi, algae, mosses, ferns, higher plants, invertebrates, fish, amphibians, reptiles, birds and mammals. They have invaded and affected native biota in virtually every ecosystem type on Earth. Hundreds of extinctions have been caused by alien invasives. The ecological cost is the irretrievable loss of native species and ecosystems.

In addition, the direct economic costs of alien invasive species run into many billions of dollars annually. Arable weeds reduce crop yields and increase costs; weeds degrade catchment areas and freshwater ecosystems; tourists and homeowners unwittingly introduce alien plants into wilderness and natural areas; pests and pathogens of crops, livestock and forests reduce yields and increase control costs. The discharge of ballast water together with hull fouling has led to unplanned and unwanted introductions of harmful aquatic organisms, including diseases, bacteria and viruses, in marine and freshwater systems. Ballast water is now regarded as the most important vector for trans-oceanic and inter-oceanic movements of shallow-water coastal organisms. Factors like environmental pollution and habitat destruction can provide conditions that favour alien invasive species.

---

[1] Definition of Terms in section 3.

The degradation of natural habitats, ecosystems and agricultural lands (e.g. loss of cover and soil, pollution of land and waterways) that has occurred throughout the world has made it easier for alien species to establish and become invasive. Many alien invasives are "colonising" species that benefit from the reduced competition that follows habitat degradation. Global climate change is also a significant factor assisting the spread and establishment of alien invasive species. For example, increased temperatures may enable alien, disease-carrying mosquitoes to extend their range.

Sometimes the information that could alert management agencies to the potential dangers of new introductions is not known. Frequently, however, useful information is not widely shared or available in an appropriate format for many countries to take prompt action, assuming they have the resources, necessary infrastructure, commitment and trained staff to do so.

Few countries have developed the comprehensive legal and institutional systems that are capable of responding effectively to these new flows of goods, visitors and 'hitchhiker' species. Many citizens, key sector groups and governments have a poor appreciation of the magnitude and economic costs of the problem. As a consequence, responses are too often piecemeal, late and ineffective. It is in this context that IUCN has identified the problem of alien invasive species as one of its major initiatives at the global level.

While all continental areas have suffered from biological alien invasions, and lost biological diversity as a result, the problem is especially acute on islands in general, and for small island countries in particular. Problems also arise in other isolated habitats and ecosystems, such as in Antarctica. The physical isolation of islands over millions of years has favored the evolution of unique species and ecosystems. As a consequence, islands and other isolated areas (e.g. mountains and lakes) usually have a high proportion of endemic species (those found nowhere else) and are centres of significant biological diversity. The evolutionary processes associated with isolation have also meant island species are especially vulnerable to competitors, predators, pathogens and parasites from other areas. It is important to turn this isolation of islands into an advantage by improving the capacity of governments to prevent the arrival of alien invasive species with better knowledge, improved laws and greater management capacity, backed by quarantine and customs systems that are capable of identifying and intercepting alien invasive species.

2. Goals and Objectives

The goal of these guidelines is to prevent further losses of biological diversity due to the deleterious effects of alien invasive species. The intention is to assist governments and management agencies to give effect to Article 8 (h) of the Convention on Biological Diversity (CBD – see Appendix 1), which states that:

"Each Contracting Party shall, as far as possible and as appropriate:
   …(h) Prevent the introduction of, control or eradicate those alien species which threaten ecosystems, habitats or species."

These guidelines draw on and incorporate relevant parts of the 1987 IUCN Position Statement on Translocation of Living Organisms although they are more comprehensive in scope than the 1987 Translocation Statement. The relationship to another relevant guideline, the IUCN Guidelines for Re-introductions, is elaborated in Section 7 (See Appendix 4).

These guidelines are concerned with preventing loss of biological diversity caused by biological invasions of alien invasive species. They do not address the issue of genetically modified organisms, although many of the issues and principles stated here could apply. Neither do these guidelines address the economic (agricultural, forestry, aquaculture), human health and cultural impacts caused by biological invasions of alien invasive species.

These guidelines address four substantive concerns of the biological alien invasion problem that can be identified from this background context. These are:

- improving understanding and awareness;
- strengthening the management response;
- providing appropriate legal and institutional mechanisms;
- enhancing knowledge and research efforts.

While addressing all four concerns is important, these particular guidelines focus most strongly on aspects of strengthening the management response. This focus reflects the urgent need to spread information on management that can quickly be put into place to prevent alien invasions and eradicate or control established alien invasives. Addressing the other concerns, particularly the legal and research ones, may require longer-term strategies to achieve the necessary changes.

These guidelines have the following seven objectives.

1. To increase awareness of alien invasive species as a major issue affecting native biodiversity in developed and developing counties and in all regions of the world.

2. To encourage prevention of alien invasive species introductions as a priority issue requiring national and international action.

3. To minimise the number of unintentional introductions and to prevent unauthorised introductions of alien species.

4. To ensure that intentional introductions, including those for biological control purposes, are properly evaluated in advance, with full regard to potential impacts on biodiversity.

5. To encourage the development and implementation of eradication and control campaigns and programmes for alien invasive species, and to increase the effectiveness of those campaigns and programmes.

6. To encourage the development of a comprehensive framework for national legislation and international cooperation to regulate the introduction of alien species as well as the eradication and control of alien invasive species.

7. To encourage necessary research and the development and sharing of an adequate knowledge base to address the problem of alien invasive species worldwide.

3. Definition of Terms[2]

"**Alien invasive species**" means an alien species which becomes established in natural or semi-natural ecosystems or habitat, is an agent of change, and threatens native biological diversity.

"**Alien species**" (non-native, non-indigenous, foreign, exotic) means a species, subspecies, or lower taxon occurring outside of its natural range (past or present) and dispersal potential (i.e. outside the range it occupies naturally or could not occupy without direct or indirect introduction or care by humans) and includes any part, gametes or propagule of such species that might survive and subsequently reproduce.

"**Biological diversity**" (biodiversity) means the variability among living organisms from all sources including, *inter alia*, terrestrial, marine and other aquatic ecosystems and the ecological complexes of which they are a part; this includes diversity within species, between species and of ecosystems.

---

[2] At the time of adoption of these Guidelines by IUCN, standard terminology relating to alien invasive species has not been developed in the CBD context. Definitions used in this document were developed by IUCN in the specific context of native biodiversity loss caused by alien invasive species.

"**Biosecurity threats**" means those matters or activities which, individually or collectively, may constitute a biological risk to the ecological welfare or to the well-being of humans, animals or plants of a country.

"**Government**" includes regional co-operating groupings of governments for matters falling within their areas of competence.

"**Intentional introduction**" means an introduction made deliberately by humans, involving the purposeful movement of a species outside of its natural range and dispersal potential. Such introductions may be authorised or unauthorised.

"**Introduction**" means the movement, by human agency, of a species, subspecies, or lower taxon (including any part, gametes or propagule that might survive and subsequently reproduce) outside its natural range (past or present). This movement can be either within a country or between countries.

"**Native species**"(indigenous) means a species, subspecies, or lower taxon, occurring within its natural range (past or present) and dispersal potential (i.e. within the range it occupies naturally or could occupy without direct or indirect introduction or care by humans.)

"**Natural ecosystem**" means an ecosystem not perceptibly altered by humans.

"**Re-introduction**" means an attempt to establish a species in an area which was once part of its historical range, but from which it has been extirpated or become extinct (from IUCN Guidelines for Re-Introductions).

"**Semi-natural ecosystem**" means an ecosystem which has been altered by human actions, but which retains significant native elements.

"**Unintentional introduction**" means an unintended introduction made as a result of a species utilising humans or human delivery systems as vectors for dispersal outside its natural range.

4. Understanding and Awareness

4.1 Guiding Principles

1. Understanding and awareness, based on information and knowledge, are essential for establishing alien invasive species as a priority issue which can and must be addressed.

2. Better information and education, and improved public awareness of alien invasive issues by all sectors of society, is fundamental to preventing or reducing the risk of unintentional

or unauthorised introductions, and to establishing evaluation and authorisation procedures for proposed intentional introductions.

3. Control and eradication of alien invasive species is more likely to be successful if supported by informed and cooperating local communities, appropriate sectors and groups.

4. Information and research findings which are well communicated are vital prerequisites to education, understanding and awareness (See Section 8).

4.2 Recommended Actions

1. Identify the specific interests and roles of relevant sectors and communities with respect to alien invasive species issues and target them with appropriate information and recommended actions. Specific communication strategies for each target group will be required to help reduce the risks posed by alien invasive species. The general public is an important target group to be considered.

2. Make easily accessible, current and accurate information widely available as a key component of awareness raising. Target different audiences with information in electronic form, manuals, databases, scientific journals and popular publications (See also Section 8).

3. Target importers and exporters of goods, as well as of living organisms as key target groups for information/education efforts leading to better awareness and understanding of the issues, and their role in prevention and possible solutions.

4. Encourage the private sector to develop and follow best practice guidelines and monitor adherence to guidelines (Refer to 5.2 and 5.3).

5. As an important priority, provide information and recommended actions to travellers, both within country and between countries, preferably prior to the start of journeys. Raising awareness of how much human travel contributes to alien invasive problems can improve behaviour and be cost-effective.

6. Encourage operators in eco-tourism businesses to raise awareness on the problems caused by alien invasive species. Work with such operators to develop industry guidelines to prevent the unintentional transport or unauthorised introduction of alien plants (especially seeds) and animals into ecologically vulnerable island habitats and ecosystems (e.g. lakes, mountain areas, nature reserves, wilderness areas, isolated forests and inshore marine ecosystems).

7. Train staff for quarantine, border control, or other relevant facilities to be aware of the larger context and threats to biological diversity, in addition to practical training for aspects like identification and regulation (See Section 5.2).

8. Build communication strategies into the planning phase of all prevention, eradication and control programmes. By ensuring that effective consultation takes place with local communities and all affected parties, most potential misunderstandings and disagreements can be resolved or accommodated in advance.

9. Include alien invasive species issues, and actions that can be taken to address them, in appropriate places in educational programmes and schools.

10. Ensure that national legislation applicable to introductions of alien species, both intentional and unintentional, is known and understood, not only by the citizens and institutions of the country concerned, but also by foreigners importing goods and services as well as by tourists.

5. Prevention and Introductions

5.1 Guiding Principles

1. Preventing the introduction of alien invasive species is the cheapest, most effective and most preferred option and warrants the highest priority.

2. Rapid action to prevent the introduction of potential alien invasives is appropriate, even if there is scientific uncertainty about the long-term outcomes of the potential alien invasion.

3. Vulnerable ecosystems should be accorded the highest priority for action, especially for prevention initiatives, and particularly when significant biodiversity values are at risk. Vulnerable ecosystems include islands and isolated ecosystems, such as lakes and other freshwater ecosystems, cloud forests, coastal habitats and mountain ecosystems.

4. Since the impacts on biological diversity of many alien species are unpredictable, any intentional introductions and efforts to identify and prevent unintentional introductions should be based on the precautionary principle.

5. In the context of alien species, unless there is a reasonable likelihood that an introduction will be harmless, it should be treated as likely to be harmful.

6. Alien invasives act as "biological pollution" agents that can negatively affect development and quality of life. Hence, part of the regulatory response to the introduction of alien invasive species should be the principle that "the polluter pays" where "pollution" represents the damage to native biological diversity.

7. Biosecurity threats justify the development and implementation of comprehensive legal and institutional frameworks.

8. The risk of unintentional introductions should be minimised.

9. Intentional introductions should only take place with authorisation from the relevant agency or authority. Authorisation should require comprehensive evaluations based on biodiversity considerations (ecosystem, species, genome). Unauthorised introductions should be prevented.

10. The intentional introduction of an alien species should only be permitted if the positive effects on the environment outweigh the actual and potential adverse effects. This principle is particularly important when applied to isolated habitats and ecosystems, such as islands, fresh water systems or centres of endemism.

11. The intentional introduction of an alien species should not be permitted if experience elsewhere indicates that the probable result will be the extinction or significant loss of biological diversity.

12. The intentional introduction of an alien species should only be considered if no native species is considered suitable for the purposes for which the introduction is being made.

5.2 Unintentional Introductions – Recommended Actions

Unfortunately, it can be very difficult to control unintentional introductions that occur through a wide variety of ways and means. They include the most difficult types of movement to identify, control and prevent. By their very nature the most practical means of minimising unintentional introductions is by identifying, regulating and monitoring the major pathways. While pathways vary between countries and regions, the best known are international and national trade and tourism routes, through which the unintentional movement and establishment of many alien species occurs.

Recommended actions to reduce the likelihood of unintentional introductions are:

1. Identify and manage pathways leading to unintentional introductions. Important pathways of unintentional

introductions include: national and international trade, tourism, shipping, ballast water, fisheries, agriculture, construction projects, ground and air transport, forestry, horticulture, landscaping, pet trade and aquaculture.

2. Contracting parties to the CBD, and other affected countries, should work with the wide range of relevant international trade authorities and industry associations, with the goal of significantly reducing the risk that trade will facilitate the introduction and spread of alien invasive species.

3. Develop collaborative industry guidelines and codes of conduct, which minimise or eliminate unintentional introductions.

4. Examine regional trade organisations and agreements to minimise or eliminate unintentional introductions that are caused by their actions.

5. Explore measures such as: elimination of economic incentives that assist the introduction of alien invasive species; legislative sanctions for introductions of alien species unless no fault can be proved; internationally available information on alien invasive species, by country or region, for use in border and quarantine control, as well as for prevention, eradication and control activities (See also Section 8).

6. Implement the appropriate initiatives to reduce the problems of alien invasives arising from ballast water discharges and hull fouling. These include: better ballast water management practices; improved ship design; development of national ballast water programmes; research, sampling and monitoring regimes; information to port authorities and ships' crews on ballast water hazards. Make available existing national guidelines and legislation on ballast water (for example Australia, New Zealand, USA). At the national, regional and international level, disseminate international guidelines and recommendations, such as the International Maritime Organisation's guidelines on ballast water and sediment discharges (See also Section 9.2.2).

7. Put in place quarantine and border control regulations and facilities and train staff to intercept the unintentional introduction of alien species. Quarantine and border control regulations should not be premised only on narrow economic grounds that primarily relate to agriculture and human health, but, in addition, on the unique biosecurity threats each country is exposed to. Improved performance at intercepting unintentional introductions that arrive via major pathways may require an expansion of the responsibilities and resourcing of border control and quarantine services (See also Section 9.2).

8. Address the risks of unintentional introductions associated with certain types of goods or packaging through border control legislation and procedures.

9. Put in place appropriate fines, penalties or other sanctions to apply to those responsible for unintentional introductions through negligence and bad practice.

10. Ensure compliance by companies dealing with transport or movement of living organisms with the biosecurity regimes established by governments in the exporting and importing countries. Provide for their activities to be subjected to appropriate levels of monitoring and control.

11. For island countries with high risks and high vulnerabilities to alien invasive species, develop the most cost-effective options for governments wanting to avoid the high costs of controlling alien invasive species. These include more holistic approaches to biosecurity threats and better resourcing of quarantine and border control operations, including greater inspection and interception capabilities.

12. Assess large engineering projects, such as canals, tunnels and roads that cross biogeographical zones, that might mix previously separated flora and fauna and disturb local biological diversity. Legislation requiring environmental impact assessment of such projects should require an assessment of the risks associated with unintentional introductions of alien invasive species.

13. Have in place the necessary provisions for taking rapid and effective action, including public consultation, should unintentional introductions occur.

5.3 Intentional Introductions – Recommended Actions

1. Establish an appropriate institutional mechanism such as a 'biosecurity' agency or authority as part of legislative reforms on invasives (Refer to Section 9). This is a very high priority, since at present the legislative framework of most countries rarely treats intentional introductions in a holistic manner, that is, considers all organisms likely to be introduced and their effect on all environments. The usual orientation is towards sectors, e.g. agriculture. Consequently the administrative and structural arrangements are usually inadequate to deal with the entire range of incoming organisms, the implication for the environments into which they are being introduced, or with the need for rapid responses to emergency situations.

2. Empower the biosecurity agency, or other institutional mechanism, to reach decisions on whether proposed introductions should be authorised, to develop import and release guidelines and to set specific conditions, where appropriate. Operational functions should reside with other agencies (See Section 9.2.1).

3. Give utmost importance to effective evaluation and decision-making processes. Carry out an environment impact assessment and risk assessment as part of the evaluation process before coming to a decision on introducing an alien species (See Appendix 5).

4. Require the intending importer to provide the burden of proof that a proposed introduction will not adversely affect biological diversity.

5. Include consultation with relevant organisations within government, with NGOs and, in appropriate circumstances, with neighbouring countries, in the evaluation process.

6. Where relevant, require that specific experimental trials (e.g. to test the food preferences or infectivity of alien species) be conducted as part of the assessment process. Such trials are often required for biological control proposals and appropriate protocols for such trials should be developed and followed.

7. Ensure that the evaluation process allows for the likely environmental impacts, risks, costs (direct and indirect, monetary and non-monetary) benefits, and alternatives, to have been identified and assessed by the biosecurity authority in the importing country. This authority is then in a position to decide if the likely benefits outweigh the possible disadvantages. The public release of an interim decision, along with related information, should be made with time for submissions from interested parties before the biosecurity agency makes a final decision.

8. Impose containment conditions on an introduction if and where appropriate. In addition, monitoring requirements are often necessary following release as part of management.

9. Regardless of legal provisions, encourage exporters and importers to meet best practice standards to minimise any invasive risks associated with trade, as well as containing any accidental escapes that may occur.

10. Put in place quarantine and border control regulations and facilities and train staff to intercept unauthorised intentional introductions.

11. Develop criminal penalties and civil liability for the consequent eradication or control costs of unauthorised intentional introductions.

12. Ensure that provisions are in place, including the ability to take rapid and effective action to eradicate or control, in the event that an unauthorised introduction occurs, or that an authorised introduction of an alien species unexpectedly or accidentally results in a potential threat of biological invasion (see Sections 6 and 9).

13. As well as taking the efforts that are required at global and regional levels to reduce the risk that trade will facilitate unintentional introductions (Section 5.2), utilise opportunities to improve international instruments and practices relating to trade that affect intentional introductions. For example, the Parties to the Convention on International Trade in Endangered Species of Wild Fauna and Flora (CITES) are addressing the implications alien invasive species may have on the operation of the Convention. Similar initiatives should be made with respect to relevant international trade authorities and industry associations.

6. Eradication and Control

When a potential or actual alien invasive species has been detected, in other words, when prevention has not been successful, steps to mitigate adverse impacts include eradication, containment and control. Eradication aims to completely remove the alien invasive species. Control aims for the long term reduction in abundance or density of the alien invasive species. A special case of control is containment, where the aim is to limit the spread of the alien invasive species and to contain its presence within defined geographical boundaries.

6.1 Guiding Principles

1. Preventing the introduction of alien invasive species should be the first goal.

2. Early detection of new introductions of potential or known alien invasive species, together with the capacity to take rapid action, is often the key to successful and cost-effective eradications.

3. Lack of scientific or economic certainty about the implications of a potential biological alien invasion should not be used as a reason for postponing eradication, containment or other control measures.

4. The ability to take appropriate measures against intentionally or unintentionally introduced alien invasive species should be provided for in legislation.

5. The best opportunities for eradicating or containing an alien invasive species are in the early stages of invasion, when populations are small and localised. These opportunities may persist for a short or long time, depending on the species involved and other local factors.

6. Eradication of new or existing alien invasive species is preferable and is more cost effective than long-term control, particularly for new cases.

7. Eradication should not be attempted unless it is ecologically feasible and has the necessary financial and political commitment to be completed.

8. A strategically important focus for eradication is to identify points of vulnerability in the major invasive pathways, such as international ports and airports, for monitoring and eradication activities.

6.2 Eradication – Recommended Actions

1. Where it is achievable, promote eradication as the best management option for dealing with alien invasive species where prevention has failed. It is much more cost effective financially than ongoing control, and better for the environment. Technological improvements are increasing the number of situations where eradication is possible, especially on islands. Eradication is likely to be more difficult in the marine environment. The criteria that need to be met for eradication to succeed are given in the Appendix (Section 13).

2. When a potentially alien invasive species is first detected, mobilise and activate sufficient resources and expertise quickly. Procrastination markedly reduces the chances of success. Local knowledge and community awareness can be used to detect new alien invasions. Depending on the situation, a country's response might be within the country, or may require a cooperative effort with other countries.

3. Give priority to eradication at sites where a new alien invasion has occurred and is not yet well established.

4. Ensure eradication methods are as specific as possible with the objective of having no long-term effects on non-target native species. Some incidental loss to non-target species may be an inevitable cost of eradication and should be balanced against the long-term benefits to native species.

5. Ensure that persistence of toxins in the environment does not occur as a result of eradication. However, the use of toxins that are unacceptable for long-term control may be justified in brief and intensive eradication campaigns. The costs and benefits of the use of toxins need to be carefully assessed in these situations.

6. Ensure that methods for removing animals are as ethical and humane as possible, but consistent with the aim of permanently eliminating the alien invasive species concerned.

7. Given that interest groups may oppose eradication for ethical or self-interest reasons, include a comprehensive consultation strategy and develop community support for any proposed eradication as an integral part of the project.

8. Give priority to the eradication of alien invasive species on islands and other isolated areas that have highly distinctive biodiversity or contain threatened endemics.

9. Where relevant, achieve significant benefits for biological diversity by eradicating key alien mammalian predators (e.g. rats, cats, mustelids, dogs) from islands and other isolated areas with important native species. Similarly, target key feral and alien mammalian herbivores (e.g. rabbits, sheep, goats, pigs) for eradication to achieve significant benefits for threatened native plant and animal species.

10. Seek expert advice where appropriate. Eradication problems involving several species are often complex, such as determining the best order in which to eradicate species. A multidisciplinary approach might be best, as recommended in the IUCN Guidelines for Re-introductions.

6.3 Defining the Desired Outcomes of Control

The relevant measure of success of control is the response in the species, habitat, ecosystem or landscape that the control aims to benefit. It is important to concentrate on quantifying and reducing the damage caused by alien invasives, not concentrating on merely reducing numbers of alien invasives. Rarely is the relationship between pest numbers and their impacts a simple one. Hence estimating the reduction in the density of the alien invasive species will not necessarily indicate an improvement in the wellbeing of the native species, habitat or ecosystem that is under threat. It can be quite difficult to identify and adequately monitor the appropriate measures of success. It is important to do so, however, if the main goal, namely preventing the loss of biodiversity, is to be achieved.

6.4 Choosing Control Methods

Control methods should be socially, culturally and ethically acceptable, efficient, non-polluting, and should not adversely affect native flora and fauna, human health and well-being, domestic animals, or crops. While meeting all of these criteria can be difficult to achieve they can be seen as appropriate goals, within the need to balance the costs and benefits of control against the preferred outcomes.

Specific circumstances are so variable it is only possible to give broad guidelines of generally favoured methods: specific methods are better than broad spectrum ones. Biological control agents may sometimes be the preferred choice compared to physical or chemical methods, but require rigorous screening prior to introduction and subsequent monitoring. Physical removal can be an effective option for clearing areas of alien invasive plants. Chemicals should be as specific as possible, non-persistent, and non-accumulative in the food chain. Persistent organic pollutants, including organochlorine compounds should not be used. Control methods for animals should be as humane as possible, consistent with the aims of the control.

6.5 Control Strategies – Recommended Actions

Unlike eradication, control is an ongoing activity that has different aims and objectives. While there are several different strategic approaches that can be adopted they should have two factors in common. First, the outcomes that are sought need to achieve gains for native species, be clearly articulated, and widely supported. Second, there needs to be management and political commitment to spend the resources required over time to achieve the outcomes. Badly focused and half-hearted control efforts can waste resources which might be better spent elsewhere.

Recommended actions are as follows:

1. Prioritise the alien invasive species problems according to desired outcomes. This should include identifying the areas of highest value for native biological diversity and those most at risk from alien invasives. This analysis should take into account advances in control technology and should be reviewed from time to time.

2. Draw up a formal control strategy that includes identifying and agreeing to the prime target species, areas for control, methodology and timing. The strategy may apply to parts of, or to a whole country, and should have appropriate standing as, for example, the requirements of Article 6 of the CBD ("General Measures for Conservation and Sustainable Use"). Such strategies should be publicly available, be open for public input, and be regularly reviewed.

3. Consider stopping further spread as an appropriate strategy when eradication is not feasible, but only where the range of the alien invasive is limited and containment within defined boundaries is possible. Regular monitoring outside the containment boundaries is essential, with quick action to eradicate any new outbreaks.

4. Evaluate whether long-term reduction of alien invasive numbers is more likely to be achieved by adopting one action or set of linked actions (multiple action control). The best examples of single actions come from the successful introduction of biological control agent(s). These are the 'classical' biological control programs. Any intentional introductions of this nature should be subject to appropriate controls and monitoring (see also Sections 5.3, 9 and Appendix 4). Exclusion fencing can be an effective single action control measure in some circumstances. An example of multiple action control is integrated pest management which uses biological control agents coupled with various physical and chemical methods at the same time.

5. Increase the exchange of information between scientists and management agencies, not only about alien invasive species, but also about control methods. As techniques are continuously changing and improving it is important to pass this information on to management agencies for use.

6.6 Game and Feral Species as Alien Invasives – Recommended Actions

Feral animals can be some of the most aggressive and damaging alien species to the natural environment, especially on islands. Despite any economic or genetic value they may have, the conservation of native flora and fauna should always take precedence where it is threatened by feral species. Yet some alien invasive species that cause severe damage to native biodiversity have acquired positive cultural values, often for hunting and fishing opportunities. The result can be conflict between management objectives, interest groups and communities. In these circumstances it takes longer to work through the issues, but resolution can often be achieved through public awareness and information campaigns about the damaging impacts of the alien invasives, coupled with consultation and adaptive management approaches that have community support. Risk analysis and environmental impact assessment may also help to develop appropriate courses of action and solutions.

Recommended actions are as follows:

1. Consider managing hunting conflicts on public land by designating particular areas for hunting while carrying out more stringent control to protect biodiversity values elsewhere. This option is limited in its application to situations where there is

high value attached to the alien species and yet biological diversity values can still be protected through localised action.

2. Evaluate the option of removal of a representative number of the feral animals to captivity or domestication where eradication in the wild is planned.

3. Strongly encourage owners and farmers to take due care to prevent the release or escape of domestic animals that are known to cause damage as feral animals, e.g. cats, goats.

4. Develop legal penalties to deter such releases and escapes in circumstances where costly economic or damaging ecological consequences are likely to follow.

7. Links to Re-introduction of Species

7.1 Guiding Principle

1. Successful eradications and some control programmes can significantly improve the likely success of re-introductions of native species, and thereby provide opportunities to reverse earlier losses of native biological diversity.

7.2 Links Between Eradication and Control Operations and Re-introductions

An eradication operation that successfully removes an alien invasive species, or a control operation that lowers it to insignificant levels, usually improves the conditions for native species that occupy or previously occupied that habitat. This is especially true on many oceanic islands. Eradications are often undertaken as part of the preparation for re-introduction(s).

The IUCN Guidelines for Re-introductions (May 1995) were developed to provide "...direct, practical assistance to those planning, approving or carrying out re-introductions." These guidelines elaborate requirements and conditions, including feasibility studies, criteria for site selection, socio-economic and legal requirements, health and genetic screening of individuals, and issues surrounding the proposed release of animals from captivity or rehabilitation centres. They should be referred to as part of the planning of eradication or control operations where re-introductions might be an appropriate and related objective. They should also be referred to if reviewing any re-introduction proposal.

The socio-economic considerations that apply to eradication and control operations largely apply to re-introductions as well, namely the importance of community and political support, financial commitment and public awareness. This makes it cost-effective to combine consultation over the eradication objective with proposals to re-introduce native species. It has the added advantage of offsetting the negative aspects of some eradications (killing valued animals) with the positive benefits of re-introducing native species (restoring heritage, recreation or economic values).

8. Knowledge and Research Issues

8.1 Guiding Principle

1. An essential element in the campaigns against alien invasive species at all levels (global, national, local) is the effective and timely collection and sharing of relevant information and experiences, which, in turn, assist advances in research and better management of alien invasive species.

8.2 Recommended Actions

1. Give urgency to the development of an adequate knowledge base as a primary requirement to address the problems of alien invasive species worldwide. Although a great deal is known about many such species and their control, this knowledge remains incomplete and is difficult to access for many countries and management agencies.

2. Contribute to the development of an easily accessible global database (or linked databases) of all known alien invasive species, including information on their status, distribution, biology, invasive characteristics, impacts and control options. It is important that Governments, management agencies and other stakeholders should all participate in this.

3. Develop "Black Lists " of alien invasive species at national, regional and global levels that are easily accessible to all interested parties. While "Black Lists" are a useful tool for focusing attention on known alien invasive species, they should not be taken to imply that unlisted alien species are not potentially harmful.

4. Through national and international research initiatives, improve knowledge of the following: ecology of the invasion process, including lag effects; ecological relationships between invasive species; prediction of which species and groups of species are likely to become invasive and under what conditions; characteristics of alien invasive species; impacts of global climate change on alien invasive species; existing and possible future vectors; ecological and economic losses and costs associated with introductions of alien invasive species; sources and pathways caused by human activity.

5. Develop and disseminate better methods for excluding or removing alien species from traded goods, packaging material, ballast water, personal luggage, aircraft and ships.

6. Encourage and support further management research on: effective, target-specific, humane and socially acceptable methods for eradication or control of alien invasive species; early detection and rapid response systems; development of monitoring techniques; methods to gather and effectively disseminate information for specific audiences.

7. Encourage monitoring, recording and reporting so that any lessons learned from practical experiences in management of alien invasive species can contribute to the knowledge base.

8. Make better use of existing information and experiences to promote wider understanding and awareness of alien invasive species issues. There need to be strong linkages between the actions taken under Sections 4 and 8.

9. Law and Institutions

9.1 Guiding Principles

1. A holistic policy, legal and institutional approach by each country to threats from alien invasive species is a prerequisite to conserving biological diversity at national, regional and global levels.

2. Effective response measures depend on the availability of national legislation that provides for preventative as well as remedial action. Such legislation should also establish clear institutional accountabilities, comprehensive operational mandates, and the effective integration of responsibilities regarding actual and potential threats from alien invasive species.

3. Cooperation between countries is needed to secure the conditions necessary to prevent or minimise the risks from introductions of potentially alien invasive species. Such cooperation is to be based on the responsibility that countries have to ensure that activities within their jurisdiction or control do not cause damage to the environment of other countries.

9.2 Recommended Actions

9.2.1 National level

1. Give high priority to developing national strategies and plans for responding to actual or potential threats from alien invasive species, within the context of national strategies and plans for the conservation of biological diversity and the sustainable use of its components.

2. Ensure that appropriate national legislation is in place, and provides for the necessary controls of intentional and non-intentional introductions of alien species, as well as for remedial action in case such species become invasive. Major elements of such legislation are identified in previous sections, particularly Sections 5 and 6.

3. Ensure that such legislation provides for the necessary administrative powers to respond rapidly to emergency situations, such as border detection of potential alien invasive species as well as to address threats to biological diversity caused by intentional or non-intentional introductions of alien species across biogeographical boundaries within one country.

4. Ensure, wherever possible, for the designation of a single authority or agency responsible for the implementation and enforcement of national legislation, with clear powers and functions. In cases where this proves impossible, ensure there is a mechanism to coordinate administrative action in this field, and set up clear powers and responsibilities between the administrations concerned. (Note: these operational roles regarding implementation and enforcement are different from, and in addition to the specific function of the 'biosecurity' agency that was recommended in Section 5.3).

5. Review national legislation periodically, including institutional and administrative structures, in order to ensure that all aspects of alien invasive species issues are dealt with according to the state of the art, and that the legislation is implemented and enforced.

9.2.2 International level

1. Implement the provisions of international treaties, whether global or regional, that deal with alien invasive species issues and constitute a compulsory mandate for respective Parties. Most prominent among these treaties is the CBD, and a number of regional accords.

2. Implement decisions taken by Parties to specific global and regional conventions, such as resolutions, codes of conduct or guidelines related to introductions of alien species, for example the International Maritime Organisation's (IMO) guidance on ballast water.

3. Consider the desirability, or as the case may be, necessity, of conducting further agreements, on a bilateral or multilateral basis, or adapting existing ones, with respect to the prevention or control of introduction of alien species. This includes, in particular, consideration of international agreements related to trade, such as those under the auspices of the World Trade Organisation (WTO).

4. For neighbouring countries, consider the desirability of cooperative action to prevent potential alien invasive species from migrating across borders, including agreements to share information, through, for example, information alerts, as well as to consult and develop rapid responses in the event of such border crossings.

5. Generally develop international cooperation to prevent and combat damage caused by alien invasive species, and provide assistance and technology transfer as well as capacity building related to risk assessment as well as management techniques.

10. Role of IUCN

1. IUCN will continue to contribute to the Global Invasive Species Programme (GISP)[3], together with CAB International, the United Nations Environment Programme (UNEP) and the Scientific Committee on Problems of the Environment (SCOPE).

[3] SCOPE, UNEP, IUCN and CABI have embarked on a programme on invasive species, with the objective of providing new tools for understanding as well as dealing with invasive species. This initiative is called the Global Invasive Species Programme (GISP). GISP engages the many constituencies involved in the issue, including scientists, lawyers, educators, resource managers and people from industry and government. GISP maintains close cooperation with the CBD Secretariat on the issue of alien species.

2. IUCN will actively participate in the processes and meetings of the CBD to implement Article 8(h) by providing scientific, technical and policy advice.

3. The components of IUCN (including its Commissions, Programmes and Regional Offices) will act together to support the IUCN Global Initiative on Invasive Species.

4. IUCN will maintain and develop links and cooperative programmes with other organisations involved in this issue, including international organisations such as the United Nations Environment Programme, Food and Agricultural Organisation (FAO), Scientific Committee on Problems of the Environment, WTO and international NGOs. IUCN will work with work with Parties to CITES, Parties to the CBD, Parties to the RAMSAR Convention, and with regional programmes such as the South Pacific Regional Environment Programme (SPREP).

5. IUCN regional networks will play a significant role in raising public awareness at all levels on the issues of alien invasive species, the various threats to native biological diversity and the economic implications, as well as options for control.

6. The IUCN Invasive Species Specialist Group (ISSG) of the Species Survival Commission (SSC) will, through its international network, continue to collect, organise and disseminate information on alien invasive species, on prevention and control methods, and on ecosystems that are particularly vulnerable to alien invasion.

7. The separate work of IUCN/SSC on identifying species threatened with extinction and areas with high levels of endemism and biodiversity will be supported. This work is valuable when assessing alien invasion risks, priority areas for action, and for practical implementation of these guidelines.

8. The ongoing work of the ISSG will be supported, including the following actions: the development and maintenance of a list of expert advisors on control and eradication of alien invasive species; expansion of the alien invasive species network; production and distribution of newsletters and other publications.

9. IUCN, in association with other cooperating organisations, will take a lead in the development and transfer of capacity building programmes (e.g. infrastructure, administration, risk and environmental assessment, policy, legislation), in support of any country requesting such assistance or wishing to review its existing or proposed alien invasive species programmes.

10. IUCN will take an active role in working with countries, trade organisations and financial institutions (e.g. WTO, World Bank, International Monetary Fund (IMF), IMO) to ensure that international trade and financial agreements, codes of practice, treaties and conventions take into account the threats posed to biological diversity and the financial costs and economic losses associated with alien invasive species.

11. The ISSG will support the work of the IUCN Environmental Law Programme in assisting countries to review and improve their legal and institutional frameworks concerning alien invasive species issues.

12. The ISSG will develop regional databases and early warning systems on alien invasive species and work with other cooperating organisations to ensure efficient and timely dissemination of relevant information to requesting parties.

11. Bibliography and Related Information

The guiding principles and text of these guidelines are partially based on, or sourced from the following important documents:

Translocation of Living Organisms. IUCN Position Statement, 1987. IUCN, Gland, Switzerland.

IUCN Guidelines for Re-introductions. 1995. IUCN, Gland, Switzerland.

Code of Conduct of the Import and Release of Exotic Biological Control Agents. United Nations Food and Agriculture Organisation, 1995. FAO, Rome, Italy.

Harmful Non-indigenous Species in the United States. U.S. Congress, Office of Technology Assessment, OTA-F-565, 1993. US Government Printing Office, Washington DC, USA

Proceedings. Norway/UN Conference on Alien Species. The Trondheim Conference on Biodiversity. 1-5 July 1996. Norwegian Institute for Nature Research, Trondheim, Norway.

Guidelines for Preventing the Introduction of Unwanted Aquatic Organisms and Pathogens from Ships' Ballast Water and Sediment Discharges. International Maritime Organisation (IMO) Resolution A.774(18)(4.11.93) (Annex).

12. Acknowledgements

IUCN gratefully acknowledges the dedication and efforts of the Invasive Species Specialist Group (ISSG) and other experts on alien invasive species whose collaborative work has made the production of these guidelines possible. Input from the IUCN Environmental Law Programme is also gratefully acknowledged.

13. Appendix

1. Environmental Impact Assessment (EIA)

Generic questions in the EIA process concerning impacts a proposed introduced species may have on the environment should include the following:

- Does the proposed introduction have a history of becoming invasive in other places? If yes, it is likely to do so again and should not be considered for introduction.
- What is the probability of the alien species increasing in numbers and causing damage, especially to the ecosystem into which it would be introduced?
- Given its mode of dispersal, what is the probability the alien species would spread and invade other habitats?
- What are the likely impacts of natural cycles of biological and climatic variability on the proposed introduction? (Fire, drought and flood can substantially affect the behaviour of alien plants.)
- What is the potential for the alien species to genetically swamp or pollute the gene pool of native species through interbreeding?
- Could the alien species interbreed with a native species to produce a new species of aggressive polyploid invasive?
- Is the alien species host to diseases or parasites communicable to native flora or fauna, humans, crops, or domestic animals in the proposed area for introduction?
- What is the probability that the proposed introduction could threaten the continued existence or stability of populations of native species, whether as a predator, as a competitor for food, cover, or in any other way?

- If the proposed introduction is into a contained area(s) with no intention of release, what is the probability of a release happening accidentally?
- What are the possible negative impacts of any of the above outcomes on human welfare, health or economic activity?

2. Risk Assessment

This refers to an approach that seeks to identify the relevant risks associated with a proposed introduction and to assess each of those risks. Assessing risk means looking at the size and nature of the potential adverse effects of a proposed introduction as well as the likelihood of them happening. It should identify effective means to reduce the risks and examine alternatives to the proposed introduction. The proposed importer often does a risk assessment as a requirement by the decision-making authority.

3. Criteria to be Satisfied to Achieve Eradication

- The rate of population increase should be negative at all densities. At very low densities it becomes progressively more difficult and costly to locate and remove the last few individuals.

- Immigration must be zero. This is usually only possible for offshore or oceanic islands, or for very new alien invasions.

- All individuals in the population must be at risk to the eradication technique(s) in use. If animals become bait- or trap-shy, then a sub-set of individuals may no longer be at risk to those techniques.

- Monitoring of the species at very low densities must be achievable. If this is not possible survivors may not be detected. In the case of plants, the survival of seed banks in the soil should be checked.

- Adequate funds and commitment must continuously exist to complete the eradication over the time required. Monitoring must be funded after eradication is believed to have been achieved until there is no reasonable doubt of the outcome.

- The socio-political environment must be supportive throughout the eradication effort. Objections should be discussed and resolved, as far as practicable, before the eradication is begun.

For more information about the IUCN Guidelines for the Prevention of Biodiversity Loss caused by Alien Invasive Species see: www.iucn.org/themes/ssc/news/invasives.htm.

For more information of the IUCN SSC Invasive Species Specialist Group see: www.issg.org/.

Species Index

General Index

Philip Island glory pea, 3
Philippines, 305
phytochoria, 71, 73
phytosanitary permits, 435
Picloram, 340
pigs, 227, 232, 233, 340
pineapple, 213
pioneer species, 154, 167
pioneers, 257, 263, 264
pirates, 325, 326
plant
   identification tools, 42
   quarantine, 277
   uses, 216
plant-animal interaction, 155
plantain, 213
Plantas do Nordeste (PNE), 216, 217, 218, 219
plantations, 33, 154, 166, 167, 168, 195, 214, 318, 397, 399, 407, 417, 452, 465
Pleistocene, 187, 398
pleonanthic flowering, 448
plywood, 385
poaching, 40, 76, 86, 374, 424, 441
poison, 233
Poisson distribution, 462
policing, 40
pollen flow, 379
pollination, 31, 202, 234, 235, 237, 311
pollinators, 33, 155, 234, 246, 264
population
   dynamics, 155, 449, 454, 456
   fragmentation, 306
   genetics, 351, 437
   growth, 3, 37, 72, 117, 190, 191, 271, 326, 343, 344, 478, 480
   modelling, 379
   viability, 234
potatoes, 326
poverty, 36, 216
prawns, 177
prayer, 142, 246
predators, 336
Preuss' guenon, 398
prickly pears, 351
primates, 156, 167, 398
privet, 228
productivity, 272
propagation, 118, 131, 146, 174, 178, 179, 230, 236, 247, 251, 277, 281, 311, 318, 355, 383–394, 397, 404, 405, 407, 409, 411, 412, 413, 416, 425, 426, 429, 430, 435, 436, 438, 440, 459, 477, 478

propagule storage, 99
prostate disorders, 385
protected areas, 16, 38, 40, 41, 45, 46, 74, 76, 77, 80, 82, 86, 87, 94, 118, 119, 120, 122, 123, 128, 129, 130, 131, 226, 228, 231, 232, 273, 277, 282, 289, 302, 306, 307, 309, 310, 311, 319, 331, 355, 362, 372, 375, 378, 380, 386, 447, 472, 473, 474, 475, 480, 481
pruning, 252, 257
public health, 174
publicity, 111
pumpkins, 326
Pygenil, 20

quarantine, 21, 332, 337, 341, 342, 343, 344
quarrying, 303, 317, 327
quinine, 341

rabbits, 235
rafts, 449, 451, 453, 463, 464
rain gauge, 253
rainfall, 156, 162, 166, 187, 227, 251, 252, 253, 254
Rapid Rural Appraisal (RRA), 471
Rapid Survey Method, 474
Rapid Vulnerability Assessment (RVA), 472, 473
rats, 227, 232, 233, 237, 326, 340
rattan, 380, 401, 404, 406, 407, 409, 411, 412, 445–466
recreation, 31, 176, 400
Red Data Book, 66, 173, 178, 229, 271, 277, 308
reforestation, 378, 415, 435
refugee movement, 72
refuges, 182
regeneration, 41, 155, 178, 192, 195, 202, 230, 236, 256, 263, 337, 338, 379, 437
Regional Centre of Endemism (RCE), 71
   Afromontane, 73, 77
   Guineo-Congolian, 71
   Indian Ocean Coastal Belt, 73
   Somali-Masai, 71, 77, 271
   Sudanian, 71
   Zambezian, 71, 73
regulation, 190
reintroduction, 32, 100, 118, 131, 153, 176, 178, 181, 230, 234, 277, 289, 303, 311, 318, 337, 380, 404, 425, 428, 429, 435, 436, 437, 439, 441
relict species, 372
religion, 137, 149
remote sensing, 277
replanting, 404
reproductive biology, 230, 235, 335